MATHEMATICS
FOR DAILY LIVING

Harry Lewis

Random House/McCormick-Mathers

International Standard Book Number: 0-800-91552-6

Manufactured in the United States of America

About the Author

Dr. Lewis has relied heavily on his many years of experience as a mathematics educator in writing this textbook. As both a teacher and a department chairman, he has taught mathematics on all grade levels and all ability levels to students in junior and senior high schools. He has taught methods courses in mathematics at New York University and prepared graduate students for the actuarial examinations given by the Actuarial Society of America. Dr. Lewis is also the author of a geometry textbook, three textbooks on business mathematics, and three textbooks in college mathematics. He has authored many articles on the teaching of mathematics and has been a contributor to one of the yearbooks of the National Council of Teachers of Mathematics. Dr. Lewis, formerly chairman of the mathematics department of East Side High School, Newark, New Jersey, and Principal of Arts High School in that city, is now serving as a consultant for Jersey City State College, where he served as a professor in the secondary education department.

Photo Credits

Page X – *Motor Vehicle Manufacturers' Association of the United States, Inc.;* Page 12 – *Shell Oil Company;* Page 20 – *The Port Authority of New York and New Jersey;* Page 38 – *Metropolitan Transit Authority;* Page 42 – *The Port Authority of New York and New Jersey;* Page 46 – *Metropolitan Transit Authority;* Page 56 – *Metropolitan Transit Authority;* Page 80 – *Bay Area Rapid Transit District;* Page 88 – *TransWorld Airlines;* Page 102 – *Acme Markets, Inc.;* Page 114 – *Acme Markets, Inc.;* Page 148 – *Metropolitan Transportation Authority;* Page 186 – *U.S. Internal Revenue Service;* Page 204 – *The Chase Manhattan Bank;* Page 224 – *The Chase Manhattan Bank;* Page 250 – The Chase Manhattan Bank; Page 280 – *The Chase Manhattan Bank;* Page 294 – *The Chase Manhattan Bank;* Page 304 – *The American Stock Exchange;* Page 350 – *U.S. Department of Housing and Urban Development;* Page 369 – *Honeywell, Inc.;* Page 380 – *Rosemary Mayer;* Page 396 – *U.S. Department of Commerce, Bureau of the Census;* Page 436 – *U.S. Department of Housing and Urban Development;* Page 462 – *U.S. Department of Housing and Urban Development;* Page 470 – *U.S. Department of Housing and Urban Development;* Page 494 – *Equitable Life Assurance Society of the United States.*

Preface

Perhaps the greatest change in this, the third edition, over the earlier editions, has taken place in the design of the presentation. Mapping diagrams have replaced the rather extensive explanations used in the past. For to paraphrase the obvious, it was felt that one picture could not only replace a thousand words, but would do so with greater clarity and greater appeal to the student.

Needless to say, the content of the second edition has been completely and thoroughly updated. Certain topics such as mutual funds, bank time loans, and penalties for tardiness appear to be passing from the scene and have been deleted from the book. On the other hand, discounts on air travel, daily compounding of interest, time accounts, the Keogh Plan, the IRA, and a number of other topics that have grown in importance in our economy are included in this edition. The events of each succeeding day appear to dramatize the need that each of us has for an understanding of the application of mathematics in the marketplace.

Suggestions for Using the Test

Basically, there is no reason why the topics must be presented in the order in which they appear in the text. Should you find that the interests of your students lean more toward purchasing consumer goods or computing earnings than toward buying a car, then by all means delay the teaching of Chapter 1 until a later date and move directly into either of the other topics. A word of caution, though, it is important that the units within a single chapter be taught in the order in which they are arranged. For, as in any area of mathematics, certain topics may build upon earlier ones.

The philosophy guiding the development of the material in this book is the conviction that each topic should be explored as fully as possible keeping in mind both the maturity level and grade level of the students who will be taking this course. In view of this, the presentation of the continuity of ideas within a single chapter might be preferable to token exposure to units randomly selected from chapters throughout the book.

To the many, many teachers who have accepted the earlier editions into their classrooms, I hope you will be as pleased with the present edition.

Harry Lewis, 1980

Contents

Chapter 1

Automobile Ownership

Since many of us anxiously look forward to owning our own car, we felt it would be best to begin this course with an examination of the cost of car ownership.

The Automobile Loan Unit 1

Few people have the cash necessary to pay the entire price of a car. So, they often borrow the money they need from one of three sources:

1. A bank
2. The automobile dealer
3. A small-loan agency

Usually, a bank will charge less for a car loan than either of the other two sources. The cost of borrowing the money is less if we purchase a new car rather than a used car.

The cost of a car loan is affected by both the state and the region in which we live. The monthly payments in the tables on pages (2) and (4) are about average for what we would find in the country as a whole on a new-car loan and a used-car loan. Notice that the column headings are:

Without Life Insurance **With Life Insurance**

Frequently, money cannot be borrowed unless the borrower purchases life insurance. The bank, or dealer, is concerned with who will pay off the loan in the event of the borrower's death. If the borrower should die, the money from the life insurance will be used to pay off the loan.

New-Car Monthly Payment Table (12.85%)

Unpaid Balance	18 Months Without Life Insurance	18 Months With Life Insurance	24 Months Without Life Insurance	24 Months With Life Insurance	30 Months Without Life Insurance	30 Months With Life Insurance	36 Months Without Life Insurance	36 Months With Life Insurance
$.10	$ .61	$ .62	$ .47	$ .48	$ .39	$ .40	$ .34	$ .34
20	1.23	1.24	.95	.96	.78	.79	.67	.68
30	1.84	1.86	1.42	1.44	1.17	1.19	1.01	1.02
40	2.46	2.47	1.90	1.92	1.57	1.59	1.34	1.37
50	3.07	3.09	2.37	2.40	1.96	1.98	1.68	1.71
60	3.68	3.71	2.85	2.88	2.35	2.38	2.02	2.05
70	4.30	4.33	3.32	3.36	2.74	2.78	2.35	2.39
80	4.91	4.95	3.80	3.84	3.13	3.17	2.69	2.73
90	5.52	5.57	4.27	4.32	3.52	3.57	3.03	3.07
100	6.14	6.19	4.75	4.80	3.92	3.97	3.36	3.42
200	12.27	12.37	9.49	9.59	7.83	7.93	6.72	6.83
300	18.41	18.56	14.24	14.39	11.75	11.90	10.09	10.25
400	24.55	24.74	18.99	19.19	15.66	15.86	13.45	13.66
500	30.69	30.93	23.74	23.98	19.58	19.83	16.81	17.08
600	36.83	37.12	28.48	28.78	23.49	23.79	20.17	20.49
700	42.96	43.30	33.23	33.58	27.41	27.76	23.54	23.91
800	49.10	49.49	37.98	38.38	31.32	31.72	26.90	27.32
900	55.24	55.67	42.72	43.17	35.24	35.69	30.26	30.74
1,000	61.37	61.86	47.47	47.97	39.15	39.66	33.62	34.15
2,000	122.76	123.72	94.94	95.94	78.30	79.31	67.24	68.30
3,000	184.13	185.58	142.41	143.91	117.45	118.97	100.87	102.45
4,000	245.51	247.44	189.89	191.88	156.60	158.62	134.49	136.61
5,000	306.89	309.30	237.36	239.85	195.75	198.28	168.11	170.76

Example 1 Fred Miller borrowed $3,000 to purchase a new car. He agreed to pay back the debt monthly over a three-year period. If no life insurance was required, what was his monthly payment?

Explanation

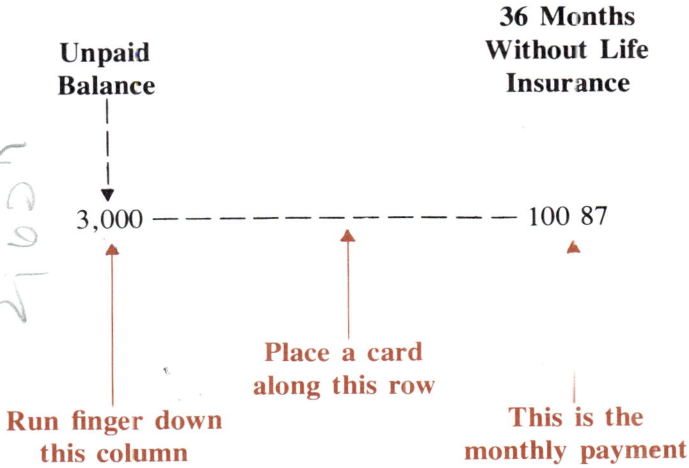

Solution

Monthly payment = $100.87

Example 2 After making a down payment on a new car George Newsome still had to borrow $4,760. He was asked to purchase life insurance to obtain this loan. If the debt was to be paid off over 24 months, how much was each payment?

Explanation

The monthly payment consists of three parts:

1. Payment on $4,000
2. Payment on $700
3. Payment on $60

By adding the three we find the total monthly payment.

Solution

Payment on $4,000 = $191.88
Payment on $700 = 33.58
Payment on $60 = 2.88
Monthly Payment on $4,760 = $228.34

Used-Car Monthly Payment Table (16.80%)

Unpaid Balance	18 Months Without Life Insurance	18 Months With Life Insurance	24 Months Without Life Insurance	24 Months With Life Insurance	30 Months Without Life Insurance	30 Months With Life Insurance	36 Months Without Life Insurance	36 Months With Life Insurance
$.10	$ .63	$ .64	$ .49	$ .50	$ .41	$ .42	$ .36	$ .36
20	1.26	1.28	.99	1.00	.82	.84	.71	.72
30	1.90	1.91	1.48	1.50	1.23	1.25	1.07	1.09
40	2.53	2.55	1.97	2.00	1.64	1.67	1.42	1.45
50	3.16	3.19	2.47	2.50	2.05	2.09	1.78	1.81
60	3.79	3.83	2.96	3.00	2.46	2.51	2.13	2.17
70	4.43	4.46	3.45	3.49	2.87	2.92	2.49	2.53
80	5.06	5.10	3.95	3.99	3.28	3.34	2.84	2.89
90	5.69	5.74	4.44	4.49	3.69	3.76	3.20	3.26
100	6.32	6.38	4.93	4.99	4.11	4.18	3.56	3.62
200	12.65	12.76	9.87	9.98	8.21	8.36	7.11	7.24
300	18.97	19.13	14.80	14.98	12.32	12.53	10.67	10.85
400	25.29	25.51	19.74	19.97	16.42	16.71	14.22	14.47
500	31.62	31.89	24.67	24.96	20.53	20.89	17.78	18.09
600	37.94	38.27	29.61	29.95	24.63	25.07	21.33	21.71
700	44.26	44.65	34.54	34.94	28.74	29.24	24.89	25.33
800	50.59	51.03	39.48	39.93	32.84	33.42	28.44	28.94
900	56.91	57.40	44.41	44.93	36.95	37.60	32.00	32.56
1,000	63.24	63.78	49.34	49.92	41.05	41.78	35.55	36.18
2,000	126.47	127.56	98.69	99.84	82.10	83.55	71.11	72.36
3,000	189.71	191.35	148.04	149.75	123.15	125.33	106.66	108.54
4,000	252.94	255.13	197.38	199.67	164.21	167.10	142.21	144.72
5,000	316.18	318.91	246.73	249.59	205.26	208.88	177.77	180.90

Example 3 An 18-month loan of $2,500 including life insurance was made on the purchase of a used car. What was the total repaid?

Explanation The first step is to determine the monthly payment. Since this payment will be made 18 times, then the total repaid will be *18 times the monthly payment*.

Solution

Payment on $2,000 = $127.56
Payment on $500 = 31.89
Monthly payment = $159.45
Total repaid = 18 times monthly payment
= 18 × $159.45
= $2,870.10

Unit 1

Exercises

A

Do you recall how to do the computation in the following exercises? This computation is the same as you will be doing when finding the cost of a car loan.

1. Find the sums:

a. $126.47 + 4.43

b. $83.55 + 25.07

c. $154.21 + 28.74

d. $95.94 + 19.19

e. $67.24 + 16.81 + 2.35

f. $63.78 + 31.89 + 4.46

g. $245.51 + 36.83 + 2.46

h. $237.36 + 42.72 + 2.85

i. $547.09 + 19.08 + 856.79

2. Find the differences:

a. $2,276.46 − 2,000.00

b. $5,872.50 − 5,000.00

c. $3,453.84 − 3,417.84

d. $4,917.96 − 4,841.64

e. $3,569.10 − 3,523.50

f. $6,051.96 − 5,696.64

3. Find the products:

a. $41.05 × 2

b. $20.53 × 4

c. $33.42 × 6

d. $42.72 × 8

e. $123.72 × 3

f. $206.89 × 8

g. $72.36 × 24

h. $41.78 × 24

i. $83.55 × 30

j. $39.66 × 30

k. $184.13 × 36

l. $102.45 × 36

m. $61.37 × 18

n. $309.30 × 18

o. $316.18 × 24

B

1. Find the monthly payment for each new-car loan.

	Unpaid Balance	Number of Months	Life Insurance	Monthly Payment
a.	$1,000	18	With	$ ______
b.	3,000	24	With	______
c.	4,000	36	Without	______
d.	800	30	Without	______
e.	500	18	Without	______
f.	2,000	24	With	______

2. Find the monthly payment for each used-car loans.

	Unpaid Balance	Number of Months	Life Insurance	Monthly Payment
a.	$3,000	18	Without	$________
b.	1,000	30	Without	________
c.	4,000	24	With	________
d.	900	18	Without	________
e.	600	24	With	________
f.	2,000	36	With	________

C

1. Determine the total monthly payment for each new-car loan. None of the loans required the purchase of life insurance.

	Unpaid Balance	Number of Months	Payments on Separate Parts			Monthly Payment
a.	$2,500	18	$________	+	$________	$________
b.	3,200	24	________	+	________	________
c.	4,600	30	________	+	________	________
d.	1,400	36	________	+	________	________
e.	5,100	30	________	+	________	________
f.	2,900	18	________	+	________	________

2. Determine the total monthly payment for each used-car loan. Each of the loans required the purchase of life insurance.

	Unpaid Balance	Number of Months	Payments on Separate Parts					Monthly Payment
a.	$2,450	18	$________	+	$________	+	$________	$________
b.	3,280	30	________	+	________	+	________	________
c.	1,370	24	________	+	________	+	________	________
d.	4,160	30	________	+	________	+	________	________
e.	5,710	36	________	+	________	+	________	________
f.	2,090	24	________	+	________	+	________	________

D

1. Determine the total repaid for each new-car loan.

	Loan	Life Insurance	Number of Months	Monthly Payment	Total Repaid
a.	$4,000	With	24	$________	$________
b.	2,000	Without	24	________	________

	Loan	Life Insurance	Number of Months	Monthly Payment	Total Repaid
c.	3,000	Without	30	$______	$______
d.	5,000	With	30	______	______
e.	1,000	With	36	______	______
f.	4,000	Without	36	______	______
g.	3,000	With	18	______	______

2. Determine the total repaid for each used-car loan. Each loan required the purchase of life insurance.

	Loan	Number of Months	Payments on Separate Parts			Monthly Payment	Total Repaid
a.	$3,500	24	$______	+	$______	$______	$______
b.	1,800	24	______	+	______	______	______
c.	2,100	30	______	+	______	______	______
d.	4,300	30	______	+	______	______	______
e.	5,400	36	______	+	______	______	______
f.	2,800	36	______	+	______	______	______
g.	950	18	______	+	______	______	______

E

1. Frank Howe borrowed $4,500 for the purchase of a new car. He agreed to pay off the debt over a 36-month period. No life insurance was required.

a. What were his monthly payments? ______

b. What was the total amount he repaid? ______

c. What was the cost of the loan; that is, how much more did he repay than he borrowed? ______

2. Janis Cleary purchased a new car priced at $5,746 by making a down payment of $746. She borrowed the remainder of the money to pay for the car. No life insurance was required.

a. How much money did she have to borrow? ______

b. If the debt is repaid over a 36-month period, what will the total repayment be? ______

c. If the debt is repaid over an 18-month period, what will the total repayment be? ______

d. How much more will the 36-month loan cost than the 18-month loan? ______

3. Four thousand dollars is borrowed for the purchase of a car. The debt is to be paid back over a 30-month period. Life insurance is required.

a. If the loan is made for the purchase of a new car, what will the total repayment be? ________

b. If the loan is made for the purchase of a used car, what will the total repayment be? ________

c. How much more is the cost of the used-car loan than the new-car loan? ________

Automobile Insurance Unit 2

Section 1 Bodily-Injury and Property-Damage Insurance

Part 1

There are four types of insurance that can be purchased to protect us if our car is involved in an accident. The two we plan to examine now are designed to pay for damages our car inflicts upon either people or property.

1. **Property-Damage Insurance** – Covers cost of damage to other people's property.
2. **Bodily-Injury Insurance** – Covers cost of injury to people other than our own family.

The basic property-damage policy is for $5,000. This means that:

a. The insurance company will pay for any damage up to $5,000.

b. We must pay for all costs over $5,000.

Example 1 Ellen Marcus was involved in an accident in which her car damaged property to the extent of $7,400. The property-damage policy she had purchased was for $5,000. In what manner was the $7,400 damage paid?

Solution

Amount paid by insurance company = $5,000
Amount paid by Ellen = $7,400 − $5,000
= $2,400

The basic bodily-injury policy is called a 15/30 policy. This means:

1. The insurance company will pay no more than $15,000 to any one person.
2. The insurance company will pay no more than $30,000 to all people injured in a single accident.
3. We must pay all damages over these amounts.

Example 2 Bert Finney's car struck another car in which three people were injured. The damages to each of these people were:

First person: $24,000
Second person: $7,000
Third person: $4,500

In what manner were these damages paid?

Solution

First person → $24,000; Second person → $7,000; Third person → $4,500

Total damages = $24,000 + $7,000 + $4,500
= $35,500

Maximum to one person → $15,000

Amount paid by insurance company = $15,000 + $7,000 + $4,500
= $26,500

Total damages → $35,500; Amount paid by ins. co. → $26,500

Amount paid by Bert = $35,500 − $26,500
= $9,000

It is possible to purchase bodily-damage insurance to protect ourself against damages greater than 15/30. Under coverage of 25/50 the insurance company will pay up to $25,000 to a single person and up to $50,000 to more than one person. This policy will cost slightly more than the 15/30 protection.

Property damage protection for more than $5,000 can also be purchased. This, too, will be somewhat more expensive than the $5,000 coverage-insurance.

The amount we pay for automobile insurance depends upon the

area in which we live. The cost is lowest in rural areas and highest in the central area of a city.

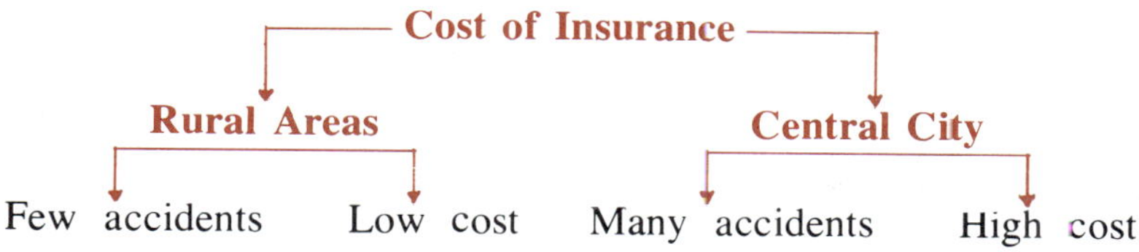

Each state is divided into many territories and a basic cost is assigned to each territory.

Semiannual Basic Cost for Bodily-Injury and Property-Damage Insurance

	Territory					
	01	02	05	06	25	26
Bodily-Injury	$45	$61	$30	$29	$23	$25
Property-Damage	$37	$46	$23	$25	$22	$20

It would appear from the figures above that territory 02 is probably within a city while territory 25 is probably a country area.

Example 3 What is the total semiannual basic premium (cost) that will have to be paid for bodily-injury and property-damage insurance in an 06 territory?

Solution

Total basic cost	=	Bodily-Injury	+	Property-Damage
	=	$29	+	$25
	=	$54		

The actual cost of automobile insurance is much more than the basic cost listed in the table on page 11. The actual cost depends on three factors:

1. **Coverage-Limit factor:** this number is determined by the amount of insurance we purchase.
2. **Driving-Record factor:** this number is determined by the number of accidents or tickets we have had.
3. **Age factor:** this number is determined by our age at the time we purchase the insurance.

Coverage—Limit Factor Bodily-Injury					
Limit:	15/30	25/50	50/100	100/200	100/300
Factor:	1.00	1.15	1.31	1.44	1.49

Property-Damage						
Limit:	$5,000	$10,000	$25,000	$50,000	$100,000	$200,000
Factor:	1.00	1.05	1.08	1.13	1.18	1.27

Unit 2

An insurance company assigns points to us in terms of our driving record. The more points we have, the higher our driving-record factor and the more we will have to pay for insurance. A sampling of the reasons for which points are assigned are:

Driving while intoxicated
Reckless driving
Accident for which we are at fault

Driving-Record Factor (Three-Year Period)					
Number of Points:	0	1	2	3	4
Factor:	.00	.40	.90	1.50	2.20

Only part of the age-factor table is shown. In this table we can see that the factor is based not only on the age of the person but also on:

1. Whether the person is a male or female.
2. Whether the person has had a driver-training course in school.
3. Whether the person has received good grades in school—a "B" average or better.
4. Whether the person uses the car for pleasure only or also for driving to work.

	Age	Age Factor: Good Student Pleasure	Good Student Business	Not Good Student Pleasure	Not Good Student Business
Unmarried	17	1.50	1.65	1.75	1.90
Female	18	1.35	1.50	1.60	1.75
No Driver	19	1.25	1.40	1.50	1.65
Training	20	1.10	1.25	1.25	1.40
Unmarried	17	1.35	1.50	1.60	1.75
Female	18	1.25	1.40	1.50	1.65
With Driver	19	1.15	1.30	1.40	1.55
Training	20	1.05	1.20	1.20	1.35
Unmarried	17	2.10	2.25	2.75	2.90
Male*	18	1.95	2.10	2.55	2.70
No Driver	19	1.85	2.00	2.40	2.55
Training	20	1.70	1.85	2.25	2.40
Unmarried	17	1.80	1.95	2.30	2.45
Male*	18	1.65	1.80	2.15	2.30
With Driver	19	1.55	1.70	2.05	2.20
Training	20	1.40	1.55	1.95	2.10

*Not car owner or principal operator

Example 4 Find the age factor of a 19-year old unmarried male who has had no driver-training course and is not eligible for good student credit. The car is used for business.

Explanation

Not good student
Age **Pleasure** **Business**

Unmarried Male
No driver training 19 - - - - - - - - - - - - → 2.55

Solution

Age Factor = 2.55

Exercises

A

1. Determine the coverage limit factor on each insurance policy.

	Type of Insurance	Coverage Limit	Factor
a.	Bodily-Injury	25/50	______
b.	Bodily-Injury	100/300	______
c.	Bodily-Injury	15/30	______
d.	Property-Damage	$25,000	______
e.	Property-Damage	$100,000	______

Unit 2

2. Determine the semiannual basic cost for each coverage.

	Coverage	Territory	Basic Cost
a.	Bodily-Injury	05	$______
b.	Bodily-Injury	26	______
c.	Property-Damage	01	______
d.	Property-Damage	25	______
e.	Bodily-Injury	02	______

3. Determine the driving-record factor for each person.

	Number of Points	Factor
a.	2	______
b.	0	______
c.	4	______

4. Determine the age-factor for each person.

	Sex	Age	Good Student	Driver Training	Use	Factor
a.	Male	17	Yes	Yes	Pleasure	______
b.	Male	18	Yes	No	Business	______
c.	Female	17	Yes	No	Business	______
d.	Female	20	No	Yes	Pleasure	______
e.	Female	19	Yes	Yes	Business	______

B

1. Roger Baker was in an automobile accident in which his car damaged property to the extent of $11,500. Roger had $5,000 property-damage insurance coverage.

a. How much of the damage was paid for by the insurance company? ______

b. How much of the cost of the damage did Roger have to pay? ______

c. If Roger had been covered by a $10,000 property-damage policy, how much of the cost of the damage would he himself have had to pay? ________

d. If Roger had been covered by a $25,000 property-damage policy, how much of the cost of the damage would he himself have had to pay? ________

2. Dorothy Lomb injured three people in an automobile accident in which she was at fault. Her automobile insurance coverage included a 25/50 bodily-injury policy.

a. If only one of the injured persons claimed any damage and that claim was $14,000, how much of this amount would the insurance company pay? Assume that the courts upheld this claim. ________

b. If the claim of the injured person in Part (a) had been $30,000, how much of this would the insurance company pay? ________

c. If the three injured persons each claimed $10,000 and were granted this by court action, what is the total amount that the insurance company would have to pay? ________

d. If the three injured persons had been granted $20,000 each, what is the total amount that the insurance company would have to pay in the settlement? ________

e. If one of the three injured persons had been granted $35,000, while the remaining two were granted $5,000 each, what is the total amount that the insurance company would have to pay in settlement? ________

f. If one of the three injured persons had been granted $35,000, while the remaining two were granted $20,000 each, what is the total amount the insurance company would have to pay? ________

Part 2

As mentioned earlier, the basic cost of bodily-injury and property-damage insurance is not the actual cost that we are charged. To determine the actual cost we must first determine the total of the three factors.

$$\text{Total factor} = \text{Coverage-limit factor} + \text{Driving-record factor} + \text{Age factor}$$

The total factor is then multiplied by the basic cost. That number, when rounded to the nearest dollar, is the actual cost of our insurance.

Example 1 Emily Greenwood, age 20, who had no driver-training course had accumulated a driver record of 1 point. She is eligible for good student credit and uses a car for pleasure only. What total factor will be used in computing the cost of a 50/100 bodily-injury policy?

Solution

Total factor	=	Coverage-limit factor	+	Driver-record factor	+	Age factor
	=	1.31	+	.40	+	1.10
	=	2.81				

Example 2 Determine the semiannual premium (actual cost) of a $25,000 property-damage policy under the following conditions.

Male:
(1) Age 18
(2) Lives in territory 02
(3) Eligible for good student credit
(4) No driver training
(5) Driver-record of 2 points
(6) Drives car for pleasure

Solution

Total factor	=	Coverage-limit factor	+	Driving-record factor	+	Age factor
	=	1.08	+	.90	+	1.95
	=	3.93				

Semiannual Premium	=	Total factor × Basic cost
	=	3.93 × $46
	=	$180.78 or $181

Explanation Notice that the $180.78 was rounded to the nearest dollar. The cost is actually $181.

Exercises A

Do you recall how to do the computation in the following exercises? This computation is the same as you will be doing when finding the cost of insurance.

1. Find the sums.

a.	c.	e.
1.25 .60 1.13	.60 2.40 1.08	.90 2.90 1.13

b.	d.	f.
1.50 1.65 1.08	1.20 1.08 1.65	1.50 1.18 .85

2. Round each number to the nearest dollar.

a. $156.30	$______	d. $184.50	$______	g. $107.49	$______
b. $127.80	______	e. $104.27	______	h. $163.59	______
c. $139.60	______	f. $176.84	______	i. $99.72	______

B

1. Determine the total factor on the purchase of a bodily-injury policy for each of the following females who are eligible for good student credit.

	Coverage Limits	Points	Age	Driver Training	Use	Coverage Factor	Points Factor	Age Factor	Total Factor
a.	15/30	0	18	Yes	Pleasure	______	______	______	______
b.	25/50	1	19	No	Pleasure	______	______	______	______
c.	50/100	2	20	No	Business	______	______	______	______
d.	15/30	3	19	Yes	Pleasure	______	______	______	______
e.	100/300	4	17	Yes	Pleasure	______	______	______	______

2. Determine the total factor on the purchase of a property-damage policy for each of the following males who have taken a driver-training course.

	Coverage Limit	Points	Age	Good Student	Use	Coverage Factor	Points Factor	Age Factor	Total Factor
a.	$ 5,000	1	18	Yes	Pleasure	______	______	______	______
b.	25,000	0	20	Yes	Business	______	______	______	______
c.	10,000	2	19	No	Pleasure	______	______	______	______
d.	50,000	4	20	No	Business	______	______	______	______
e.	100,000	3	19	Yes	Pleasure	______	______	______	______

3. Determine the semiannual premium for each of the following bodily-injury policies. The total factor has been determined for you in each exercise.

	Total Factor	Territory	Basic Cost	Semiannual Premium
a.	3.65	02	$______	$______
b.	3.70	06	______	______
c.	4.26	25	______	______
d.	4.05	01	______	______
e.	5.18	26	______	______

Unit 2

C

1. Determine the semiannual premium on a property-damage policy for each of the following females who have not taken a driver-training course but are eligible for good student credit. Each of them uses the car for pleasure.

	Coverage Limit	Points	Age	Terri-tory	Basic Cost	Coverage Factor	Points Factor	Age Factor	Total Factor	Premium
a.	$ 5,000	1	19	06	$____	____	____	____	____	$____
b.	25,000	0	17	01	____	____	____	____	____	____
c.	10,000	2	18	25	____	____	____	____	____	____
d.	50,000	4	20	26	____	____	____	____	____	____
e.	100,000	0	17	02	____	____	____	____	____	____
f.	25,000	3	19	05	____	____	____	____	____	____

2. Determine the semiannual premium on a bodily-injury policy for each of the following males who have taken a driver-training course but are not eligible for good student credit. Each of them uses the car for business.

	Coverage Limits	Points	Age	Terri-tory	Basic Cost	Coverage Factor	Points Factor	Age Factor	Total Factor	Premium
a.	15/30	0	17	05	$____	____	____	____	____	$____
b.	50/100	3	20	01	____	____	____	____	____	____
c.	25/50	1	19	25	____	____	____	____	____	____
d.	100/300	2	18	02	____	____	____	____	____	____
e.	25/50	0	17	06	____	____	____	____	____	____
f.	50/100	4	20	26	____	____	____	____	____	____

D

1. Peter Ryan lives in an 02 territory and drives a car that has a 50/100 bodily-injury coverage. He is 17 years old, has no points against his record, and uses the car for pleasure only.

a. What will the semiannual premium be if Peter took a driver-training course and is eligible for good student credit? ________

b. What will the semiannual premium be if Peter did not take a driver-training course and is not eligible for good student credit? ________

c. How much can be saved on the semiannual premium if Peter did take the driver-training course and was eligible for good student credit? ________

2. Sarah Channing who is 19 drives a car that is insured for $25,000 property-damage. Sarah lives in an 01 territory, has had a driver-training course, is eligible for good student credit, and uses the car for business.

a. What will the semiannual premium be if Sarah has a record of 0 points? ________

b. What will the semiannual premium be if Sarah has a record of 4 points? ________

c. How much more would the semiannual premium be if Sarah has 4 points rather than 0 points? ________

Unit 2

Section 2 Collision and Comprehensive Insurance

Bodily-injury and property-damage insurance protects us in the event our car inflicts injury upon some other person or upon another person's property. The two policies we will examine in this section protect us in the event of damage to our own car.

1. **Collision Insurance** - covers damage to our car that was caused by collision
2. **Comprehensive Insurance** - covers damage to our car from any cause other than collision. (Fire, theft, hail, falling objects, etc.)

Collision coverage (insurance) and most comprehensive coverage is of a deductible variety. As an example, the basic collision coverage is the *$200-deductible policy*. This means that we must pay for the first $200 in the cost of the damages while the insurance company will pay the remainder. Similarly, the basic comprehensive coverage is the *$50-deductible policy*. In the event a rock destroys the windshield of our car and the replacement costs $175, then:

1. We must pay the first $50 of the cost.
2. The insurance company will pay the remaining $125.

$$\$175 - \$50 = \$125$$

Example 1 A car is insured under a $200-deductible collision policy. The damage resulting from an accident amounts to $795. In what manner will these damages be paid?

Solution

Amount owner pays = $200

Deduct owner's share

Amount insurance company pays = $795 − $200
= $595

Example 2 Under a $50-deductible comprehensive policy, how much will the insurance company pay for the replacement of a slashed tire worth $35?

Explanation Since $35 is less than $50, the owner must pay the entire $35.

Solution

Amount owner pays = $35
Amount insurance company pays = $0

The basic cost of collision and comprehensive insurance depends upon three elements:

1. The territory in which we live.
2. The price of the car when it was new.
3. The present age of the car.

Semiannual Basic Cost for
A: $200-deductible collision insurance
B: $50-deductible comprehensive insurance

Territory	Cost When New	Age 1 A	Age 1 B	Age 2,3 A	Age 2,3 B	Age 4,5 A	Age 4,5 B	Age 6 A	Age 6 B
	$2,751-$3,700	$128	$ 73	$109	$62	$90	$55	$ 70	$40
	$3,701-$5,000	141	88	120	74	99	66	78	48
02	$5,001-$6,500	166	120	142	102	116	91	92	66
	$6,501-$8,000	192	150	164	127	134	112	106	82
	$2,751-$3,700	118	40	100	34	83	30	65	22
	$3,701-$5,000	130	48	111	41	91	36	72	26
05	$5,001-$6,500	153	66	131	36	107	50	85	36
	$6,501-$8,000	177	82	151	70	124	62	98	45
	$2,751-$3,700	102	26	87	22	71	20	57	14
	$3,701-$5,000	113	31	96	27	79	23	62	17
25	$5,001-$6,500	124	43	106	36	87	32	69	24
	$6,501-$8,000	147	53	125	45	103	40	81	29

Example 3 Determine the semiannual basic cost for $200-deductible collision insurance on a 4-year old car that was worth $7,000 when new. The car is garaged in an 05 territory.

Explanation

Territory	**Cost when new**	**Age 4,5**
05	$6,501-$8,000	$124

Solution

Semiannual basic cost = $124

Collision insurance can be purchased for deductible amounts other than $200. The basic cost of a $250-deductible policy is only 90% of $200-deductible basic cost. As we can see in the table below, as the deductible amount increases, the basic cost decreases.

COLLISION INSURANCE

Deductible Amount	Percent of $200-deductible
$250	90%
500	65%
1,000	50%

Comprehensive insurance can also be purchased such that the deductible amount is $100 rather than $50. Here, too, the greater the deductible amount the less the cost.

COMPREHENSIVE INSURANCE

Deductible Amount	Percent of $50-deductible
$100	80%

Example 4 Determine the semiannual basic cost for $500-deductible collision insurance on a 3-year old car that was worth $3,900 when new. The car is garaged in a 25 territory.

Explanation Just a reminder on how a percent numeral is renamed as a decimal:

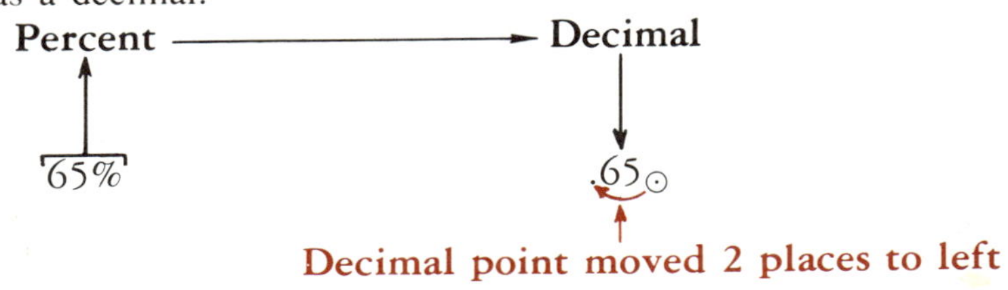

Solution

Basic cost for \$200-deductible = \$96
Basic cost for \$500-deductible = 65% of \$96
= .65 × \$96
= \$62.40 or \$62

The actual premium on both collision and comprehensive insurance is determined in the same manner as on property-damage and bodily-injury insurance.

Exercises

A

Do you recall how to do the computation in the following exercises? This computation is the same as you will be doing when finding the cost of insurance.

1. Rename each percent numeral as a decimal.

a.	25%	______	e.	65%	______
b.	16%	______	f.	50%	______
c.	90%	______	g.	6%	______
d.	80%	______	h.	8%	______

2. Find the products.

a.	50% × \$83	\$______	e.	80% × \$118	\$______
b.	50% × \$79	______	f.	80% × \$106	______
c.	90% × \$109	______	g.	65% × \$152	______
d.	90% × \$125	______	h.	65% × \$177	______

B

1. In each of the following exercises the car is covered by collision insurance. Determine the amount the owner must pay and the amount the insurance company must pay for the following accidents.

	Deductible Coverage	Damage	Owner's Share	Ins. Company's Share
a.	$200	$150	$______	$______
b.	200	500	______	______
c.	500	2,400	______	______
d.	1,000	800	______	______
e.	1,000	4,600	______	______
f.	250	700	______	______
g.	500	700	______	______

Unit 2

2. Determine the semiannual basic cost for the coverage in each of the following.

	Territory	Cost of Car When New	Age	Coverage	Basic Cost
a.	02	$4,000	1	$200-deduct. col.	$______
b.	05	3,000	4	$200-deduct. col.	______
c.	25	6,200	3	$50-deduct. comp.	______
d.	05	5,000	6	$50-deduct. comp.	______
e.	25	2,950	2	$200-deduct. col.	______
f.	02	6,500	5	$50-deduct. comp.	______

3. In each of these exercises the semiannual basic cost of a $200-deductible collision policy is shown. Determine the semiannual basic cost of the collision policy purchased.

	Basic Cost of $200-Deductible	Policy Purchased	Basic Cost of Policy Purchased
a.	$ 63	$500-deductible	$______
b.	74	$500-deductible	______
c.	127	$1,000-deductible	______
d.	143	$1,000-deductible	______
e.	119	$250-deductible	______
f.	102	$250-deductible	______

4. Determine the semiannual basic cost of each collision insurance policy.

	Territory	Cost of Car When New	Age	Basic Cost of $200-Deduct.	Deductible Amount	Basic Cost
a.	05	$3,400	3	$______	$1,000	$______
b.	25	4,100	4	______	1,000	______
c.	02	6,200	1	______	500	______
d.	05	7,600	2	______	500	______
e.	02	5,460	5	______	250	______
f.	25	6,900	3	______	250	______

5. The age factor and points factor are given in each exercise.

Determine the semiannual premium for each $50-deductible comprehensive policy.

	Territory	New-Car Cost	Age	Basic Cost	Age Factor	Point Factor	Total Factor	Premium
a.	02	$5,200	1	$____	1.20	.40	____	$____
b.	05	6,100	3	____	1.65	.00	____	____
c.	25	7,900	5	____	1.75	1.50	____	____
d.	25	3,800	2	____	1.90	.90	____	____
e.	02	4,690	4	____	2.25	2.20	____	____

C

1. A car that cost $6,375 when new is garaged in an 05 territory.

a. What is the semiannual basic cost of $50-deductible comprehensive insurance during the first year? ________

b. What is the semiannual basic cost of $50-deductible comprehensive insurance during the sixth year? ________

c. How much more is the semiannual basic cost during the first year than during the sixth year? ________

2. When computing the premium for a $200-deductible collision insurance policy the following was determined:

Age Factor: 1.80 **Points Factor: .30**
Present Age of Car: 3 **New-Car Value: $6,575**

a. What would the premium be if the car is garaged in an 02 territory? ________

b. What would the premium be if the car is garaged in a 25 territory? ________

c. How much less is the premium in a 25 territory than in an 02 territory? ________

Unit 3 Depreciation on a Car

Perhaps the greatest expense associated with owning a car relates to the depreciation (decrease) in the value of the car. As new models are manufactured each year, the value of older cars fall rapidly. No matter how little a car may be used, its value decreases year after year.

The rate at which a car depreciates depends not only on the make of that car but also on the particular model of the car. The table below is not exact for any particular make car but does show how an average car might depreciate. Each of the rates in this table shows the percent of decrease based on the car's value during the previous year.

Average Automobile Depreciation Rates					
Year	1st	2nd	3rd	4th	5th
Rate	35%	25%	20%	30%	35%

Unit 3

Example 1 Paul Elliot purchased a car for $5,840. Should he decide to trade in the car after one year, approximately how much will the car be worth?

Explanation To determine the value of the car after one year:

1. We must first determine the decrease in value during the year.
2. Then, we subtract the decrease in value from the original price to give us the car's value after one year.

Solution

Depreciation during 1st year = 35% of $5,840
= .35 × $5,840
= $2,044

Value after 1 year = $5,840 - $2,044
= $3,796

Example 2 Marie DeMarco paid $6,250 for her new car. What will the approximate trade-in value of the car be after three years?

Explanation The solution to this problem requires computing the depreciation for each of the first three years. To simplify the numbers, round each depreciation to the nearest dollar.

Solution

Year	Depreciation	Value
1	.35 × \$6,250 = \$2,188	\$6,250 − \$2,188 = \$4,062
2	.25 × \$4,062 = \$1,016	\$4,062 − \$1,016 = \$3,046
3	.20 × \$3,046 = \$609	\$3,046 − \$ 609 = \$2,437

Trade-in value after 3 years—↑

Exercises A

1. Determine the approximate depreciation for each car during the first year. Round each answer to the nearest dollar.

	New-Car Value	1st Year Depreciation		New-Car Value	1st Year Depreciation
a.	\$6,000	\$________	d.	\$5,780	\$________
b.	4,600	________	e.	6,390	________
c.	3,900	________	f.	7,985	________

2. Determine the approximate value for each car at the end of the first year. Round each depreciation to the nearest dollar.

	New-Car Value	1st Year Depreciation	Value After 1 Year
a.	\$7,000	\$________	\$________
b.	5,000	________	________
c.	4,200	________	________
d.	5,600	________	________
e.	6,700	________	________
f.	7,350	________	________
g.	8,940	________	________
h.	8,325	________	________

3. Determine the approximate value for each car at the end of the second year. Round each depreciation to the nearest dollar.

	New-Car Value	1st Year Depreciation	Value After 1st Year	2nd Year Depreciation	Value After 2nd Year
a.	\$8,000	\$________	\$________	\$________	\$________
b.	6,100	________	________	________	________
c.	5,400	________	________	________	________
d.	3,500	________	________	________	________
e.	4,800	________	________	________	________
f.	8,350	________	________	________	________

B

1. A new car is priced at $5,642. Determine its approximate value at the end of each of the first three years. Round each depreciation to the nearest dollar.

Year	1st	2nd	3rd
Depreciation	$______	$______	$______
Value	$______	$______	$______

2. Judy Bailey paid $4,954 for her car five years ago. What is the approximate trade-in value of the car at the present time? Round each depreciation to the nearest dollar.

Year	1st	2nd	3rd	4th	5th
Depreciation	$____	$____	$____	$____	$____
Value	____	____	____	____	____

Operating Costs Unit 4

The cost of the car and the cost of the insurance are but the beginning of the expenses that are involved in owning a car. Among others are:

1. Cost of gasoline
2. Cost of oil
3. Cost of repairs
4. Cost of tires

In this unit we will examine the costs of both gasoline and oil when operating a car.

Example 1 During a year Joe purchased 1,150 gallons of gasoline. The state tax on gasoline where he lives is 8.5¢. What was the total state gasoline tax that Joe paid that year?

Explanation Change cents to dollars:

8.5¢ ⟶ \$.08.5

Decimal point moved 2 places to left.

The tax on 1 gallon is \$.085
The tax on 2 gallons is 2 times \$.085
The tax on 1,150 gallons is 1,150 times \$.085

Solution

$$\begin{aligned}\text{Tax on 1,150 gallons} &= 1{,}150 \times 8.5¢ \\ &= 1{,}150 \times \$.085 \\ &= \$97.75\end{aligned}$$

Example 2 Kim drove her car 15,000 miles last year. Her car averages 22 miles to the gallon of gasoline. To the nearest gallon, how many gallons of gasoline did she buy?

Explanation

For 22 miles she used 1 gallon: $22 \div 22 = 1$
For 44 miles she used 2 gallons: $44 \div 22 = 2$
For 66 miles she used 3 gallons: $66 \div 22 = 3$

Therefore, for 15,000 miles the number of gallons used will be:

$$15{,}000 \div 22$$

Solution

$$\begin{aligned}\text{Number of gallons} &= 15{,}000 \div 22 \\ &= 681.8 \text{ or } 682\end{aligned}$$

Example 3 Carlos averages 850 miles to a quart of oil. Last year he drove his car 24,600 miles. If he purchased oil at 85¢ a quart, what was his total expense for oil last year?

Explanation The solution to this problem requires two steps:

1. First we must find the number of quarts used:

$$24{,}600 \div 850$$

2. The number of quarts must then be multiplied by the price of each quart to find the total cost.

Solution

Number of quarts used = 24,600 ÷ 850
= 28.9 or 29

Total cost = 29 × 85¢
= 29 × $.85
= $24.65

Unit 4

Exercises A

Do you recall how to do the computation in the following exercises? This computation is the same as you will be doing when finding the cost of operating a car.

1. Change each from cents to dollars.

a.	7.5¢	$ ______	e.	11¢	$ ______
b.	9.5¢	______	f.	10¢	______
c.	6.58¢	______	g.	8¢	______
d.	7.75¢	______	h.	7¢	______

2. Round each to the nearest whole number.

a.	69.2	______	d.	753.9	______
b.	84.7	______	e.	829.6	______
c.	561.5	______	f.	909.8	______

3. Round each amount to the nearest cent.

a.	$42.673	$ ______	d.	$8.695	$ ______
b.	$59.846	______	e.	$187.493	______
c.	$14.061	______	f.	$689.718	______

4. Find the products. (Round your answer to the nearest cent.)

a.	8 × $.625	$ ______	d.	23 × $.609	$ ______
b.	14 × $.639	______	e.	720 × $.689	______
c.	16 × $.589	______	f.	1,050 × $.679	______

5. Find the quotients.

a.	8,600 ÷ 20	________	e.	14,400 ÷ 600	________
b.	18,000 ÷ 30	________	f.	21,000 ÷ 750	________
c.	8,424 ÷ 24	________	g.	46,800 ÷ 2,600	________
d.	31,104 ÷ 36	________	h.	27,600 ÷ 1,200	________

6. Determine the quotients to the nearest whole number.

a.	2,500 ÷ 24	________	d.	16,210 ÷ 850	________
b.	950 ÷ 32	________	e.	21,940 ÷ 2,500	________
c.	3,450 ÷ 650	________	f.	36,500 ÷ 1,600	________

B

1. The Federal tax on gasoline is 4¢ per gallon. Determine the tax on these purchases made during one year.

	Number of Gallons	Federal Tax		Number of Gallons	Federal Tax
a.	1,100	$________	d.	1,480	$________
b.	950	________	e.	1,780	________
c.	1,240	________	f.	2,045	________

2. The following exercises show the number of gallons of gasoline purchased during the period of one year. Also shown is the State Tax per gallon. What is the total State and Federal Tax for each purchase?

	Number of Gallons	State Tax Per Gallon	State Tax	Federal Tax	Total Tax
a.	1,400	8¢	$________	$________	$________
b.	900	7¢	________	________	________
c.	1,800	10¢	________	________	________
d.	1,910	5¢	________	________	________
e.	1,500	9.5¢	________	________	________
f.	1,250	7.75¢	________	________	________
g.	2,100	6.58¢	________	________	________

3. To the nearest cent, what is the cost of each of the following gasoline purchases?

	Number of Gallons	Price Per Gallon	Exact Cost	Cost to the Nearest Cent
a.	8	63.9¢	$________	$________
b.	12	67.5¢	________	________
c.	75	64.6¢	________	________
d.	950	72.9¢	________	________
e.	1,625	68.7¢	________	________

4. Determine the number of gallons of gasoline used during each of the following trips.

	Miles Traveled	Miles Per Gallon	Number of Gallons
a.	3,000	15	______
b.	7,200	25	______
c.	4,420	26	______
d.	8,550	38	______

5. Determine the number of liters of oil used under each of the following conditions.

	Kilometers Traveled	Kilometers Per Liter	Number of Liters
a.	11,200	1,400	______
b.	51,300	2,700	______
c.	53,200	1,900	______
d.	54,150	950	______

6. Determine the cost of gasoline consumed during each trip.

	Miles Traveled	Miles Per Gallon	Number of Gallons Consumed	Cost Per Gallon	Total Cost
a.	6,000	12	______	70.9¢	$______
b.	4,480	16	______	65.8¢	______
c.	3,082	23	______	72.7¢	______
d.	6,324	34	______	68.9¢	______

7. Determine the cost of gasoline consumed during each trip.

	Kilometers Traveled	Kilometers Per Liter	Number of Liters	Cost Per Liter	Total Cost
a.	3,250	10	______	16.5¢	$______
b.	2,676	12	______	17.1¢	______
c.	7,875	9	______	16.8¢	______
d.	3,157	11	______	17.5¢	______
e.	5,096	14	______	18.3¢	______

C

1. The Gavins took a 12,600 mile trip around the country. On the average they spent 68.9¢ per gallon for gasoline.

a. If their car averaged 30 miles per gallon, how many gallons of gasoline were purchased? ______

b. What was the total cost of the gasoline? ______

c. Had the car averaged 21 miles per gallon, how many gallons would they have purchased? ______

d. What would the total cost have been at 21 miles per gallon? ________

e. How much was saved by the fact that the car averaged 30 miles per gallon rather than 21? ________

2. As part of his work Gene Bradley drives his car approximately 45,000 miles each year. He averages 25 miles per gallon of gasoline and 1,500 miles per quart of oil.

a. If the average price he spends for gasoline is 63.4¢ per gallon, what does the gasoline cost him for the year? ________

b. If the average price he spends for oil is 87¢ per quart, what does the oil cost him for the year? ________

c. What is his total expenditure for both gasoline and oil during a period of one year? ________

3. Jim Butler owns two cars that require a total of 36 liters of oil each year.

a. If he purchased the oil at a gasoline station he would have to pay $1.15 per liter. What would the total cost of oil be? ________

b. If he purchased the oil by the case at an auto store, he would pay 72¢ per liter. What would the total cost of oil be? ________

c. How much can he save each year by purchasing the oil by the case? ________

Unit 5 Traveling by Road Map

One of the reasons many of us purchase a car is as a means of transportation at the time of our vacation. During these trips we rely on road maps to help us locate the roads we will use to reach our destination. These maps also help us learn in advance both the quality of these roads and the distance we will travel.

The major highways in the country and the drawings used to designate these roads are shown on page 35.

National Interstate Highways

United States Highways

State Highways

Of the three, the interstate highways are usually the best.

The distance between points on the map is frequently indicated with two colors.

1. The black numbers name very short distances such as:

a. Between two towns

b. Between a town and an intersection of two roads

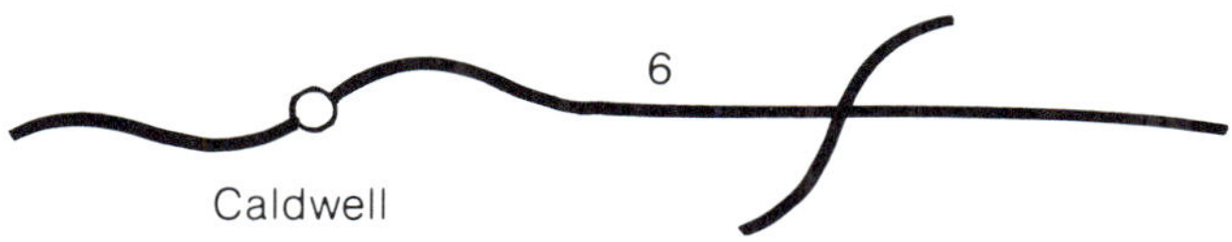

2. The red numbers name the distance between two red markers such as this ▲.

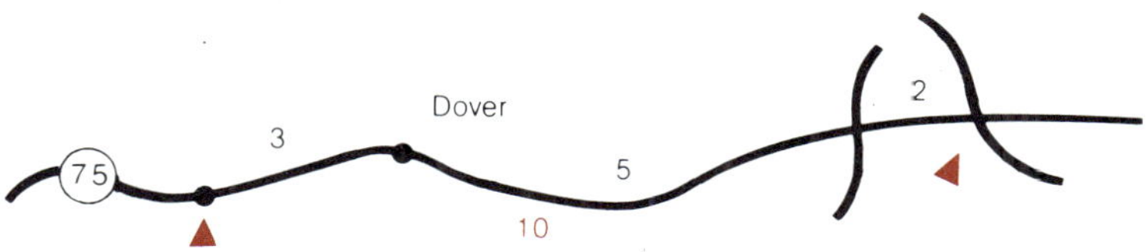

In the situation above we would simply read the red number, 10 miles, rather than add 3 + 5 + 2.

Example 1 Using U.S. 20, what is the distance from E. Avon to the point where U.S. 20 meets Alternate U.S. 20?

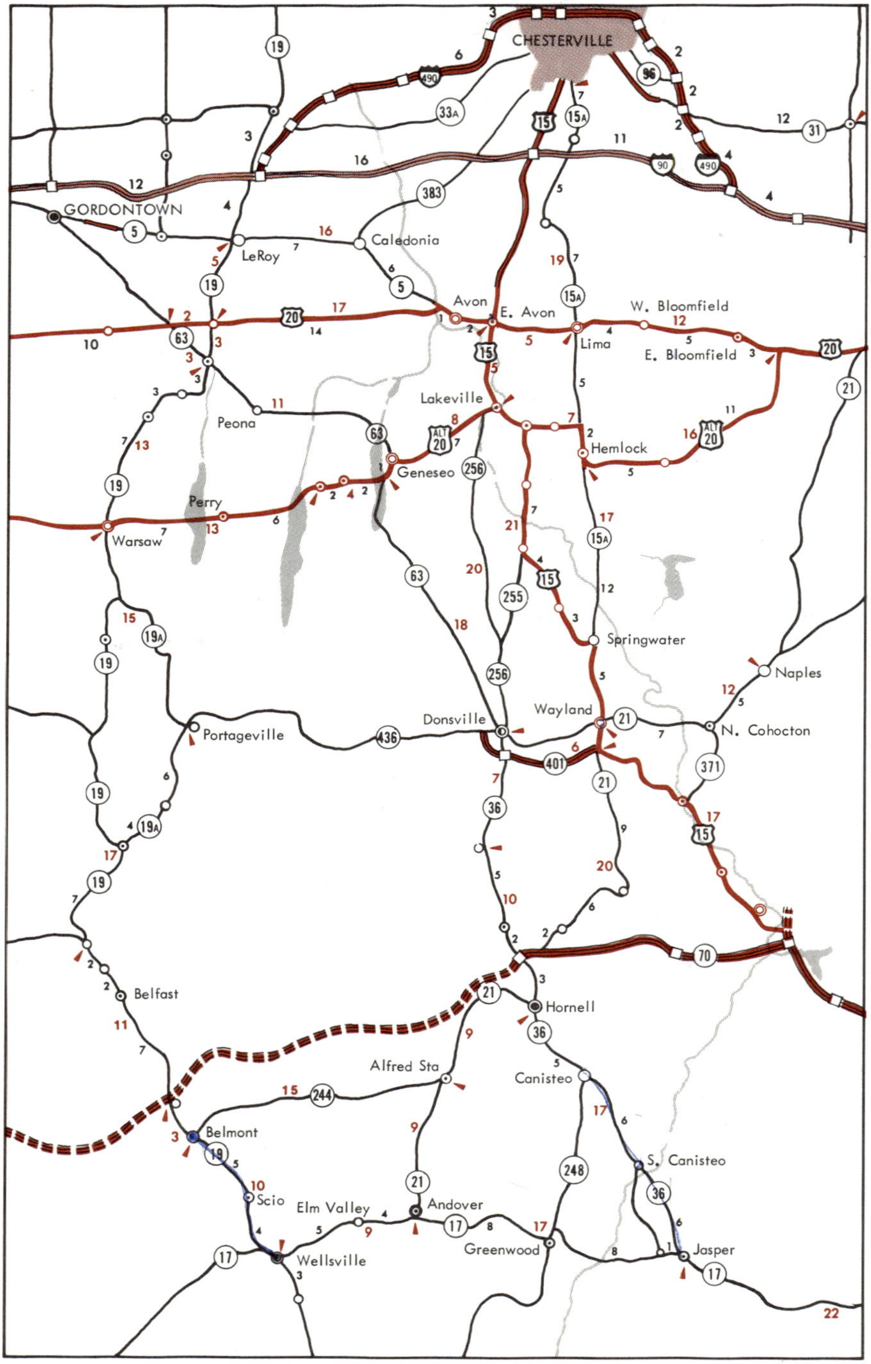
CHESTERVILLE
GORDONTOWN
LeRoy
Caledonia
Avon
E. Avon
W. Bloomfield
Lima
E. Bloomfield
Lakeville
Peona
Hemlock
Geneseo
Perry
Warsaw
Springwater
Naples
Wayland
Donsville
N. Cohocton
Portageville
Belfast
Hornell
Alfred Sta
Canisteo
Belmont
S. Canisteo
Scio
Elm Valley
Andover
Jasper
Greenwood
Wellsville

Solution

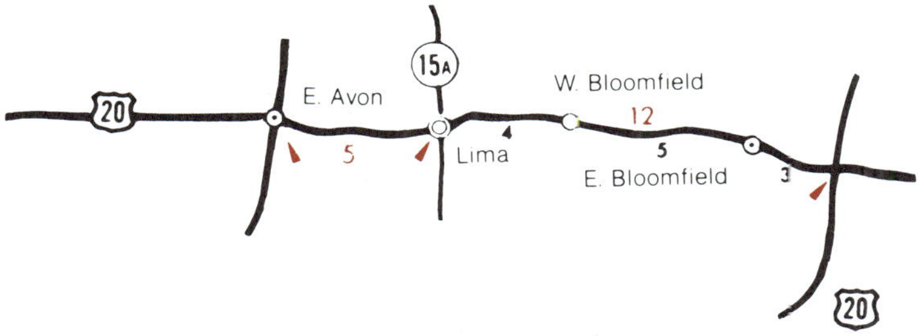

Distance from E. Avon to Lima = 5 miles
Distance from Lima to intersection = 12 miles
Total distance = 5 + 12
= 17 miles

One concern when traveling is to try to judge the distance we might cover during a certain period of time. If we plan to drive the car at an average speed of 50 mph (miles per hour), then in 1 hour we will travel 50 miles.

Distance traveled in 1 hour = 50 × 1 = 50 miles

Similarly, in 2 hours we will travel 2 times that distance. In 2 hours we will travel 100 miles.

Distance traveled in 2 hours = 50 × 2 = 100 miles

And in a period of 8 hours we will travel 8 times the distance traveled in 1 hour.

Distance traveled in 8 hours = 50 × 8 = 400 miles

This can be generalized as follows:

Distance traveled = 50 × 8
↓ ↓
Distance traveled = Average speed × Number of hours

This principle is shortened by using letters to replace these words.

Distance traveled is replaced by D
Average speed is replaced by R
Number of hours is replaced by T

Therefore, the principle becomes the formula:

$$D = R \times T$$

Example 2 On the first day of his trip Charles McCabe drove his car at an average speed of 52 mph for 7 1/2 hours. How far did he travel?

Solution

$$
\begin{aligned}
D &= R \times T \\
&\;\;\;\downarrow \quad\; \downarrow \\
&= 52 \times 7\tfrac{1}{2} \\
&= 52 \times 7.5 \\
\text{Distance traveled} &= 390 \text{ miles}
\end{aligned}
$$

Example 3 The driving time over several days was 27 hours. The average speed of the car during this trip was 48 mph. If the car averaged 26 miles per gallon of gasoline, to the nearest gallon, how many gallons were consumed?

Explanation This solution requires two steps:

1. We must first determine the distance traveled by using the distance formula,

$$D = R \times T$$

2. Then, using the information we learned in the previous section, we can determine the number of gallons used.

Solution

$$D = R \times T$$
$$= 48 \times 27$$
$$= 1{,}296 \text{ miles}$$
$$\text{Number of gallons used} = 1{,}296 \div 26$$
$$= 49.8 \text{ or } 50 \text{ gallons}$$

Exercises

A

Do you recall how to do the computation in the following exercises? This computation is the same as you will be doing when finding the distance a car travels.

1. Rename each mixed number as a decimal.

a. $6\frac{1}{2}$ ________
b. $10\frac{1}{2}$ ________
c. $18\frac{1}{2}$ ________
d. $29\frac{1}{2}$ ________

2. Find the products.

a. $18 \times 5\frac{1}{2}$ ________
b. $57 \times 8\frac{1}{2}$ ________
c. $62 \times 4\frac{1}{2}$ ________
d. $86 \times 21\frac{1}{2}$ ________

B

Use the map on page 36 when doing the following exercises.

1. Each of the highways listed below passes through Chesterville. Identify each as to whether it is an interstate highway, United States highway, or state highway.

a.	15	________	d.	96	________
b.	31	________	e.	490	________
c.	33A	________	f.	383	________

2. How far is it between the following towns?

a. From Belmont to Alfred Station along highway 244. ________

b. From Wellsville to Belmont along highway 19. ________

c. From Wayland to Naples along highway 21. ________

d. From East Avon to Lima along highway 20. ________

e. From Canisteo to South Canisteo along highway 36. ________

f. From LeRoy to Caledonia along highway 5. ________

g. From Geneseo to Lakeville along alternate highway 20. ________

3. How far is it between the following points?

a. From Hemlock to highway 20 along alternate highway 20. ________

b. From Warsaw to LeRoy along highway 19. ________

c. From Lima to Chesterville along highway 15A. ________

d. From Avon to East Bloomfield along highway 20. ________

e. From Belfast to Portageville along highways 19 and 19A. ________

4. Gloria Tremarco drove from her home in Wellsville to her friend's home in Alfred Station.

a. If she used routes 17 and 21, how far did she travel? ________

b. If she used routes 19 and 244, how far did she travel? ________

c. How many miles less would she travel if she uses routes 17 and 21? ________

C

1. Determine the number of miles traveled for each exercise.

	Miles Per Hour	Number of Hours	Miles		Miles Per Hour	Number Of Hours	Miles
a.	40	5	________	e.	42	$6\frac{1}{2}$	________
b.	50	3	________	f.	54	$9\frac{1}{2}$	________
c.	39	23	________	g.	37	$12\frac{1}{2}$	________
d.	48	18	________	h.	49	$17\frac{1}{2}$	________

2. The speedometers of the cars below registered the speed in kilometers per hour (kph). Determine the number of kilometers traveled for each exercise.

	KPH	Number of Hours	Kilo- meters		KPH	Number of Hours	Kilo- meters
a.	65	4	________	d.	86	$18\frac{1}{2}$	________
b.	72	15	________	e.	61	$7\frac{1}{2}$	________
c.	74	$5\frac{1}{2}$	________	f.	89	$26\frac{1}{2}$	________

3. To the nearest gallon, determine the number of gallons of gasoline consumed during each trip.

	Miles Per Hour	Number of Hours	Miles Traveled	Miles Per Gallon	Number of Gallons
a.	40	5	________	20	________
b.	52	12	________	18	________
c.	48	26	________	24	________
d.	53	$6\frac{1}{2}$	________	27	________
e.	54	$9\frac{1}{2}$	________	31	________
f.	51	$12\frac{1}{2}$	________	36	________

4. To the nearest liter, determine the number of liters of gasoline that was consumed during each trip.

	KPH	Number of Hours	Kilometers Traveled	Kilometers Per Liter	Number of Liters
a.	50	3	________	6	________
b.	70	4	________	7	________
c.	84	5	________	8	________
d.	78	16	________	9	________
e.	89	27	________	14	________

D

1. Last summer, Mr. and Mrs. Norris took a trip around the country. During a period of 8 weeks, they traveled 13,200 miles. They kept an accurate record of the gas they purchased and they found that their car averaged 23 miles to a gallon of gasoline.

a. How many gallons of gasoline did they buy on the trip? Find your answer to the nearest gallon. ________

b. If the average cost of the gasoline they purchased was 69.7¢ per gallon, what was the total cost of gasoline? ________

Unit 5

2. On the third day of his trip, Roland Green put in a full 9 1/2 hours of driving on throughways. He averaged 52 miles per hour.

a. How far did Roland travel that day? ________

b. If his car averages 28 miles to the gallon of gasoline, how many gallons did the car consume that day? Find your answer to the nearest gallon. ________

c. If gas cost him 73.9¢ per gallon, what was the total cost of gasoline for that day? ________

3. The Foleys purchased a trailer which they towed around behind their car during their 6-week summer vacation. They found that they traveled about 5 hours each day at the average rate of 35 miles per hour.

a. How far did they travel during the 6-week period? ________

b. With the trailer attached, the car averaged only 15 miles to the gallon of gasoline. How many gallons of gas did they purchase on the trip? ________

c. If the trailer were not attached, the car would have averaged 24 miles to the gallon of gasoline. How many gallons of gas would the Foleys have purchased if they did not have the trailer? Find your answer to the nearest gallon. ________

d. The average cost of gasoline was 69.3¢ per gallon. How much would the Foleys have saved on the cost of gasoline if they had not been towing a trailer? ________

Determining Average Speed Unit 6

Unit 6

Before setting out on a trip we frequently wonder how many hours the trip will take. To determine this, let's consider the following situations:

1. If we travel at a speed of 80 km per hour, then it will take 1 hour to travel 80 km:

80 km ÷ 80 kph = 1 hour

2. To travel 160 km at 80 kph, it will take 2 hours:

160 km ÷ 80 kph = 2 hours

3. Similarly, to find the time it will take to travel 720 km at 80 kph, we simply divide 720 by 80

720 km ÷ 80 kph = 9 hours

Using the information above we can derive the following principle for finding the number of hours of traveling time:

720 kilometers	÷	80 kph	=	9 hours
↓		↓		↓
Distance	÷	Rate	=	Time

And this can be expressed by a formula:

$$D \div R = T$$

Example 1 Carol plans to travel 375 miles to visit her parents. If she drives at an average speed of 45 miles per hour, to the nearest tenth of an hour how long will the trip take her?

Solution

$$T = D \div R$$
$$= 375 \div 45$$
$$= 8.33 \text{ or } 8.3 \text{ hours}$$

One thing that many of us are concerned about is how fast we can stop the car in the event the need arises. Experiments were conducted to determine the distance a car will travel from the moment a person realizes the need to brake the car until the moment the car comes to rest.

It was found that this distance is composed of two parts:

1. **Reaction distance -** this is the distance our car will travel from the instant we realize we must stop the car until the instant we place our foot on the brake pedal.
2. **Braking distance -** this is the distance our car will travel after we have applied pressure to the brake pedal.

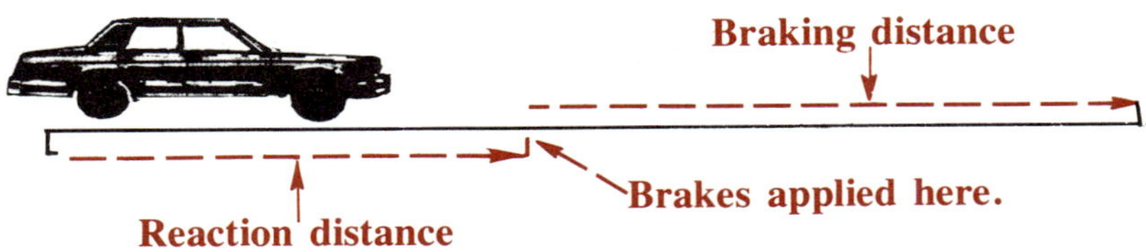

The studies have shown that by the time our foot reaches the brake pedal the car will travel approximately 1 foot for each mile per hour that the car is traveling.

Example 2 A car is traveling at a speed of 37 miles per hour. What is the reaction distance at this speed?

Solution

Speed = 37 mph
Reaction distance = 37 feet

The studies on braking have also shown that the braking distance can be found by using the following formula:

$$\text{Braking distance} = \left(\tfrac{1}{10} \times \text{Speed}\right)^2 \times 5$$

The small 2 that appears in the formula above means that the number is to be multiplied by itself. Thus,

8^2 means 8×8;
17^2 means 17×17.

Example 3 What is the braking distance of a car traveling 48 miles per hour?

Solution

$$\text{Braking distance} = (\tfrac{1}{10} \times 48)^2 \times 5$$
$$= (4.8)^2 \times 5$$
$$= 4.8 \times 4.8 \times 5$$
$$= 23.04 \times 5$$
$$= 115.2 \text{ or } 115 \text{ feet}$$

Example 4 A car is traveling at the rate of 52 miles per hour. Approximately how far will the car travel from the moment the driver realizes he must apply the brakes until the moment the car comes to rest?

Explanation The actual stopping distance is found by adding the reaction distance to the braking distance.

Stopping distance = Reaction distance + Braking distance

Solution

Reaction distance = 52 feet (1 ft per mile of speed)

$$\text{Braking distance} = (\tfrac{1}{10} \times 52)^2 \times 5$$
$$= (5.2)^2 \times 5$$
$$= 5.2 \times 5.2 \times 5$$
$$= 27.04 \times 5$$
$$= 135.2 \text{ or } 135 \text{ feet}$$

$$\text{Stopping distance} = 52 + 135$$
$$= 187 \text{ feet}$$

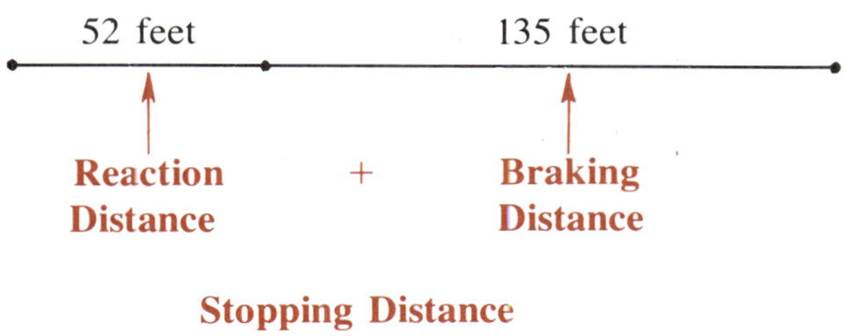

Exercises A

Do you recall how to do the computation in the following exercises? This computation is the same as you will be doing when finding the traveling time.

1. Round each number to the nearest tenth.

a.	4.72	______	d.	12.35	______
b.	6.87	______	e.	14.08	______
c.	3.59	______	f.	17.96	______

B

1. Determine the approximate reaction distance for each speed.

	Miles Per Hour	Reaction Distance		Miles Per Hour	Reaction Distance
a.	60	______ft	c.	35	______ft
b.	40	______ft	d.	72	______ft

2. Determine the approximate braking distance for each speed.

	Miles Per hour	Braking Distance		Miles Per Hour	Braking Distance
a.	40	______ft	e.	35	______ft
b.	50	______ft	f.	65	______ft
c.	80	______ft	g.	25	______ft
d.	20	______ft	h.	42	______ft

3. Determine the approximate stopping distance for each speed.

	Miles Per Hour	Reaction Distance	Braking Distance	Stopping Distance
a.	10	______ft	______ft	______ft
b.	30	______ft	______ft	______ft
c.	70	______ft	______ft	______ft
d.	25	______ft	______ft	______ft
e.	55	______ft	______ft	______ft
f.	85	______ft	______ft	______ft

C

1. To the nearest tenth of an hour determine the number of hours it will take to travel the distance shown.

	Miles Traveled	Speed of Car	Number of Hours
a.	200	40 mph	______
b.	322	46 mph	______
c.	325	53 mph	______
d.	530	42 mph	______
e.	1,400	51 mph	______

2. To the nearest tenth of an hour determine the number of hours it will take to travel the distance shown.

	Kilometers Traveled	Speed of Car	Number of Hours
a.	224	56 kph	________
b.	806	62 kph	________
c.	600	74 kph	________
d.	1,280	85 kph	________
e.	2,450	105 kph	________

D

1. In order to visit their daughter, the Vail's have to travel a distance of 548 miles. Because of traffic and road conditions they will probably average about 35 miles per hour. To the nearest tenth of an hour, how many hours of driving time should they plan for? ________

2. On a wet road, the reaction time of the driver is the same as on a dry road. However, the braking distance is twice as great. A car is traveling on a wet road at the rate of 52 miles per hour.

a. What is the reaction distance? ________

b. What is the braking distance? ________

c. What is the total distance a driver will need to stop a car that is traveling at 52 miles per hour on a wet road? ________

3. Joseph Gomez is traveling at the rate of 41 miles per hour.

a. What is the total distance he will need to stop the car if he is traveling on a dry road? ________

b. What is the total distance he will need to stop the car if he is traveling on a wet road? ________

c. How much more distance is needed if the road is wet than if it is dry? ________

4. How much more distance is needed to stop a car on a dry road if it is traveling at 60 miles per hour than if it is traveling at 30 miles per hour? ________

5. Phyllis Trent is driving her car on a clear day at the rate of 60 miles per hour. She rounds a bend in the road and spots an accident pileup 200 feet ahead. Assuming she has good brakes, will she be able to stop in time? ________________________________

____

Travel Costs Unit 7

In this unit we want to examine some of the travel costs other than those concerned with our car.

To relieve themselves of concern over theft during a trip many people will carry traveler's checks rather than actual cash. Should these checks be stolen, the company that issues them will cover the loss. There are one or two companies that issue traveler's checks at no charge whatsoever. The more widely-known companies have a fee of 10¢ for each $10 in the value of the check.

Example 1 Before Edwina Battle set out on her trip she purchased $850 worth of traveler's checks. What fee did she pay for these checks?

Explanation To determine the charge for these checks we must know the number of $10 amounts in $850. To understand how this is found, consider the following pattern:

In $20 there are 2 $10 amounts: $20 ÷ $10 = 2
In $30 there are 3 $10 amounts: $30 ÷ $10 = 3
In $40 there are 4 $10 amounts: $40 ÷ $10 = 4

Therefore, to find the number of $10 amounts in $850, we would divide $850 by $10.

Solution

Number of $10 amounts in $850 = 850 ÷ 10
= 85

Fee for $850 worth of traveler's checks = 85 × 10¢
= 85 × $.10
= $8.50

Rather than rely on traveler's checks many people prefer to charge their bills using credit cards. Two of the better known credit companies are Master Charge and VISA. Neither of these companies require a membership fee. However, if we do not pay the entire bill when it is due, an interest charge is added to the following month's bill.

Below is a copy of the form we sign at a motel, gasoline station, restaurant, or any other place where we charge our purchase. This form is sent to the credit company which eventually bills us.

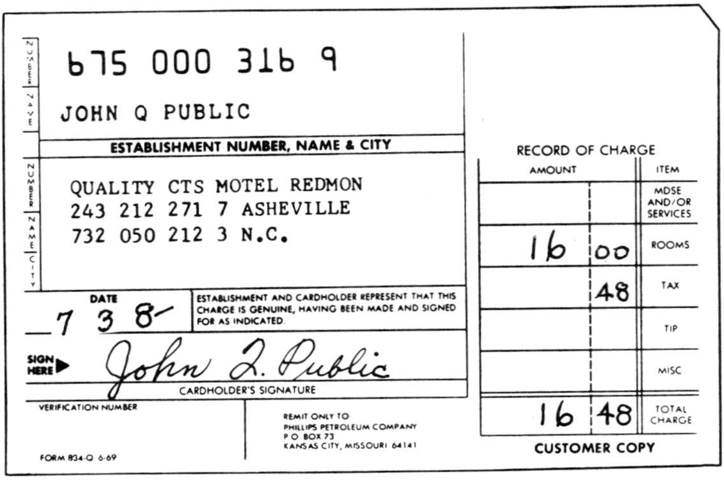
675 000 316 9
JOHN Q PUBLIC
ESTABLISHMENT NUMBER, NAME & CITY
QUALITY CTS MOTEL REDMON
243 212 271 7 ASHEVILLE
732 050 212 3 N.C.
DATE 7 3 8-
ESTABLISHMENT AND CARDHOLDER REPRESENT THAT THIS CHARGE IS GENUINE, HAVING BEEN MADE AND SIGNED FOR AS INDICATED.
SIGN HERE ▶ John Q. Public
CARDHOLDER'S SIGNATURE
VERIFICATION NUMBER
REMIT ONLY TO PHILLIPS PETROLEUM COMPANY P O BOX 73 KANSAS CITY, MISSOURI 64141
FORM 834-Q 6-69

RECORD OF CHARGE

AMOUNT		ITEM
		MDSE AND/OR SERVICES
16	00	ROOMS
	48	TAX
		TIP
		MISC
16	48	TOTAL CHARGE

CUSTOMER COPY

Notice that there is a line provided on this form for a Tip. Were this a restaurant, the amount we would very likely write on this line would be 15% of the cost. This percent will vary, however, depending upon how pleased we are with the service of the waiter.

Example 2 The cost of food at a restaurant amounted to \$23.40. To this was added a 6 1/2% state sales tax and a 15% tip for the waiter. What was the total cost of the meal?

Solution

$$\begin{aligned} \text{State sales tax} &= 6\tfrac{1}{2}\% \text{ of } \$23.40 \\ &= 6.5\% \text{ of } \$23.40 \\ &= .065 \times \$23.40 \\ &= \$1.521 \text{ or } \$1.52 \end{aligned}$$

$$\begin{aligned} \text{Tip} &= 15\% \text{ of } \$23.40 \\ &= .15 \times \$23.40 \\ &= \$3.51 \end{aligned}$$

$$\begin{aligned} \text{Total cost} &= \$23.40 + \$1.52 + \$3.51 \\ &= \$28.43 \end{aligned}$$

Exercises

A

Do you recall how to do the computation in the following exercises? This computation is the same as you will be doing when finding travel costs.

1. Rename each percent numeral as a decimal.

a.	$4\frac{1}{2}\%$	________	c.	$12\frac{1}{2}\%$	________
b.	$5\frac{1}{2}\%$	________	d.	$10\frac{1}{2}\%$	________

Unit 7

2. Find the products. Round each answer to the nearest cent.

a.	$5.80 × $5\frac{1}{2}\%$	$________	d.	$26.75 × $7\frac{1}{2}\%$	$________
b.	$9.50 × $6\frac{1}{2}\%$	________	e.	$48.55 × $6\frac{1}{2}\%$	________
c.	$18.20 × $4\frac{1}{2}\%$	________	f.	$134.96 × $8\frac{1}{2}\%$	________

B

1. Determine the cost of each traveler's check purchase.

	Total Check Purchase	Fee		Total Check Purchase	Fee
a.	$ 80	$________	d.	$ 580	$________
b.	700	________	e.	4,600	________
c.	350	________	f.	2,370	________

2. Determine the change a person will receive after paying each bill with a traveler's check.

	Amount of Bill	Size of Traveler's Check	Change
a.	$7.50	$10	$________
b.	6.95	10	________
c.	8.50	20	________
d.	12.45	20	________
e.	37.25	50	________
f.	29.30	100	________
g.	64.19	100	________

3. Determine the waiter's tip on each meal if 15% is used as the tipping rate.

	Cost of Food	Tip		Cost of Food	Tip
a.	$ 4.00	$________	e.	$14.60	$________
b.	6.00	________	f.	27.50	________
c.	3.50	________	g.	31.95	________
d.	5.70	________	h.	46.25	________

4. Determine the total restaurant bill including state tax and tip for each meal. A 15% tip was given in each situation.

	Cost of Food	State Sales Tax Rate	State Tax	Tip	Total Cost
a.	$5.00	6%	$______	$______	$______
b.	17.50	7%	______	______	______
c.	24.75	6%	______	______	______
d.	19.85	5%	______	______	______
e.	37.45	6%	______	______	______

5. Determine the state sales tax that had to be paid on each motel bill.

	Motel Bill	State Sales Tax Rate	Sales Tax
a.	$18.00	5%	$______
b.	26.00	7%	______
c.	22.50	8%	______
d.	34.00	$5\frac{1}{2}$%	______
e.	76.50	$6\frac{1}{2}$%	______
f.	134.75	$7\frac{1}{2}$%	______

6. Determine the total motel cost including tax.

	Cost Per Night	Number of Nights	Total Motel Charge	State Tax Rate	State Tax	Total Cost Including Tax
a.	$15.00	2	$______	5%	$______	$______
b.	17.00	3	______	6%	______	______
c.	18.50	4	______	6%	______	______
d.	21.50	3	______	$4\frac{1}{2}$%	______	______
e.	27.50	7	______	$6\frac{1}{2}$%	______	______
f.	34.50	6	______	$5\frac{1}{2}$%	______	______

C

1. Janice Evans purchased three $100 traveler's checks, eight $50 checks, ten $20 checks, and twenty $10 checks. What was the fee for these checks? ______

2. Vivian Spellman and her family spent ten days at a shore motel at a cost of $43.50 per day. The sales tax rate in that state is 6%. In traveling to the motel and returning home they purchased gasoline on four occasions at a cost of $10.25, $8.90, $7.65, and $11.40.

a. What was the total cost for gasoline? ______

b. Including tax, what was their total motel bill? ______

c. If both were charged against her credit card, for what amount will she be billed by the credit company? ________

3. Ronald McEvory used his car to travel from his home in Omaha, Nebraska to the home of a friend in San Francisco, California – a distance of 1,725 miles. He plans to drive for 8 hours each day at a rate of 50 miles per hour.

a. How many miles will he drive each day? ________

b. How many full days will he spend driving? ________

c. If he averages $21.50 per night as the cost for a motel room and $15.75 per day for the cost of food, what will be the total of these two expenses? In computing your answer, include the cost of food for the last partial day of traveling, but do not include the cost of a motel room for that night. ________

Review

Review of Chapter 1

When doing many of the exercises in this review, you will have to refer to the tables in this chapter.

1. Find the monthly payment for each car loan.

	Type of Loan	Unpaid Balance	Number of Months	Life Insurance	Monthly Payment
a.	New-Car	$4,000	18	Without	$________
b.	Used-Car	4,000	18	Without	________
c.	New-Car	5,600	24	With	________
d.	Used-Car	2,950	36	With	________

2. Determine the total repaid for each car loan.

	Type of Loan	Unpaid Balance	Number of Months	Life Insurance	Monthly Payment	Total Repaid
a.	New-Car	$5,000	30	Without	$________	$________
b.	Used-Car	740	18	Without	________	________
c.	New-Car	4,300	24	With	________	________

3. Determine the coverage factor for each insurance policy.

	Type of Insurance	Coverage Limit	Factor
a.	Bodily-Injury	100/200	________
b.	Property-Damage	$200,000	________

4. Determine the semiannual basic cost for each coverage.

	Type of Insurance	Territory	Basic Cost
a.	Bodily-Injury	05	$________
b.	Property-Damage	25	________

5. Determine the semiannual basic cost of a $500-deductible collision insurance policy under the following conditions.

Territory	Cost of Car When New	Age	Basic Cost of $200-Deduct.	Basic Cost of $500-Deduct.
25	$6,950	2	$________	$________

6. Determine the total factor on a 25/50 bodily-injury policy for the male in the exercise below. He is not eligible for good student credit.

Points	Age	Driver Training		Coverage Factor	Points Factor	Age Factor	Total Factor
2	19	No	Business	____	____	____	____

7. Tom Willard was at fault in an automobile accident. As a result of court action, the driver of the other car was granted $30,000 for the injuries he suffered. If Tom carried a 25/50 bodily-injury insurance policy, how much did the insurance company pay the injured party? ________

8. When Joe Fry applied his brakes on an icy road, his car slid into a tree. The extent of damage to the car came to $435. How much of the cost of repairing the car did the insurance company pay if Joe had purchased a $200-deductible collision insurance policy? ________

9. Reggie Moran purchased a car for $6,290. If this car depreciates 35% during the first year, what will the trade-in value of the car be at the end of the first year? ________

10. What is the cost of 10.6 gallons of gasoline at 72.9¢ per gallon? Find your answer to the nearest cent. ________

11.

a. How many gallons of gasoline were used on a trip of 12,500 miles if the car averaged 23 miles to the gallon of gasoline? Find your answer to the nearest gallon. ________

b. How much did the driver of this car have to pay for gasoline if the average price per gallon was 68.9¢? Find your answer to the nearest cent. ________

12. According to the road map on page 36, how far is it from Springwater to Lima along state highway 15A? ________

13. How far will a car travel in 8 hours if it is moving at an average speed of 90 kilometers per hour? ________

Review

14. During a trip, Cheryl Bell's total driving time was 54 hours. She found that she had averaged 50 miles per hour.

a. How many miles did Cheryl travel? ________

b. If the car averaged 28 miles to the gallon of gasoline, determine, to the nearest gallon, how many gallons were consumed. ________

c. If the average price of gasoline was 67.7¢ per gallon, what was the total cost of the gasoline consumed? ________

15. At an average speed of 72 kilometers per hour, how long will it take a car to travel 840 kilometers? ________

16. A car is traveling at the rate of 70 miles per hour.

a. What is the approximate reaction distance at this speed? ________

b. What is the approximate braking distance at this speed? ________

c. How far will the car travel before coming to a complete stop after the driver has seen the need to apply his brakes? ________

17. What is the cost of $350 worth of traveler's checks? ________

18. Jim Clemens and his wife stayed at a motel for a period of 5 nights at a cost of $28.50 per night.

a. What was the cost of lodging for the 5 nights? ________

b. If the state tax rate was 6%, how much did Jim Clemens have to pay? ________

c. What was the total bill for these 5 nights? ________

Ft. Wash-
ington Av-
Broadway-
Riverside Dr-
110 St-5 Av
M4
TO 34 St-Pennsylvania Sta
NEW YORK CITY
M
SURFACE
Nightcoach Ticket, $1
LEYLAND

Chapter 2

Commercial Transportation

Prior to now we have been concerned with the cost of transportation as it is related to a car of our own. In this chapter we plan to take a look at what is in store for us should we prefer to travel by

1. Rented Car
2. Bus
3. Train
4. Airplane

Renting a Car Unit 1

The cost of car rentals, even for the very same model car, varies widely from city to city. In fact, renting a car at an airport outside a city will cost more than renting the same model car from an agency of that company within the city.

However, car rentals within the same areas vary slightly with different agencies. If the basic charge at a particular company is smaller, we will probably find that other costs are higher.

The rate table on page 58 is typical of the manner in which car-rental rates are quoted.

CAR-RENTAL RATES

Type of Car	Daily Rates	Weekly Rates
Compact	$15.95 + 13¢/mile	$ 99.95 + 13¢/mile
Mid-size	18.95 + 14¢/mile	119.95 + 14¢/mile
Full-size	20.95 + 15¢/mile	129.95 + 15¢/mile
Luxury	23.95 + 17¢/mile	149.95 + 17¢/mile

Daily Rates: No mileage charge for first 100 miles per day
Weekly Rates: No mileage charge for first 700 miles per week
All rates include $300,000 bodily-injury and $25,000 property-damage insurance
$3.50 per day for $250-deductible collision insurance
Rental does not include gas

Example 1 Determine the basic charge for the rental of a mid-size car for a period of 4 days.

Solution

Number of days → 4; **Basic charge per day** → $18.95

Basic charge = 4 × $18.95
= $75.80

Example 2 A person rented a car for three days and drove 896 miles. For how many of these miles must he pay an additional fee?

Solution

Number of days → 3; **Free mileage per day** → 100

Mileage at no charge = 3 × 100
= 300 miles

Excess mileage = 896 − 300
= 596 miles

Example 3 A full-size car is rented for one day and driven 317 miles. Determine the excess mileage charge.

Solution

Free mileage for 1 day

$$\text{Excess mileage} = 317 - 100 = 217 \text{ miles}$$

Number of miles; Charge per mile

$$\begin{aligned}\text{Excess mileage charge} &= 217 \times 15¢ \\ &= 217 \times \$.15 \\ &= \$32.55\end{aligned}$$

Unit 1

Example 4 Fred Morel rented a full-size car for three days and drove 572 miles. How much rental did he pay for the three days?

Solution

$$\begin{aligned}\text{Basic charge for 3 days} &= 3 \times \$20.95 \\ &= \$62.85\end{aligned}$$

Free mileage for 3 days

$$\begin{aligned}\text{Excess mileage} &= 572 - 300 \\ &= 272 \text{ miles} \\ \text{Excess mileage charge} &= 272 \times 15¢ \\ &= 272 \times \$.15 \\ &= \$40.80\end{aligned}$$

$$\begin{aligned}\textbf{Total rental} &= \textbf{Basic charge} + \textbf{Excess mileage charge} \\ &= \$62.85 + \$40.80 \\ &= \$103.65\end{aligned}$$

Example 5 Joyce Thomson rented a compact car for a period of one week. During that time she drove 1,850 miles. The car averaged 28 miles to a gallon of gasoline and the average price paid for the gasoline was 64.3¢ per gallon. Including the cost of the gasoline, what was the total cost for the use of the car for the week?

Solution

$$\begin{aligned}\text{Basic charge for 1 week} &= 1 \times \$99.95 \\ &= \$99.95\end{aligned}$$

No charge for 700 miles

$$\begin{aligned}\text{Excess mileage} &= 1{,}850 - 700 \\ &= 1{,}150 \text{ miles} \\ \text{Mileage charge} &= 1{,}150 \times 13¢ \\ &= 1{,}150 \times \$.13 \\ &= \$149.50\end{aligned}$$

Number of miles **Miles per gallon**

$$\begin{aligned}\text{Gasoline used} &= 1{,}850 \div 28 \\ &= 66 \text{ gallons}\end{aligned}$$

Number of gallons **Cost per gallon**

$$\begin{aligned}\text{Gasoline cost} &= 66 \times 64.3¢ \\ &= 66 \times \$.643 \\ &= \$42.438 \text{ or } \$42.44\end{aligned}$$

$$\begin{aligned}\textbf{Total cost} &= \textbf{Basic charge} + \textbf{Mileage charge} + \textbf{Gasoline cost} \\ &= \$99.95 + \$149.50 + \$42.44 \\ &= \$291.89\end{aligned}$$

Exercises A

1. Determine the basic charge for each car rental.

	Type of Car	Base Rate Per Day	Number of Days	Basic Charge
a.	Mid-size	$_____	1	$_____
b.	Compact	_____	3	_____
c.	Luxury	_____	2	_____
d.	Full-size	_____	4	_____

2. Determine the excess mileage a person must pay if the car is rented on a daily basis and driven the following number of miles during the rental period.

	Miles Driven	Rental Period	Free Mileage	Excess Mileage
a.	135	1 day	_____	_____
b.	358	2 days	_____	_____
c.	275	4 days	_____	_____
d.	923	5 days	_____	_____

3. Determine the excess mileage charge on each of these one day rentals.

	Type of Car	Miles Driven	Excess Mileage	Cost Per Mile	Excess Mileage Charge
a.	Full-size	160	______	______¢	$______
b.	Compact	280	______	______¢	______
c.	Mid-size	59	______	______¢	______
d.	Luxury	325	______	______¢	______

4. Determine the excess mileage charge on each of these weekly rentals.

Unit 1

	Type of Car	Miles Driven	Number of Weeks	Free Mileage	Excess Mileage	Cost Per Mile	Excess Mileage Charge
a.	Mid-size	840	1	______	______	______¢	$______
b.	Luxury	1,960	2	______	______	______¢	______
c.	Compact	4,675	4	______	______	______¢	______

5. Determine the car rental.

	Type of Car	Basic Charge	Miles Driven	Excess Mileage	Cost Per Mile	Excess Mileage Charge	Car Rental
a.	Full-size	$______	49	______	______¢	$______	$______
b.	Mid-size	______	125	______	______¢	______	______
c.	Compact	______	264	______	______¢	______	______
d.	Luxury	______	358	______	______¢	______	______

6. Determine the car rental.

	Type of Car	Number of Days	Basic Charge	Miles Driven	Excess Mileage Charge	Car Rental
a.	Compact	3	$______	651	$______	$______
b.	Full-size	2	______	486	______	______
c.	Luxury	5	______	947	______	______

7. In the following situations a mid-size car was rented on a weekly basis. Including gasoline, what was the total cost for each rental?

	Number of Weeks	Basic Charge	Miles Driven	Excess Mileage Charge	Miles Per Gallon	Gasoline Cost/Gal.	Gasoline Cost	Total Cost
a.	2	$______	1,240	$______	20	72.8¢	$______	$______
b.	1	______	960	______	24	65.7¢	______	______
c.	3	______	3,200	______	25	64.9¢	______	______
d.	2	______	3,780	______	28	68.5¢	______	______

B

1. How much can be saved by renting a compact car for one week at the weekly rate rather than at the daily rate? ________

2. Sheila Hagel rented a car for 1 day and drove it 375 miles.

a. If she rented a compact car, what would the rental charge have been? ________

b. If she rented a luxury car, what would the rental charge have been? ________

c. How much less is the rental charge on a compact car than on a luxury car? ________

3. The car rental agency whose rates are shown on page 58 advertises a special in which a compact car can be rented for $129.60 for one week. Under this arrangement the car can be driven an unlimited number of miles at no charge.

a. If a compact car is rented at the normal weekly rate and driven 1,500 miles during the week, what will the cost be? ________

b. How much can be saved by renting the car under the advertised special? ________

Unit 2 Traveling By Bus

Section 1 Local Bus Transportation
Part 1

Public transportation involving either a bus, a train, or an airplane must operate according to published timetables. The table on page 63 is the weekly schedule of a bus line whose route is between Butler and Newark.

TO NEWARK								
BUTLER Bus Station	POMPTON LAKES Bus Station	MOUNTAIN VIEW Center	SINGAC Four Corners	LITTLE FALLS Main & Center Sts.	CEDAR GROVE Bowden Road	MONTCLAIR D.L. & W. Station	BLOOMFIELD Bloomfield Center	NEWARK P.R.R. Station
—	5.32	5.39	5.44	5.47	5.54	6.05	6.10	6.29
—	—	6.05	6.10	6.13	6.20	6.31	6.36	6.55
—	6.10	6.30	6.35	—	—	6.48	6.53	7.13
—	—	—	6.35	6.38	6.47	7.01	7.07	7.34
—	—	6.52	6.57	7.00	7.09	7.23	7.29	7.51
6.20	6.40	7.00	7.05	—	—	7.23	7.29	7.49
—	—	—	7.12	7.15	7.24	7.38	7.44	8.04
—	—	7.20	7.25	7.28	7.37	7.51	7.57	8.17
—	7.03	7.25	7.30	—	—	7.48	7.54	8.14
—	—	7.35	7.40	7.43	7.52	8.06	8.12	8.32
—	—	—	8.00	8.03	8.12	8.26	8.32	8.59
7.30	7.50	8.12	8.17	8.20	8.29	8.43	8.49	9.09
8.10	8.25	8.47	8.52	8.55	9.04	9.17	9.22	9.42
—	9.00	9.22	9.27	9.30	9.39	9.52	9.57	10.17
—	10.00	10.22	10.27	10.30	10.39	10.52	10.57	11.17
—	11.00	11.22	11.27	11.30	11.39	11.52	11.57	**12.17**
11.40	**12.00**	**12.22**	**12.27**	**12.30**	**12.39**	**12.52**	**12.57**	**1.17**
—	**1.00**	**1.22**	**1.27**	**1.30**	**1.39**	**1.52**	**1.57**	**2.17**
—	**1.55**	**2.17**	**2.22**	**2.25**	**2.34**	**2.47**	**2.52**	**3.12**
—	—	—	—	**2.54**	**3.03**	**3.16**	**3.22**	**3.49**
2.50	**3.10**	**3.32**	**3.37**	**3.40**	**3.49**	**4.02**	**4.08**	**4.35**
—	—	—	—	**4.10**	**4.19**	**4.32**	**4.38**	**5.05**
—	**4.20**	**4.42**	**4.47**	**4.50**	**4.59**	**5.12**	**5.18**	**5.45**
—	—	—	—	**5.00**	**5.09**	**5.22**	—	—
4.30	**5.15**	**5.35**	**5.40**	**5.43**	**5.49**	**6.00**	**6.04**	**6.23**
—	**6.45**	**7.05**	**7.10**	**7.13**	**7.19**	**7.30**	**7.34**	**7.53**
8.13	**8.25**	**8.45**	**8.50**	**8.53**	**8.59**	**9.10**	**9.14**	**9.33**
—	—	—	—	12.00	12.06	12.17	—	—

The numbers in light-face type,

8.30 ← Light-face type

show the time of day from 12.00 midnight until 12.00 noon. The numbers in bold-face type,

4.47 ← Bold-face type

show the time of day from 12.00 noon until 12.00 midnight.

The decimal point in each of the numbers in this table is used to separate the hours from the minutes. Thus,

7.25 means 7 hours 25 minutes.

However, we usually read this as 25 minutes past 7.
This same hour can also be written as 7:25 or 7 25.

Example 1 If a person leaves Pompton Lakes at 8.25 AM, at what time will he arrive in Bloomfield?

Explanation

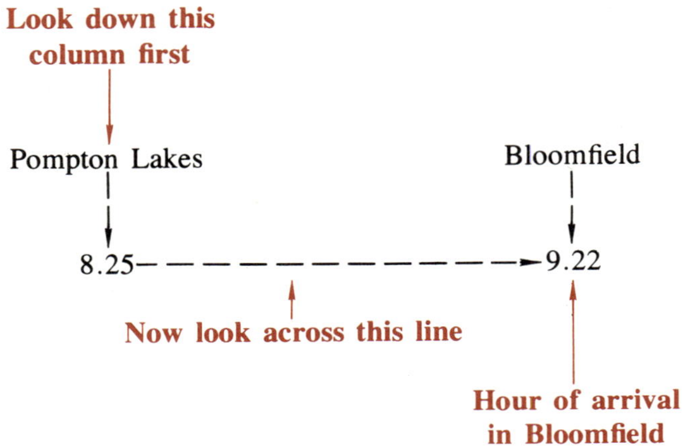

Solution

The person arrives in Bloomfield at 9.22 AM.

Example 2 At what time must a person leave Butler if he wants to arrive in Cedar Grove by 9.04 AM?

Explanation

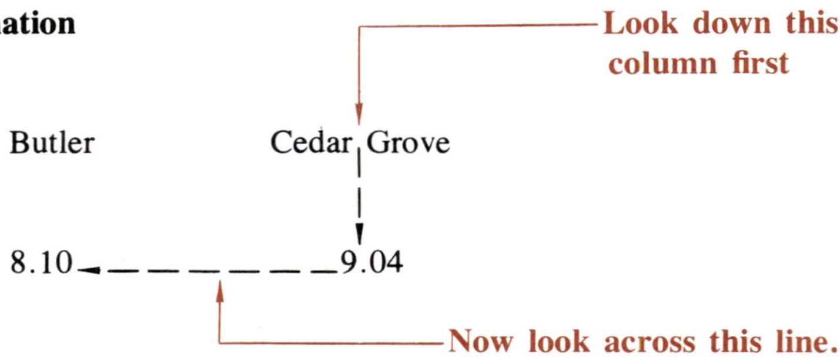

Solution

The person must leave Butler at 8.10 AM.

Exercises

A

1. Each bus leaves Butler at the time indicated. At what time will it arrive at its destination?

	Departure Time	Destination	Arrival Time
a.	6.20 AM	Singac	______
b.	7.30 AM	Montclair	______
c.	8.10 AM	Bloomfield	______
d.	11.40 AM	Newark	______
e.	2.50 PM	Mountain View	______
f.	4.30 PM	Little Falls	______

Unit 2

2. Each bus leaves the bus stop shown at the time indicated. At what time will it reach each destination?

	Bus Stop	Departure Time	Destination	Arrival Time
a.	Pompton Lakes	9.00 AM	Montclair	______
b.	Singac	9.27 AM	Bloomfield	______
c.	Cedar Grove	3.49 PM	Newark	______
d.	Little Falls	5.43 PM	Cedar Grove	______

3. At what time must a person leave each bus stop in order to arrive at the destination at the time indicated?

	Bus Stop	Destination	Arrival Time at Destination	Departure Time
a.	Butler	Bloomfield	12.57 PM	______
b.	Singac	Montclair	5.12 PM	______
c.	Cedar Grove	Newark	8.32 AM	______
d.	Pompton Lakes	Little Falls	8.55 AM	______
e.	Mountain View	Cedar Grove	11.39 AM	______

B

1.

a. When does the first bus leave Butler for Newark each weekday? ______

b. When does the last bus leave Mountain View for Newark each weekday? ______

c. If we miss the 6.40 AM bus out of Pompton Lakes, until when must we wait for the next bus? ______

d. How many bus trips out of Butler to Newark are there each day? ______

2. Paul Phillips lives in Butler but works in Montclair. He must be at his place of work at 9.00 AM. There is a 10-minute walk after leaving the bus. Can he take the 7.30 bus out of Butler or must he take the 6.20 bus? ________

____

Part 2

Another thing we can determine by examining a timetable is travel time. As an example, consider the situation in which we board the bus at 7.35 AM and leave it at 8.52 AM. The time spent traveling can be found by subtracting minutes from minutes and hours from hours.

Subtract hours → | ← **Subtract minutes**

	hours	minutes
	8 hours	52 minutes
−	7 hours	35 minutes
	1 hour	17 minutes

Example 1 How long will it take the 8.25 AM bus out of Pompton Lakes to arrive in Newark?

Solution

Time of leaving Pompton Lakes: 8.25 AM
Time of arrival in Newark: 9.42 AM
Travel time:

	hours	minutes
	9 hours	42 minutes
−	8 hours	25 minutes
	1 hour	17 minutes

There are times when it is not possible to find the difference between the minutes. This occurs when trying to determine the travel time from 6.42 AM to 9.27 AM.

	hours	minutes
	9 hours	27 minutes
−	6 hours	42 minutes

42 minutes cannot be subtracted from 27 minutes.

In order to complete the computation it is necessary to rename 1 hour as 60 minutes. For the situation above the 9 hours 27 minutes is renamed as follows:

9 hrs 27 min = 8 hrs 1 hr 27 min = 8 hrs 60 min 27 min = 8 hrs 87 min

We can now determine the travel time from 6.42 to 9.27.

	8 hours 87 minutes
−	6 hours 42 minutes
	2 hours 45 minutes

Example 2 How long will it take the 5.35 PM bus out of Mountain View to arrive in Newark?

Solution

Time of leaving Mountain View: 5.35 PM
Time of arrival in Newark: 6.23 PM
Travel time: 6 hours 23 minutes = 5 hours 83 minutes
5 hours 35 minutes = 5 hours 35 minutes
48 minutes

There is one more difficulty relating to travel time that requires investigation. This difficulty arises when determining the travel time of a person who boards the bus in the morning and leaves it in the afternoon.

Boarding time: 11:57 AM
Leaving time: 2:31 PM

One of the ways to overcome this problem is to think of a clock that is marked off into 24 hours rather than into 12 hours.

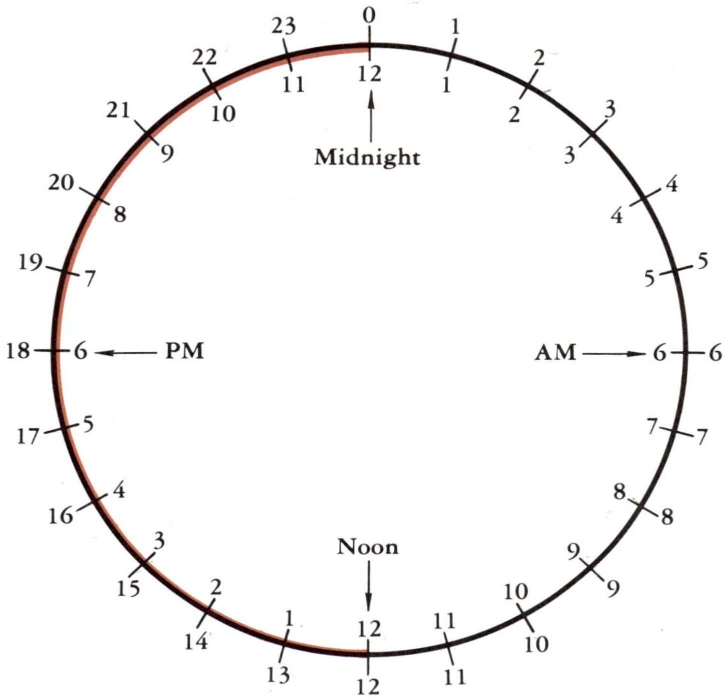

By examining the diagram above we can see that the morning hours or AM hours are identically the same for the 12-hour clock as they are for the 24-hour clock. The PM hours, however, differ:

1:00 PM	becomes	13 hours
2:00 PM	becomes	14 hours
3:00 PM	becomes	15 hours

and similarly for other PM hours.

A close examination of these numbers reveals that those PM hours starting at 1:00 PM on the 12-hour clock can easily be changed to hours on the 24-hour clock. This is done simply by adding 12 to the PM hours.

12-hour Clock Time		**24-hour Clock** Time
5:00 PM	⟶	5 + 12 or 17 hours
9:00 PM	⟶	9 + 12 or 21 hours
11:00 PM	⟶	11 + 12 or 23 hours

Example 3 Rename 6:23 PM by using the 24-hour clock.

Solution

	6:00 PM →	6 + 12 = 18 hours
Therefore,	6:23 PM →	18:23 hours

Example 4 A person boards a bus at 11:57 AM and leaves it at 2:31 PM. What was the travel time?

Solution

12-hour Clock	24-hour Clock		
2:31 PM →	14 hours 31 minutes	=	13 hours 91 minutes
− 11:57 AM →	11 hours 57 minutes	=	−11 hours 57 minutes
			2 hours 34 minutes

Travel time = 2 hours 34 minutes

Example 5 How long will it take the 12:22 AM bus out of Mountain View to reach its station in Newark?

Solution

Time of leaving Mountain View:	12:22 PM
Time of arrival in Newark:	1:17 PM
Travel time:	

12-hour Clock	24-hour Clock		
1:17 PM →	13 hours 17 minutes	=	12 hours 77 minutes
− 12:22 PM →	12 hours 22 minutes	=	− 12 hours 22 minutes
			55 minutes

Travel time = 55 minutes

Exercises A

1. The time below is named in terms of the 12-hour clock. Rename each in terms of the 24-hour clock.

	12-hour clock	24-hour clock		12-hour clock	24-hour clock
a.	2:00 AM	________	b.	5:00 AM	________
c.	3:30 AM	________	d.	9:45 AM	________
e.	10:17 AM	________	f.	11:23 AM	________
g.	4:00 PM	________	h.	3:00 PM	________

Unit 2

	12-hour clock	24-hour clock		12-hour clock	24-hour clock
i.	1:20 PM	________	j.	5:19 PM	________
k.	11:34 PM	________	l.	6:42 PM	________
m.	9:56 PM	________	n.	12:35 PM	________

2. Rename each in terms of the number of hours and minutes indicated. The first one is completed for you.

a.	5 hr 20 min	=	4 hr	80	min
b.	7 hr 20 min	=	6 hr	________	min
c.	6 hr 0 min	=	5 hr	________	min
d.	10 hr 15 min	=	9 hr	________	min
e.	17 hr 0 min	=	16 hr	________	min
f.	15 hr 36 min	=	14 hr	________	min
g.	22 hr 7 min	=	21 hr	________	min
h.	19 hr 42 min	=	18 hr	________	min

3. Find the differences.

a. 7 hr 25 min − 4 hr 12 min

f. 15 hr − 13 hr 35 min

b. 5 hr 47 min − 2 hr 24 min

g. 17 hr 10 min − 10 hr 50 min

c. 18 hr 31 min − 12 hr 19 min

h. 14 hr 5 min − 12 hr 20 min

d. 21 hr 55 min − 9 hr 38 min

i. 23 hr 21 min − 20 hr 25 min

e. 8 hr − 5 hr 20 min

j. 14 hr 7 min − 12 hr 53 min

4. Find the travel time for each.

	From	To	Travel Time		From	To	Travel Time
a.	3:12 PM	5:18 PM	______	h.	10:00 AM	2:00 PM	______
b.	2:15 AM	7:25 AM	______	i.	2:00 AM	5:00 PM	______
c.	9:34 AM	11:48 AM	______	j.	9:20 AM	1:30 PM	______
d.	5:07 PM	11:32 PM	______	k.	8:25 AM	3:20 PM	______
e.	5:40 PM	6:10 PM	______	l.	7:45 AM	2:18 PM	______
f.	9:30 PM	11:20 PM	______	m.	5:37 AM	7:10 PM	______
g.	2:53 AM	9:05 AM	______	n.	6:52 AM	9:07 PM	______

B

For this set of exercises you will have to refer to the bus schedule on page 63.

1. Determine the travel time for each trip out of Pompton Lakes at the hour shown.

	Departure Time	Destination	Arrival Time	Travel Time
a.	9.00 AM	Bloomfield	______	______
b.	10.00 AM	Newark	______	______
c.	7.03 AM	Montclair	______	______
d.	7.50 AM	Cedar Grove	______	______
e.	3.10 PM	Montclair	______	______
f.	12.00 PM	Newark	______	______

Unit 2

2. Determine the travel time from each stop.

	Bus Stop	Departure Time	Destination	Arrival Time	Travel Time
a.	Singac	8.17 AM	Bloomfield	______	______
b.	Little Falls	1.30 PM	Montclair	______	______
c.	Mountain View	7.35 AM	Newark	______	______
d.	Butler	2.50 PM	Bloomfield	______	______
e.	Butler	11.40 AM	Newark	______	______

3.

a. Determine the travel time for the 6.20 AM bus out of Butler to Newark. ______

b. Determine the travel time for the 4.30 PM bus out of Butler to Newark. ______

c. Why should the afternoon trip take so much longer than the morning trip? ______

Section 2 Interstate Bus Transportation

The timetables of interstate buses are often arranged vertically. For the route shown in the schedule on page 72, the bus departs from the Port Authority Terminal in New York. Its first stop is Jersey City, then Newark, then Bellmawr and on to the other cities named in the first column.

Each trip out of New York will not always stop at every point

NEW YORK—BALTIMORE—WASHINGTON

Trip Numbers		7051	9	2029	117	2033	7121	173	2057	2037	2053	115	2039
New York, NY Port Authority Term	Lv	12 01	2 00	3 30	5 30	7 45	8 30	9 30	10 30	10 45	12 30	1 30	1 45
Jersey City, NJ	Lv	↓				8 10		↓		11 10		↓	↓
Newark, NJ	Lv	12 35				8 30		10 05		11 30		2 05	2 20
Bellmawr, NJ						10 00							↓
Wilmington, DE	Lv	↓		↓		10 50				1 45			4 35
State Road, DE	Ar	2 45	NON-STOP	5 50	NON-STOP	11 05	NON-STOP		NON-STOP	2 00	NON-STOP	NON-STOP	4 50
State Road, DE	Lv	2 45		6 05		11 15				2 10			5 00
Elkton Jct., MD				6 25		11 35				2 30			5 20
Northeast Jct., MD		↓		6 30		11 40		↓		2 35			5 25
Perryville, MD		3 20		6 40		11 50		12 50		2 45			5 35
Aberdeen, MD		3 30		6 50		12 01		1 00		2 55			5 45
Edgewood Road, MD		↓	↓	7 00	↓	12 10	↓	↓	↓	3 05	↓	↓	5 55
Baltimore, MD	Ar	4 15	5 30	7 45	9 05	12 50	11 59	1 45	2 00	3 55	4 00	5 30	6 40
Baltimore, MD	Lv	4 20	5 35	7 50	9 10	12 55		1 50		4 00			6 45
Washington, DC	Ar	5 15	6 30	8 45	10 05	1 50		2 45		5 00			7 40

along its route. As an example, the first stop that is made by Trip Number 9 after leaving New York is Baltimore, Maryland.

Trip Numbers	**9**
New York, N.Y.	2 00
↓	↓ Non-Stop
Baltimore, Md	5 30

As is usual with timetables, light face type indicates AM time and bold face type indicates PM time.

Therefore,

9 45	means	9:45 AM,
5 27	means	5:27 PM.

Travel time is computed as we learned earlier.

Example 1 Determine the travel time from New York to Washington, DC on trip number 2037.

Solution

Departure time: 10:45 AM
Arrival time: 5:00 PM

12-hour clock		24-hour clock		
5:00 PM	→	17 hours 0 minutes	=	16 hours 60 minutes
− 10:45 AM	→	10 hours 45 minutes	=	− 10 hours 45 minutes
				6 hours 15 minutes

Travel time = 6 hours 15 minutes

Airlines publish small booklets showing the costs of their flights between various points. Railways do much the same. However, since interstate buses stop at so many different places along their route it would be very difficult to set up a price schedule showing each separate charge.

The fares, though, can easily be determined by phoning any

Unit 2

ticket agency of the company. A few sample fares are shown below.

One-Way Bus Fares Between New York and Selected Cities

Albany	$12.70	Harrisburg	$16.40
Annapolis	24.55	Los Angeles	172.30
Atlanta	63.05	Philadelphia	9.40
Baton Rouge	90.45	Tampa	78.30
Boston	20.40	Washington D.C.	20.95
Columbus	50.05	Williamsburg	31.25

In order to encourage people to use the same bus line both traveling from their home and returning to it, bus companies offer a 5% discount on round-trip fares.

Example 2 Determine the round-trip fare from New York to Annapolis.

Solution

One-way fare to Annapolis = \$24.55

Cost going → first \$24.55; Cost returning → second \$24.55

Two-way fare = \$24.55 + \$24.55
= \$49.10

Discount = 5% of Two-Way Fare
= .05 × \$49.10
= \$2.455 or \$2.46

Round-trip fare = Two-way Fare − Discount
= \$49.10 − \$2.46
= \$46.64

In competing with airlines and also with each other, interstate bus lines have been offering travelers some greatly reduced rates. As an example, the company whose sample fares are given in the earlier tables advertises the following travel bargains.

Bargain Fares

If the normal one-way fare is:	You pay: One-way	Round-trip
\$49 through \$58.95	\$49	\$98
\$59 through \$74.95	\$59	\$118
\$75 and above	\$75	\$150

To qualify for these bargain fares, two conditions must be met:

1. The trip must be to a point out of state.
2. The round trip must be completed within 30 days.

Unit 2

Example 3 The normal cost of a one-way fare from New York to Las Vegas is \$161.75. How much can a person save on a round trip by using the bargain fares?

Solution

One-way fare = \$161.75
Two-way fare = \$161.75 + \$161.75
= \$323.50
Discount = 5% of Two-way fare
= .05 × \$323.50
= \$16.175 or \$16.18
Round-trip fare = Two-way fare − discount
= \$323.50 − \$16.18
= \$307.32
Saving on bargain fare = Round-trip fare − bargain fare
= \$307.32 − \$150
= \$157.32

Exercises A

Do you recall how to do the computation in the following exercises? This computation is the same as you will be doing when finding interstate bus fares.

1. Rename each percent numeral as a decimal.

a. 2% __________ **c.** 3% __________
b. 7% __________ **d.** 5% __________

2. Determine the product to the nearest cent.

a.	5% × $62.00	$________	d.	5% × $176.60	$________
b.	5% × $25.40	________	e.	5% × $159.90	________
c.	5% × $40.80	________	f.	5% × $187.90	________

B

1. At what time will each trip arrive in Baltimore?

a.	2029	________	c.	173	________
b.	115	________	d.	2039	________

2. Determine the time spent on each of the following trips.

	Trip Number	Bus Stop	Destination	Departure Time	Arrival Time	Travel Time
a.	7051	Baltimore	Washington D.C.	______	______	______
b.	2037	Wilmington	Baltimore	______	______	______
c.	117	New York	Washington D.C.	______	______	______
d.	2033	Jersey City	Aberdeen	______	______	______
e.	2033	Wilmington	Washington D.C.	______	______	______
f.	2057	New York	Baltimore	______	______	______

C

1. Determine the one-way fare from New York to each of the following destinations.

a. Atlanta ________ b. Harrisburg ________

2. Determine the round-trip fare from New York to each of the following destinations.

Destination	One-way Fare	Two-way Fare	Discount	Round-trip Fare
a. Boston	$________	$________	$________	$________
b. Philadelphia	________	________	________	________
c. Albany	________	________	________	________
d. Harrisburg	________	________	________	________
e. Washington D.C.	________	________	________	________

3. Determine the bargain fare saving over the one-way bus fare from New York to each destination.

	Destination	One-way Normal Fare	One-way Bargain Fare	Saving
a.	Atlanta	$________	$________	$________
b.	Tampa	________	________	________
c.	Williamsburg	________	________	________
d.	Los Angeles	________	________	________

4. Determine the bargain fare saving over the round trip bus fare from New York to each destination.

Destination	One-way Fare	Two-way Fare	Discount	Round-trip Fare	Round-trip Bargain Fare	Saving
a. Baton Rouge	$_____	$_____	$_____	$_____	$_____	$_______
b. Atlanta	_____	_____	_____	_____	_____	_______
c. Columbus	_____	_____	_____	_____	_____	_______
d. Tampa	_____	_____	_____	_____	_____	_______

D

1. James Nazario is planning to take the bus from Newark, N.J., to Aberdeen, Md.

a. How much time will it take the bus on Trip Number 2033? _________

b. How much time will it take the bus on Trip Number 7051? _________

c. How much more time will James spend on the 2033 trip? _________

2. How much more time will it take to travel between New York and Baltimore by bus on Trip Number 2037 than on Trip Number 173? _________

3. Use the bus schedule on page 72 to answer each of the following questions.

a. How long does it take the bus on Trip Number 2033 to make the trip from New York to Newark? _________

b. How long does it take the bus on Trip Number 173 to make the trip from New York to Newark? _________

c. Why should more time be needed for Trip Number 2033?

____

Traveling by Train Unit 3

Train timetables are much the same as the timetables of interstate buses. The schedule on page 78 is a small section of one of them. As usual, the numbers across the top line designate the trip numbers.

Trip Numbers 867 874 903 945

Also, the names in the first column are the stations along the route.

Arlington
Rowe
Glen Ridge
⋮

Toward the bottom of the table we find the word

Connections.

This word implies that we will probably have to leave the train that brought us to Denville and board a second train that will take us on to Netcong.

Mondays through Fridays except Holidays

	867	874	903	945
From Clarkstown	7.12	9.56	10.47	4.58
to				
Arlington	7.31	10.15	11.06	5.17
Rowe	7.56	10.40	11.31	5.42
Glen Ridge	8.23	11.07	11.58	6.09
Watchung	8.49	—	—	6.35
Great Notch	9.31	12.15	—	7.54
Lincoln Park	—	12.31	—	8.10
Boonton	10.03	12.47	1.33	8.22
Ar Denville	10.46	1.30	2.16	9.09
Connections				
Lv Denville	11.12	1.54	3.07	10.01
Netcong	12.01	2.43	3.56	10.50

Example 1 How long is the waiting time at Denville on trip number 903?

Solution

Time of arrival in Denville: 2.16 PM
Time of departure from Denville: 3.07 PM

Waiting time 3 hours 7 minutes ⟶ 2 hours 67 minutes
− 2 hours 16 minutes ⟶ − 2 hours 16 minutes
51 minutes

Waiting time = 51 minutes

Example 2 Determine the travel time from Clarkstown to Boonton on trip number 903.

Solution

Departure time: 10.47 AM
Arrival time: 1.33 PM

Travel time:

12-hour clock	24-hour clock	
1.33 PM ⟶	13 hours 33 minutes =	12 hours 93 minutes
− 10.47 AM ⟶	− 10 hours 47 minutes =	− 10 hours 47 minutes
		2 hours 46 minutes

Travel time = 2 hours 46 minutes

Unit 3

At present, the great majority of passenger trains are commuter lines. That is, they carry large numbers of people from their homes in the suburbs into the city each morning. And at the end of the work-day, they carry these same people in the reverse direction.

Railways offer commuters an opportunity to save a good deal of money by purchasing commutation tickets. The railway whose fare schedule is shown below sells two different commutation tickets:

Monthly commutation **Weekly commutation**

By purchasing a monthly commutation ticket we can take as many trips as we desire during the month. Similarly, a weekly commutation ticket enables us to use the train as often as we want during the week.

Fares To and From Salem

From Salem to	One-way	1-Day Round Trip	Monthly Commutation	Weekly Commutation
Medway	$.70	$1.10	$21.00	$6.50
Arcadia	.90	1.40	27.00	8.00
West Valley	1.10	1.70	30.00	9.00
Fairfield	1.20	1.80	34.00	10.00
Knox	1.40	2.10	38.00	11.50
Drake	1.80	2.70	42.00	12.50
Hopewell	1.90	2.90	44.00	13.00
Pine Bush	2.10	3.20	45.00	13.50
Fremont	2.70	4.10	51.00	15.50
Campbell	3.80	5.70	58.00	17.50

Example 3 What is the cost of a monthly commutation ticket from Hopewell to Salem?

Explanation

Monthly commutation

Salem

Read Down

Hopewell — — — — — — — → $44.00

Read Across

Solution Cost of monthly ticket = $44.00

Example 4 Sylvia Kerner makes 10 trips each week in traveling from her home in Fremont to Salem where she works. How much can she save each week by purchasing a weekly commutation ticket than by purchasing 10 one-way tickets?

Explanation The saving can be found by subtracting this cost ⟶ from this cost

1 weekly ticket ↓ 10 one-way tickets ↓

Solution

Number of tickets ⟶ 10; **Cost of each ticket** ⟶ \$2.70

Cost of 10 one-way tickets = 10 × \$2.70
= \$27.00

Cost of the weekly ticket = \$15.50

Saving = Cost of 10 one-way tickets − Cost of weekly ticket
= \$27.00 − \$15.50
= \$11.50

Unit 3

Exercises A

1. Determine the time at which each train arrives at the station below.

	Trip Number	Station	Time		Trip Number	Station	Time
a.	874	Rowe	________	c.	945	Boonton	________
b.	903	Netcong	________	d.	867	Watchung	________

2. Determine the waiting time at Denville for each trip.

	Trip	Waiting Time		Trip	Waiting Time		Trip	Waiting Time
a.	867	________	b.	874	________	c.	945	________

3. Determine the travel time for each trip.

	Trip Number	From	To	Departure Time	Arrival Time	Travel Time
a.	867	Clarkstown	Great Notch	________	________	________
b.	874	Arlington	Boonton	________	________	________
c.	874	Glen Ridge	Denville	________	________	________
d.	903	Rowe	Denville	________	________	________
e.	945	Clarkstown	Netcong	________	________	________
f.	874	Boonton	Netcong	________	________	________

B

1. Determine the cost of a one-way train ticket between Salem and each of the following stations.

a.	Fairfield	$________	c.	Campbell	$________
b.	Drake	________	d.	Arcadia	________

2. Determine the cost of a monthly commutation ticket between Salem and each of the following stations.

a.	Knox	$________	c.	Pine Bush	$________
b.	Fremont	________	d.	Hopewell	________

3. Each of the following commuters purchased 50 weekly commutation tickets during the year. What is each person's annual commuting cost?

	Station	Weekly Commutation	Annual Cost
a.	Arcadia	$________	$________
b.	West Valley	________	________
c.	Drake	________	________
d.	Campbell	________	________

4. Each of the following commuters purchase 12 monthly commutation tickets during the period of one year. What is the annual commuting cost of each?

	Station	Monthly Commutation	Annual Cost
a.	Fairfield	$________	$________
b.	Hopewell	________	________
c.	Fremont	________	________
d.	West Valley	________	________

5. During the period of one week (10 trips), how much can be saved by the purchase of a weekly-commutation ticket over the one-way tickets?

Station	One-way Ticket	10 One-way Tickets	Weekly Ticket	Saving
a. Medway	$________	$________	$________	$________
b. Knox	________	________	________	________
c. Drake	________	________	________	________
d. Arcadia	________	________	________	________
e. Fremont	________	________	________	________

6. During the period of one month (44 trips), how much can be saved by the purchase of a monthly-commutation ticket over the one-way tickets?

Station	One-way Ticket	44 One-way Tickets	Monthly Ticket	Saving
a. West Valley	$________	$________	$________	$________
b. Hopewell	________	________	________	________
c. Pine Bush	________	________	________	________
d. Campbell	________	________	________	________

Unit 3

C

1. The 1-day round trip ticket can not be used during the rush hours. It is designed to attract people to use the trains after the commuter rush hour.

a. What is the cost of two one-way tickets to Campbell? ________

b. How much can be saved by purchasing a round-trip ticket to Campbell rather than 2 one-way tickets? ________

2. During the year Carmine DeNoia travels to work by train 237 days. His home is in Hopewell but his office is in Salem.

a. If he purchased one-way tickets each day, what would his total travel cost be for the year? ________

b. If he purchased monthly commutation tickets, what would his total cost be for the year? ________

c. How much can be saved each year by purchasing monthly commutation tickets rather than the one-way tickets? ________

3.

a. Determine the travel time from Clarkstown to Denville on trip number 874. ________

b. Determine the travel time from Clarkstown to Denville on trip number 903. ________

c. How do you account for the difference in the answers above? ________________

Unit 4 Traveling by Airplane

Section 1 Flight Schedules

Airlines publish flight schedules which contain a great deal more information than either the bus or train schedules we have been examining. The table on pages 86 and 87 shows a number of the flights out of Los Angeles made by a single airline.

We will begin by using the first flight to Chicago to point out the nature of the information given.

Time when plane leaves Los Angeles –8:45 A.M.

Time when plane arrives in Chicago –2:25 PM

8:45a **2:25p** 318 NON-STOP Daily

This is the flight number

This flight is made every day

This implies that there are no stops along the route

There are only two other points of information that are of interest to us at this time. Both of these points occur in the last flight to Chicago.

The plane will land at Dallas and another plane will continue to Chicago.

12:40a **9:52p** 262/238 Dallas Daily

Flight number from Los Angeles to Dallas

Flight number from Dallas to Chicago

Determining the travel time when traveling by air involves but one step more than what we had been doing in the past. Our country is divided into four time zones. When we pass from one zone to the next we will either lose an hour of time or gain an hour of time.

The map below shows the different time zones.

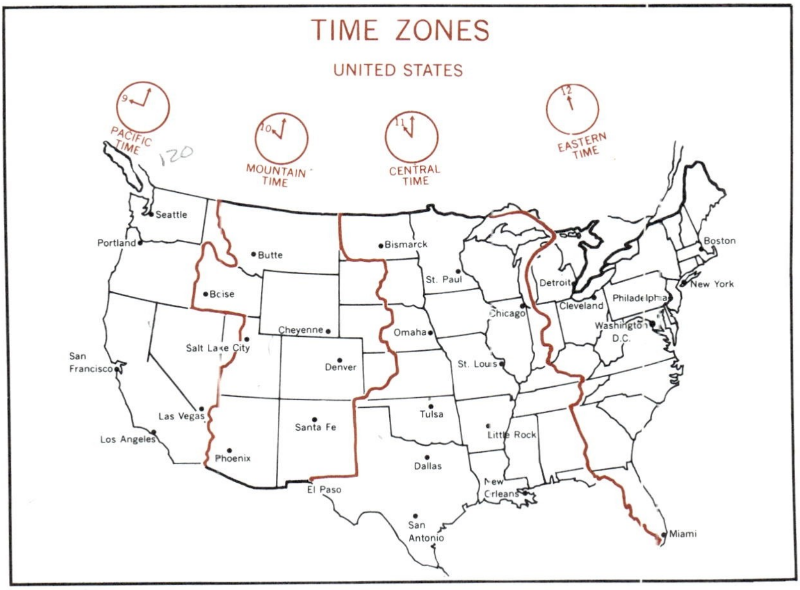

By examining the clock directly above each time zone, we can see that when it is

	9 AM Pacific Time
it is,	10 AM Mountain Time,
it is,	11 AM Central Time,
and	12 Noon Eastern Time.

Immediately following the name of the city in the flight schedule we will find letters such as PDT, MDT, CDT, EDT. These letters are an abbreviation of the time zone in which these cities are located.

Eastern Daylight Savings Time
↓
Cleveland (EDT)

Perhaps one of the better ways of computing travel time between different time zones is by not moving our watch when in flight. The examples below will clarify this point.

FROM LOS ANGELES (PDT)

Leave	Arrive	Flight	Stops or Connecting City	Operates
TO: CHICAGO (CDT)				**F$215 Y$165 FN$165 YN$132**
8:45a	**2:25p**	318	NON-STOP	Daily
10:00a	**3:38p**	184	NON-STOP	Daily
12:00p	**5:43p**	194	NON-STOP	Daily
3:30p	**9:10p**	192	NON-STOP	Daily
12:20a	5:55a	196	NON-STOP	Daily
12:40a	8:56a	262/206*	Dallas	Ex Su
12:40a	8:56a	262/206*	Dallas	Su
12:40a	9:52a	262/238*	Dallas	Daily
TO: CINCINNATI (EDT)				**F$230 Y$177 FN$177 YN$142**
7:40a	**3:52p**	200/342	Dallas	Sa
7:40a	**3:52p**	200/342	Dallas	Ex Sa
8:30a	**5:25p**	250/514	Memphis	Daily
12:00p	**7:40p**	390	One-Stop	Daily
1:00p	**9:13p**	20/394	Dallas	Daily
12:40a	9:10a	262/612*	Dallas	Daily
TO: CLEVELAND (EDT)				**F$244 Y$188**
9:15a	**4:19p**	74	NON-STOP	Daily
12:45p	**7:45p**	72	NON-STOP	Daily
11:00p	6:00a	36	NON-STOP	Daily
TO: COLUMBUS, Ohio (EDT)				**F$239 Y$184**
7:40a	**4:48p**	200/424	Dallas	Daily
1:00p	**10:26p**	20/120	Dallas	Daily
TO: DALLAS FT WORTH (CDT)				**F$164 Y$126 FN$126 YN$101**
7:40a	**12:22p**	200	NON-STOP	Daily
10:00a	**2:40p**	122	NON-STOP	Daily
1:00p	**5:40p**	20	NON-STOP	Daily
2:30p	**7:10p**	324	NON-STOP	Daily
3:45p	**8:25p**	78	NON-STOP	Sa
5:10p	**9:50p**	306	NON-STOP	Daily
9:30p	2:04a	314*	NON-STOP	Daily
10:35p	3:12a	378*	NON-STOP	Daily
12:40a	5:17a	262*	NON-STOP	Daily

*Nightcoach Fares, FN, YN

Leave	Arrive	Flight	Stops or Connecting City	Operates
TO: LOUISVILLE (EDT)				**F$225 Y$173**
7:40a	**3:40p**	200/424	Dallas	Daily
8:30a	**4:17p**	250	One-Stop	Daily
1:00p	**9:03p**	20	One-Stop	Daily
12:40a	10:29a	262/344★	Dallas	Daily
TO: MEMPHIS (CDT)				**F$203 Y$156**
8:30a	**1:45p**	250	NON-STOP	Daily
10:00a	**5:27p**	122	Two-Stop	Daily
1:35p	**6:50p**	528	NON-STOP	Daily
5:10p	**11:33p**	306/264	Dallas	Daily
12:49a	7:58a	262/180	Dallas	Daily
TO: NEW YORK NEWARK (EDT)				**F$286 Y$220 FN$220 YN$176**
8:45a	**4:39p**	2	NON-STOP	Daily
8:45a	**5:44p**	318/220	Chicago	Daily
9:00a	**5:53p**	68/418	Detroit	Ex Sa
11:00a	**6:47p**	8	NON-STOP	Daily
12:00p	**9:10p**	390/146	St Louis	Daily
1:00p	**8:52p**	32	NON-STOP	Daily
4:30p	12:23a	4	NON-STOP	Ex Sa
10:30p	6:20a	10	NON-STOP	Daily
10:35p	8:03a	378/428★	Dallas	Daily
12:20a	9:43a	196/358	Chicago	Daily
12:20a	9:50a	196/246	Chicago	Daily
12:40a	11:13a	262/84★	Dallas	Daily
TO: OKLAHOMA CITY (CDT)				**F$159 Y$122 FN$122 YN$98** **Via Dallas** **F$164 Y$127 FN$127 YN$103**
9:25a	**1:57p**	382	NON-STOP	Daily
12:40p	**5:10p**	178	NON-STOP	Daily
2:30p	**8:58p**	324/69	Dallas	Daily
5:10p	**11:05p**	306	One-Stop	Daily
12:40a	7:40a	262/238★	Dallas	Daily

★Nightcoach Fares, FN, YN

Example 1 If the time is 6:25 AM PDT, what is the CDT?

Explanation

	PDT	MDT	CDT
Start here →	6:25 AM	7:25 AM	8:25 AM
	+ 1 hr	+ 1 hr	

Solution CDT: 8:25 AM

Example 2 If the time is 1:50 PM EDT, what is the PDT?

Explanation

PDT	MDT	CDT	EDT	
10:50a	11:50a	12:50p	1:50p	← Start here
− 1 hr	− 1 hr	− 1 hr		

Solution PDT: 10:50 AM

Example 3 A plane leaves Kansas City at 5:35p CDT and arrives in San Francisco at 8:15p PDT. What is the flight time?

Explanation Had the watch not been moved in flight, then the arrival time of 8:15p PDT will be recorded on the person's watch as 10:15p CDT.

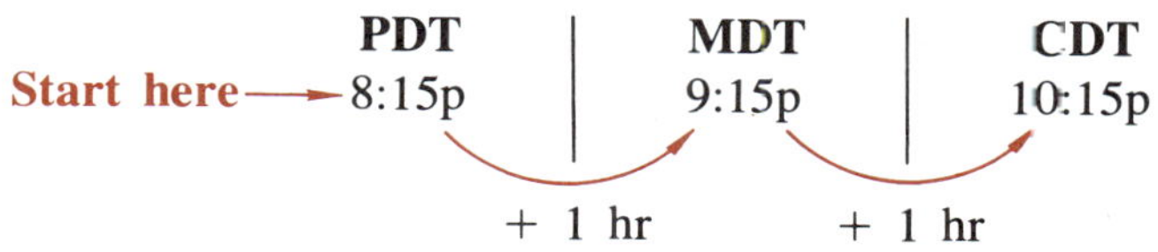

Solution

Departure time: 5:35p CDT
Arrival time: 8:15p PDT = 10:15p CDT

Unit 4

Travel time:

10:15p=	9 hours 75 minutes	
− 5:35p=	−5 hours 35 minutes	
	4 hours 40 minutes	

Travel time = 4 hours 40 minutes

Example 4 Anna Caruso took flight number 250 out of Los Angeles to Memphis. What was the travel time of this flight?

Explanation

Arrival time: 1:45p CDT

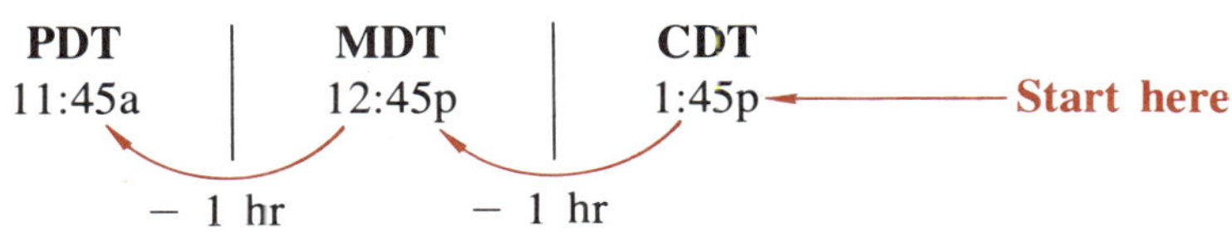

Solution

Departure time: 8:30a PDT
Arrival time: 1:45p CDT = 11:45a PDT

Travel time:	11:45a	=	11 hours 45 minutes
	−8:30a	=	−8 hours 30 minutes
			3 hours 15 minutes

Travel time: 3 hours 15 minutes

Exercises A

1. The following travelers do not change the setting of their watches while in flight. What time will each watch read at the time of arrival after traveling between the time zones shown?

	Zone Traveling From	Zone Traveling To	Arrival Time	Watch Time
a.	Pacific	Central	5:45p	________
b.	Pacific	Eastern	7:25p	________
c.	Pacific	Mountain	2:30a	________
d.	Pacific	Eastern	9:30p	________
e.	Mountain	Central	6:40a	________
f.	Mountain	Eastern	9:43a	________
g.	Central	Eastern	8:53p	________
h.	Pacific	Central	1:10p	________
i.	Pacific	Eastern	2:04p	________

2. The following travelers do not change the setting of their watches while in flight. What time will each watch read at the time of arrival after traveling between the time zones shown?

	Zone Traveling From	Zone Traveling To	Arrival Time	Watch Time
a.	Eastern	Central	2:35p	________
b.	Eastern	Pacific	4:27p	________
c.	Eastern	Mountain	5:54a	________
d.	Central	Pacific	2:15a	________
e.	Central	Mountain	10:36p	________
f.	Mountain	Pacific	6:42a	________
g.	Eastern	Pacific	11:25a	________
h.	Eastern	Mountain	12:14a	________
i.	Central	Pacific	11:53a	________

3. Determine the travel time for each flight.

	Departure Time	Arrival Time	Watch Time at Arrival	Flight Time
a.	1:20p PDT	3:40p MDT	________	________
b.	4:10p PDT	8:30p CDT	________	________
c.	10:45a PDT	7:40p EDT	________	________
d.	10:30a MDT	5:04p EDT	________	________
e.	12:45p MDT	3:56p CDT	________	________

4. Determine the travel time for each flight.

	Departure Time	Arrival Time	Watch Time at Arrival	Flight Time
a.	5:30p EDT	5:50p CDT	________	________
b.	4:25p EDT	7:02p MDT	________	________
c.	5:50p CDT	7:37p MDT	________	________
d.	11:45a CDT	2:04p PDT	________	________
e.	9:00a CDT	11:20a PDT	________	________

B

Unit 4

1. At what time will the following flights out of Los Angeles arrive at their destination?

	Flight Number	City of Arrival	Time of Arrival
a.	184	Chicago	________
b.	72	Cleveland	________
c.	20	Louisville	________
d.	528	Memphis	________
e.	306	Oklahoma City	________

2. A traveler wants to leave Los Angeles at approximately the time indicated. Which flight should be taken?

	Approx. Time of Leaving	Destination	Flight Number
a.	10:00 A.M.	Chicago	________
b.	7:30 A.M.	Columbus	________
c.	1:30 P.M.	Memphis	________
d.	5:00 P.M.	Dallas	________
e.	9:30 A.M.	Oklahoma City	________

3. How many hours will it take each of the following flights out of Los Angeles to reach its destination?

	Flight Number	Destination	Departure Time	Arrival Time	Watch Time at Arrival	Flight Time
a.	192	Chicago	____	____	____	______
b.	20	Louisville	____	____	____	______
c.	200	Dallas	____	____	____	______
d.	20/120	Columbus	____	____	____	______
e.	74	Cleveland	____	____	____	______
f.	8	New York	____	____	____	______
g.	196	Chicago	____	____	____	______
h.	72	Cleveland	____	____	____	______

C

1. A traveler takes the 5:10 PM flight out of Los Angeles for Memphis.

a. What is the flight number for the plane leaving Los Angeles? ________

b. At what city will the plane make a stop? ________

c. What is the flight number for the plane that the traveler will take from that city to Memphis? ________

d. At what time will the traveler land in the stop-over city? ________

2. Travelers fly from Los Angeles to Oklahoma City.

a. If they leave on the 12:40 AM flight, how long will it take them? ________

b. If they leave on the 2:30 PM flight, how long will it take them? ________

c. How much traveling time can they save by taking the earlier flight? ________

Section 2 Cost of Air Transportation

The cost of traveling by air depends on the service we select:

1. First Class Service
2. Coach Service

The cost of First Class service is the greater of the two. However, in return for the extra money the quality of the food and the seating comfort is superior to that of the Coach service.

There are some airlines that offer discounts on some of their night-flights. The service on these flights are also separated into two groups of First Class and Coach.

The cost of each of these services is usually shown on the flight schedule. As an example, in the schedule on pages 86 and 87, the very top line in the section of the flights to Chicago shows the following:

First Class cost → F$215

Coach cost → Y$165

F$215 Y$165 FN$132 YN$106

FN$132 ← **First Class night cost**

YN$106 ← **Coach night cost**

These prices are what we might call the basic cost. However, in order to encourage people to travel by plane, many airlines offer the traveler discount prices. Within recent years, among the more popular of the discount arrangements are the two plans that follow.

TOUR AMERICA PLAN

Peak Season	**Off Peak Season**
June 1 - Sept. 14	Sept. 15 - May 31
15%	20%

The above plan is for coach round-trip service only and carries with it two conditions:

1. The tickets must be purchased at least 14 days before the departure date.

2. The return flight must be made at some time between 8 days and 30 days after the departure date.

Unit 4

Example 1 Determine the fare on a Tour America Plan round trip from Los Angeles to Cleveland. The trip is to begin on November 5.

Solution

One-way coach fare = \$188

Cost going ↓   ↓ **Cost returning**

Round trip coach fare = \$188 + \$188
= \$376

Off-peak discount rate ↓

Tour America Plan discount = 20% of Round-trip fare
= .20 × \$376
= \$75.20

Tour America Plan fare = Round-trip fare − Discount
= \$376.00 − \$75.20
= \$300.80

With the second plan, there is greater saving. However, the ticket must be purchased much further in advance than under the previous plan.

1. Coach tickets have to be purchased at least 30 days before the departure date.
2. The return flight has to be made at some time between 8 days and 45 days after the departure date.

DOLLAR SAVER PLAN

Peak Days Saturday, Sunday	Shoulder Days Monday, Friday	Off Peak Days Tuesday, Wednesday, Thursday
35%	40%	45%

Example 2 Determine the fare on a Dollar Saver Plan round trip from Los Angeles to New York if the day of departure is a Saturday.

Solution

$$\begin{aligned}\text{One-way coach fare} &= \$220\\ \text{Round-trip coach fare} &= \$220 + \$220\\ &= \$440\end{aligned}$$

Saturday discount rate

$$\begin{aligned}\text{Dollar Saver Plan discount} &= 35\% \text{ of Round-trip fare}\\ &= .35 \times \$440\\ &= \$154\\ \text{Dollar Saver Plan fare} &= \text{Round-trip fare} - \text{Discount}\\ &= \$440 - \$154\\ &= \$286\end{aligned}$$

Exercises A

Do you recall how to do the computation in the following exercises? This computation is the same as you will be doing when finding flight fares.

1. Rename each as a decimal.

a. 35% __________ d. 45% __________
b. 20% __________ e. 40% __________
c. 60% __________ f. 10% __________

2. Find the products.

a.	20% × $150	$__________	d.	45% × $236	$__________
b.	40% × $175	__________	e.	45% × $273	__________
c.	35% × $167	__________	f.	35% × $297	__________

B

Unit 4

1. Find the one-way fare from Los Angeles to each of the cities indicated.

	Destination	Flight Service	Fare
a.	Chicago	First Class	$__________
b.	Columbus	Coach	__________
c.	Cincinnati	Night First Class	__________
d.	Memphis	Coach	__________
e.	New York	Night Coach	__________

2. Determine the fare for the following Tour America Plan round-trip flights.

	Normal Round-Trip Coach Fare	Departure Date	Discount Rate	Discount	Tour America Plan Fare
a.	$120	Feb. 4	__________%	$__________	$__________
b.	250	July 9	__________%	__________	__________
c.	172	Oct. 16	__________%	__________	__________
d.	348	Jan. 12	__________%	__________	__________
e.	297	Aug. 23	__________%	__________	__________

3. Determine the fare for the following Dollar Saver Plan round-trip flights.

	Normal One-way Coach Fare	Normal Round-Trip Coach Fare	Departure Day	Discount Rate	Discount	Dollar Saver Plan Fare
a.	$150	$__________	Monday	__________%	$__________	$__________
b.	87	__________	Friday	__________%	__________	__________
c.	123	__________	Sunday	__________%	__________	__________
d.	186	__________	Tuesday	__________%	__________	__________
e.	239	__________	Thursday	__________%	__________	__________

4. Determine the fare for the following Dollar Saver Plan round-trip flights out of Los Angeles.

Destination	Normal One-way Fare	Normal Round-trip Fare	Departure Day	Discount Rate	Discount	Dollar Saver Plan Fare
a. Cincinnati	$_____	$_____	Friday	_____%	$_____	$_____
b. Memphis	_____	_____	Saturday	_____%	_____	_____
c. New York	_____	_____	Tuesday	_____%	_____	_____
d. Oklahoma City	_____	_____	Sunday	_____%	_____	_____
e. Louisville	_____	_____	Thursday	_____%	_____	_____

C

1. John Vaughn bought a first-class ticket from Los Angeles to Oklahoma City.

a. If the plane stops off at Dallas, what will the cost be? ________

b. If the flight is non-stop, what will the cost be? ________

c. How much can John save by traveling non-stop? ________

2. Bert Phillips made a round-trip flight from Los Angeles to Cleveland. The departure date was October 5.

a. If he traveled coach service, what would the flight have cost him? ________

b. If he was able to fly under the Tour America Plan, what would the flight have cost him? ________

c. How much would he have been able to save by flying under the Tour America Plan? ________

3. Alice Bradley who lives in Los Angeles decided to spend her vacation in New York. Her departure date was Tuesday, May 3.

a. If she was able to fly under the Tour America Plan, what would the flight have cost her? ________

b. If she was able to fly under the Dollar Saver Plan, what would the flight have cost her? ________

c. How much would she have been able to save by flying under the Dollar Saver Plan? ________

Section 3 Flight Insurance

People who travel by plane are often concerned about the possibility of an accident. Although plane accidents receive a great deal of publicity, the risk of being killed is rather small. In fact, insurance companies will pay as much as $150,000 for a premium of $3 in the event of death as a result of a commercial airline accident.

RATE SCHEDULE

Principal Sum	Premium
$150,000	$3.00
125,000	2.50
100,000	2.00
87,500	1.75
62,500	1.25
50,000	1.00
37,500	.75
25,000	.50

Rates are for one-way or round-trip.

Unit 4

Should death occur, then the insurance company will pay the Principal Sum to the person whom we have named. On the other hand, if we are injured, the amount we receive will depend upon the extent of the injury.

The principal sum will be paid for loss of any of the following:

1. Both feet
2. Both hands
3. Both eyes
4. One hand and one foot
5. One hand and one eye
6. One foot and one eye

50% of the principal sum will be paid for loss of any of the following:

1. One foot
2. One hand

25% of the principal sum will be paid for loss of the following:

1. One eye

Example 1 At the start of an air trip, Sam Traymore purchased flight insurance at the cost of $1.00. As a result of a forced landing, one of Sam's eyes was so badly injured that he lost the sight of it. How much did Sam receive from the insurance company?

Solution

Principal sum for a $1 premium = $50,000

Coverage for loss of one eye = 25% of principal sum

= 25% × $50,000

= .25 × $50,000

= $12,500

Although we stand little risk of personal injury during a trip by plane, our luggage stands a much greater risk of damage or loss. Included in the price of the airline ticket is coverage on our luggage to the extent of \$750. Should we care for greater coverage, some airlines will permit this at a cost for the additional insurance of

10¢ per \$100.

Example 2 A person wanted his baggage insured to the extent of \$9,000 during a plane trip. What charge did he have to pay on the excess coverage?

Explanation The charge will depend on the number of \$100-amounts in the excess coverage. To understand how this is determined, consider the following pattern:

In \$200 there are 2 \$100-amounts: \$200 ÷ \$100 = 2
In \$300 there are 3 \$100-amounts: \$300 ÷ \$100 = 3
In \$400 there are 4 \$100-amounts: \$400 ÷ \$100 = 4

Therefore, to find the number of \$100-amounts in the excess coverage, simply *divide excess coverage by \$100*.

Solution

No charge for \$750

Excess coverage = \$9,000 − \$750
= \$8,250

Number of \$100s in \$8,250 = \$8,250 ÷ \$100
= 82.5

Number of 100s **Cost for each 100**

Premium for \$8,250 coverage = 82.5 × 10¢
= 82.5 × \$.10
= \$8.25

Exercises

A

Do you recall how to do the computation in the following exercises? This computation is the same as you will be doing when finding the cost of baggage insurance.

1. Determine the number of $100-amounts in each exercise.

a.	$800	________	d.	$4,640	________
b.	$1,500	________	e.	$9,350	________
c.	$6,000	________	f.	$14,650	________

B

1. Determine the flight insurance principal sum that can be purchased for each premium.

	Premium	Principal Sum		Premium	Principal Sum
a.	$2.00	$________	c.	$1.25	$________
b.	75¢	________	d.	50¢	________

Unit 4

2. Flight insurance was purchased for each of the following premiums. Determine the payment that will be made in the event of an accident.

	Premium	Principal Sum	Extent of Injury	Payment Made
a.	$1.25	$________	Death	$________
b.	1.00	________	Death	________
c.	3.00	________	Loss of one hand	________
d.	.75	________	Loss of one foot	________
e.	1.75	________	Loss of both hands	________
f.	2.00	________	Loss of one eye	________
g.	2.50	________	Loss of both eyes	________

3. The cost of the flight ticket includes baggage insurance to the extent of $750. Additional insurance can be purchased at 10¢ per $100 for excess coverage. Determine the cost of the following coverages.

	Coverage	Excess Above $750	Cost of Insurance
a.	$2,750	$________	$________
b.	3,850	________	________
c.	7,200	________	________
d.	18,400	________	________
e.	23,500	________	________

4. The cost of the flight ticket on overseas travel includes baggage insurance to the extent of $400. Additional insurance can be purchased at 15¢ per $100 on the excess above $400. Determine the cost of the following coverages for overseas flights.

	Coverage	Excess Above $400	Cost of Insurance
a.	$3,400	$______	$______
b.	6,700	______	______
c.	12,100	______	______
d.	15,300	______	______

Review of Chapter 2

When doing the computation for many of the following problems it will be necessary to refer to the tables in this chapter.

1. Determine the basic charge of the following car rentals. (Table, page 58)

	Type of Car	Base Rate Per Day	Number of Days	Basic Charge
a.	Compact	$______	2	$______
b.	Full-size	______	5	______

2. Determine the *mileage charge* of the following *one* day car rentals. (Table, page 58)

	Type of Car	Miles Driven	Excess Mileage	Cost Per Mile	Mileage Charge
a.	Mid-size	120	______	$______	$______
b.	Luxury	256	______	______	______

3. Determine the total car rental for the following situations. (Table, page 58)

	Type of Car	Number of Days	Basic Charge	Miles Driven	Mileage Charge	Car Rental
a.	Compact	2	$______	310	$______	$______
b.	Full-size	4	______	785	______	______

4. The time below is named in terms of the 12-hour clock. Rename each in terms of the 24-hour clock.

	12-hour clock	24-hour clock		12-hour clock	24-hour clock
a.	5:12 AM	______	c.	11:17 PM	______
b.	2:50 PM	______	d.	12:15 PM	______

5. Determine the travel time for each bus trip.

	From	To	Travel Time		From	To	Travel Time
a.	4:20 AM	6:35 AM	______	c.	8:30 AM	1:40 PM	______
b.	2:45 PM	8:12 PM	______	d.	10:58 AM	5:23 PM	______

6. Determine the travel time for the 2:50 PM bus out of Butler to Newark. (Table, page 63) ________

7. Determine the travel time for Trip Number 2037 out of Jersey City for Washington, D.C. (Table, page 72) ________

8. Determine the round-trip bus fare from New York to Harrisburg. (Table, page 74) ________

9. How much can a person save by taking advantage of the bargain one-way fare from New York to Baton Rouge when traveling by bus? (Table, page 75) ________

Review

10. A person boards train number 874 in Rowe and leaves the train in Denville. How long did this trip take? (Table, page 78) ________

11. How much money can be saved by purchasing a monthly commutation ticket from Knox to Salem rather than 42 one-way tickets? (Table, page 79) ________

12. Determine the travel time of the following flights.

	Departure Time	Arrival Time	Watch Time at Arrival	Travel Time
a.	2:30p PDT	6:45p CDT	________	________
b.	3:50a EDT	9:27a MDT	________	________
c.	7:14a PDT	3:41p EDT	________	________

13. Determine the travel time for flight number 122 out of Los Angeles for Memphis. (Table, page 87) ________

14. Janet Richards made a round-trip flight from Los Angeles to Dallas, departing on February 17. How much money was she able to save by taking advantage of the Tour America Plan? (Tables, pages 86 and 93) ________

15. As the result of a plane accident Arthur Jamison lost one foot. Prior to the flight he purchased flight-insurance at a cost of $2.50. How much will he receive for the loss of his foot? (Table, page 97) ________

You're going to like it here!
You're going to like our super Grocery savings!
3
2
CLOSED

Chapter 3

Purchasing Consumer Goods

During recent years we have been made very much aware of the fact that the cost of almost everything is going up very rapidly. In view of this, most people are making a determined attempt to shop more wisely than they had in the past. The object of this chapter is to point out the arithmetic that might aid us in making decisions before making a purchase.

Shopping at a Supermarket Unit 1

Section 1 Unit Pricing
Part 1

Some confusion exists in shopping at a supermarket over the manner in which articles are priced. Only recently have several states passed laws that require markets to post unit prices. Thus, on each shelf are signs containing information such as,

Unit price - 2.6¢ per ounce
Unit price - 54.8¢ per pound
Unit price - 29.4¢ per quart

By comparing unit prices of the same article in two different size cans we can immediately determine the better buy. As an example,

Two Different Size Jars of Jelly

12-ounce jar	18-ounce jar
Unit price - 5.2¢ per ounce	Unit price - 5.6¢ per ounce

Better buy (12-ounce jar, 5.2¢ per ounce)

To show how the unit price is determined let us examine the following pattern:

If 2 ounces cost 10¢, then each ounce costs 5¢:

10¢ ÷ 2 = 5¢

Total cost (10¢) — **Number of ounces** (2)

Similarly, if 5 ounces cost 15¢, then each ounce costs 3¢:

15¢ ÷ 5 = 3¢

Total cost (15¢) — **Number of ounces** (5)

The *cost per ounce* is found by *dividing the total cost by the number of ounces*. This principle can be expressed in the following manner:

Cost per ounce = Total cost ÷ Number of ounces

For many articles at a supermarket, unless the unit price is expressed to the nearest tenth of a cent, comparisons between prices can not be made easily.

Example 1 Determine the unit price on a 14-ounce can of merchandise that sells for 87¢.

Solution

Cost per ounce = Total cost ÷ Number of ounces

= 87¢ ÷ 14

= 6.2¢ ← **Unit price rounded to nearest tenth of a cent.**

Unit prices are also given in terms of measures other than the ounce. However, the principle for determining the unit price is the same no matter what the unit may be.

Cost per pound = Total cost ÷ Number of pounds

Cost per quart = Total cost ÷ Number of quarts

Unit cost = Total cost ÷ Number of units

Example 2 Determine the unit price (cost per pound) on a 25 pound bag of birdseed that sells for $2.87.

Solution

Cost per pound = Total cost ÷ Number of pounds
= $2.87 ÷ 25
= 287¢ ÷ 25
= 11.5¢

Exercises A

Do you recall how to do the computation in the following exercises? This computation is the same as you will be doing when finding the unit price.

1. Round each number to the nearest tenth of a cent.

a.	6.34¢	____¢	c.	5.37¢	____¢	e.	42.96¢	____¢
b.	47.82¢	____¢	d.	32.85¢	____¢	f.	29.97¢	____¢

2. Rename each number in cents rather than dollars.

a.	$2.43	____¢	c.	$1.98	____¢	e.	$.47	____¢
b.	$5.29	____¢	d.	$3.04	____¢	f.	$.05	____¢

3. Find the quotients. Round your answer to the nearest tenth of a cent.

a.	53¢ ÷ 7	________¢	d.	65¢ ÷ 18	________¢
b.	74¢ ÷ 12	________¢	e.	89¢ ÷ 32	________¢
c.	92¢ ÷ 14	________¢	f.	95¢ ÷ 46	________¢

4. Find the quotients. Round your answer to the nearest tenth of a cent.

a.	$1.25 ÷ 16	________¢	c.	$4.95 ÷ 32	________¢
b.	$2.47 ÷ 18	________¢	d.	$6.09 ÷ 46	________¢

B

1. Determine the cost per ounce for each. Round each answer to the nearest tenth of a cent.

	Total Cost	Number of Ounces	Cost per Ounce		Total Cost	Number of Ounces	Cost per Ounce
a.	65¢	8	________¢	d.	$1.54	9	________¢
b.	59¢	14	________¢	e.	$1.89	12	________¢
c.	86¢	16	________¢	f.	$2.15	18	________¢

2. Determine each unit price. Round each answer to the nearest tenth of a cent.

	Total Cost	Number of Units	Unit Price
a.	37¢	7 ounces	________¢ per ounce
b.	51¢	9 ounces	________¢ per ounce
c.	95¢	3 quarts	________¢ per quart
d.	75¢	4 pints	________¢ per pint
e.	87¢	5 pounds	________¢ per pound
f.	$1.23	10 ounces	________¢ per ounce
g.	$1.37	6 pounds	________¢ per pound
h.	$1.96	8 quarts	________¢ per quart
i.	$2.57	15 pounds	________¢ per pound
j.	$7.95	65 ounces	________¢ per ounce

3. Prices in a supermarket are stamped on cans as 2/47¢, or 5/83¢, and so on. These prices are read as 2 cans for 47¢, or 5 cans for 83¢. If a person purchases 1 can, the price is rounded to the next higher cent. How much will have to be paid for the purchase of 1 can of the following items?

	Marking on Can	Cost Per Can		Marking on Can	Cost Per Can
a.	2/15¢	________¢	f.	2/$1.15	________¢
b.	3/37¢	________¢	g.	5/$1.67	________¢
c.	4/85¢	________¢	h.	9/$3.46	________¢
d.	5/96¢	________¢	i.	7/$2.98	________¢
e.	3/89¢	________¢	j.	6/$5.95	________¢

Part 2

Soon all measurements in the United States will be done with the metric system. Even now, most canned goods purchased in a supermarket show their size in metric measure.

There are four metric measures normally found on items in a market. Two of these measures appear on containers whose contents are liquids such as milk, soda, or juice. The other two are used for measuring the weight of items such as potatoes, canned fruit, frozen vegetables, and the like.

Unit 1

Liquid Measure		Dry Measure	
liter	(l)	kilogram	(kg)
milliliter	(ml)	gram	(g)

As these units are new to many of us we tend to have difficulty picturing their size. The following relations might help:

A liter is just slightly larger than a quart.
A kilogram is just slightly larger than two pounds.

More important are the relations between the liter and the milliliter, and also between the gram and the kilogram.

1,000 ml = 1 l
1,000 g = 1 kg

Considering the size of the liter, it is quite apparent that the milliliter must be very, very small. Therefore, it would be best to compute all unit prices in terms of the liter. And the same comments can be made for the gram. In view of this, we will find it necessary to rename milliliters as liters and grams as kilograms. To do this, let us examine the following pattern. The pattern begins by making use of the relation between the milliliter and the liter.

1,000 ÷ 1,000 = 1

1,000 ml = 1 liter

2,000 ÷ 1,000 = 2

2,000 ml = 2 liters

3,000 ÷ 1,000 = 3

3,000 ml = 3 liters

Therefore, we can say,

Number of liters = Number of milliliters ÷ 1,000 .

In exactly the same manner we can show that,

Number of kilograms = Number of grams ÷ 1,000 .

Example 1 Determine the number of liters in 325 milliliters.

Solution

Number of liters = Number of milliliters ÷ 1,000
= 325 ÷ 1,000
= .325

You may recall that a simple device for dividing a number by 1,000 is to merely move the decimal point three places to the left in the number. Thus,

325 ÷ 1,000
= 325. ÷ 1,000

Supply missing decimal point

= .325.

Decimal point moved three places to left.

Example 2 Determine the number of kilograms in 695 grams.

Solution

Number of kilograms = Number of grams ÷ 1,000
= 695. ÷ 1,000
= .695

Example 3 A 680 gram container of an iced tea mix costs \$1.67. To the nearest cent, what is the unit price?

Explanation To find the unit price, we must first determine the number of kilograms in 680 grams.

Solution

$$\begin{aligned}\text{Number of kilograms} &= \text{Number of grams} \div 1{,}000\\ &= 680. \div 1{,}000\\ &= .680\end{aligned}$$

$$\begin{aligned}\text{Cost per kilogram} &= \text{Total cost} \div \text{Number of kilograms}\\ &= \$1.67 \div .68\\ &= \$2.46\end{aligned}$$

Example 4 A 532 ml can of tomato juice costs 37¢. What is the unit price?

Unit 1

Solution

$$\begin{aligned}\text{Number of liters} &= \text{Number of milliliters} \div 1{,}000\\ &= 532. \div 1{,}000\\ &= .532\end{aligned}$$

$$\begin{aligned}\text{Cost per liter} &= \text{Total cost} \div \text{Number of liters}\\ &= 37¢ \div .532\\ &= 69.5¢\end{aligned}$$

Exercises A

Do you recall how to do the computation in the following exercises? This computation is the same as you will be doing when finding the unit price.

1. Find the quotients.

a. 527 ÷ 1,000 ________ d. 1,465 ÷ 1,000 ________
b. 346 ÷ 1,000 ________ e. 3,540 ÷ 1,000 ________
c. 250 ÷ 1,000 ________ f. 4,200 ÷ 1,000 ________

2. Find the quotients. Round your answer to the nearest cent.

a. 43¢ ÷ .52 ________¢ d. 75¢ ÷ .31 ________¢
b. 57¢ ÷ .78 ________¢ e. 92¢ ÷ .82 ________¢
c. 28¢ ÷ .64 ________¢ f. 128¢ ÷ .47 ________¢

3. Find the quotients. Round your answer to the nearest cent.

a.	\$1.46 ÷ 1.20	\$________	d.	\$3.59 ÷ 2.75	\$________
b.	\$1.58 ÷ 2.35	________	e.	\$3.04 ÷ 1.58	________
c.	\$2.74 ÷ 2.46	________	f.	\$5.29 ÷ 3.15	________

B

1. Rename each as liters.

	Milliliters	Liters		Milliliters	Liters
a.	265	________	d.	2,540	________
b.	372	________	e.	1,075	________
c.	690	________	f.	3,090	________

2. Rename each as kilograms.

	Grams	Kilograms		Grams	Kilograms
a.	425	________	d.	1,754	________
b.	695	________	e.	2,650	________
c.	250	________	f.	8,300	________

3. Determine the cost per liter. Round each answer to the nearest cent.

	Total Cost	Number of Milliliters	Number of Liters	Cost per Liter
a.	53¢	250	________	________¢
b.	48¢	560	________	________¢
c.	61¢	480	________	________¢
d.	27¢	720	________	________¢
e.	86¢	1,240	________	________¢
f.	75¢	2,370	________	________¢

4. Determine each unit price. Round each answer to the nearest cent.

	Total Cost	Number of Units	Unit Price
a.	\$2.47	5 kilograms	________¢ per kilogram
b.	1.69	3 liters	________¢ per liter
c.	3.74	6 kilograms	________¢ per kilogram
d.	4.59	7 kilograms	________¢ per kilogram
e.	8.54	8 liters	________¢ per liter

5. Determine each unit price. Round each answer to the nearest cent.

	Total Cost	Number of Units	Number of Units	Unit Price
a.	$1.85 ÷	250 grams	______ kilograms	$______ per kilogram
b.	.75	470 grams	______ kilograms	______ per kilogram
c.	.98	560 grams	______ kilograms	______ per kilogram
d.	1.09	740 milliliters	______ liters	______ per liter
e.	1.59	1,250 milliliters	______ liters	______ per liter
f.	3.42	2,560 grams	______ kilograms	______ per kilogram

C

1. A 454 gram jar of beets can be purchased for 49¢. To the nearest cent, what is the unit price per kilogram of these beets? ________

2. A 1,360 milliliter can of fruit juice costs 79¢. To the nearest cent, what is the unit price (price per liter) of this juice? ________

3. A jar containing 1,640 milliliters of ammonia costs $1.07. To the nearest cent, what is the unit price of this brand of ammonia? ________

4. Two jars of peanut butter each weighing 170 grams can be purchased for $1.03. To the nearest cent, what is the unit price of this brand of peanut butter? ________

5. Three 340-gram boxes of corn flakes can be purchased at a special sale for $1.39. To the nearest cent, what is the unit price on this purchase? ________

Section 2 Comparison Shopping

It is one short step from finding the unit price of a purchase to determining which is the better buy between two purchases. The better buy is the one with the lower unit price. It is important to remember that price comparisons can be made only between articles having the same brand name.

Example 1 Fruit cocktail can be purchased in either of the following weights: 850 grams for 59¢ or 625 grams for 47¢. Which is the better buy?

Solution

Unit price for first can = Total cost ÷ Number of *kilograms*
= 59¢ ÷ .85
= 69¢ ⟵ Nearest cent

Unit price for second can = Total cost ÷ Number of *kilograms*
= 47¢ ÷ .625
= 75¢ ⟵ Nearest cent

The better buy is the 850 gram can.

Having determined which is the better buy, we would now want to know how much can be saved by making this purchase. Finding that saving requires the following steps:

1. The saving on one unit is determined first.

Saving per unit = Larger unit price − Smaller unit price

2. That saving is then multiplied by the number of units purchased.

Total Saving = Number of units × saving per unit

Example 2 Apple pie-filling can be purchased in either of the following size cans: 850 grams for 69¢ or 1,400 grams for 83¢. How much can be saved by purchasing the better buy?

Solution

Unit price for first can = Total cost ÷ Number of *kilograms*
= 69¢ ÷ .85
= 81¢ ⟵ Nearest cent

Unit price for second can = Total cost ÷ Number of *kilograms*
= 83¢ ÷ 1.4
= 59¢ ⟵ Nearest cent

Saving per kilogram = 81¢ − 59¢
= 22¢

Total saving = Number of kg purchased × Saving per kg
= 1.4 × 22¢
= 31¢ ⟵ Nearest cent

Exercises

A

Do you recall how to do the computation in the following exercises? This computation is the same as you will be doing when finding the saving on a purchase at a supermarket.

1. Find the products. Round your answer to the nearest cent.

a.	.24 × 14¢	________¢	d.	1.25 × 29¢	________¢
b.	.51 × 23¢	________¢	e.	1.03 × 36¢	________¢
c.	.67 × 18¢	________¢	f.	2.52 × 84¢	________¢

2. Find the quotients. Round your answer to the nearest cent.

a.	36¢ ÷ .84	________¢	c.	87¢ ÷ .48	________¢
b.	52¢ ÷ .73	________¢	d.	95¢ ÷ 1.67	________¢

Unit 1

B

1. In these exercises the same item is sold in two different size cans. Which of the two is the better buy?

Weight and Cost of Smaller Can	Smaller Can Unit Price	Weight and Cost of Larger Can	Larger Can Unit Price	Better Buy
a. 270 g for 67¢	_____¢	420 g for 79¢	_____¢	_____
b. 350 ml for 25¢	_____¢	530 ml for 41¢	_____¢	_____
c. 560 ml for 47¢	_____¢	870 ml for 83¢	_____¢	_____
d. 475 g for 36¢	_____¢	750 g for 52¢	_____¢	_____
e. 620 ml for $1.43	$_____	825 ml for $1.68	$_____	_____
f. 740 g for $1.79	$_____	960 g for $2.09	$_____	_____
g. 1,200 ml for $1.48	$_____	1,500 ml for $2.19	$_____	_____
h. 1,450 g for $3.09	$_____	2,600 g for $4.95	$_____	_____

2. Determine the amount of money saved by purchasing the better buy.

Weight and Cost of Smaller Can	Smaller Can Unit Price	Weight and Cost of Larger Can	Larger Can Unit Price	Saving Per Unit	Total Saving
a. 310 g for 29¢	_____¢	520 g for 39¢	_____¢	_____¢	_____¢
b. 230 ml for 36¢	_____¢	720 ml for 85¢	_____¢	_____¢	_____¢
c. 580 g for 48¢	_____¢	810 g for 79¢	_____¢	_____¢	_____¢
d. 1.3 l for $1.86	$_____	2.5 l for $3.29	$_____	$_____	$_____
e. 1,250 ml for $2.09	$_____	3.6 l for $4.25	$_____	$_____	$_____

3. Determine the amount of money saved by purchasing the better buy.

a. One 227 g box of cornflakes for 43¢
One 340 g box of cornflakes for 55¢ __________

b. One 335 g jar of strawberry jelly for 63¢
One 500 g jar of strawberry jelly for \$1.19 ________

c. Four 336 ml cans of vegetable juice for 99¢
One 1.3 l can of vegetable juice for 63¢ ________

d. One 700 g jar of applesauce for 59¢
Two 420 g jars of applesauce for 79¢ ________

e. One 960 ml jar of mouthwash for \$1.59
One 540 ml jar of mouthwash for \$1.09 ________

f. Two 210 ml jars of shampoo for \$1.12
One 750 ml jar of shampoo for \$2.15 ________

g. Three 75 g cans of soup for \$1.00
One 312 g can of soup for 65¢ ________

h. Two 235 g bags of popcorn for \$1.35
Three 140 g bags of popcorn for \$1.09 ________

Counting Change Unit 2

It would be unfortunate that after spending so much time searching out the best buy, we did not spend another moment to count our change at the checkout counter. The cashiers usually have little or nothing to gain by giving us the wrong change. However they can make errors and we would not want to lose the saving we had gained by careful shopping.

Perhaps the best way to count the change is to add the total bill to the amount we receive from the cashier. The sum of the two should be the money we gave her. As an example, consider the fact that a *$5.00 bill is given in payment of a $2.36 purchase.*

1. Begin by **adding** 4¢ to the total purchase:
$2.36 + 4¢ = $2.40

2. Now add 10¢ to the sum just found:
$2.40 + 10¢ = $2.50

3. To that sum add either 2 quarters or 1 halfdollar:
$2.50 + 50¢ = $3.00

4. Finally, 2 $1 bills are added to the $3.00:
$3.00 + $2.00 = $5.00

$2.40 $2.50 $3.00 $5.00

$2.36 + 4¢ + 10¢ + 50¢ + $2

Unit 2

Example 1 Determine the change that will be received if a $10-bill is given in payment of a $3.78 purchase.

Solution

$3.80 $3.90 $4.00 $5.00 $10.00

$3.78 + 2¢ + 10¢ + 10¢ + $1 + $5

Change = 2¢ + 10¢ + 10¢ + $1 + $5

There are times when we have a great deal of change. In order to avoid accumulating more change on a purchase costing \$3.07, we begin by giving the cashier 7¢. After doing this, we can then consider the cost of the purchase to be \$3.00 rather than \$3.07.

Example 2 Determine the change that will be received if \$10 and 3¢ is given in payment of a \$6.68 purchase.

Explanation Deduct 3¢ from \$6.68 and consider the purchase to be *\$6.65* rather than \$6.68.

Solution

	\$6.75	\$7.00	\$8.00	\$9.00	\$10.00
\$6.65	+ 10¢	+ 25¢	+ \$1	+ \$1	+ \$1

CHANGE

1¢	5¢	10¢	25¢	50¢	\$1	\$5	\$10
		1	1		3		

Exercises

1. Draw a table similar to the following and indicate how much change will be returned if a \$1 bill is given for each of the purchases shown.

Amount of Purchase	Change 1¢	5¢	10¢	25¢	50¢
85¢					
75¢					
67¢					
54¢					
36¢					
28¢					
19¢					
6¢					

2. Draw a table similar to the following and indicate how much change will be returned if a $5-bill is given for each of the purchases shown.

Amount of Purchase	Change					
	1¢	5¢	10¢	25¢	50¢	$1
$4.50						
$4.16						
$4.02						
$3.78						
$2.53						
$1.89						
98¢						
59¢						
25¢						

Unit 2

3. Draw a table similar to the following and indicate how much change will be returned if the amount shown is given for each of the purchases listed.

Amount of Purchase	Payment Made	Change							
		1¢	5¢	10¢	25¢	50¢	$1	$5	$10
62¢	$1.02								
83¢	$1.03								
54¢	$1.04								
26¢	$1.01								
$1.08	$5.08								
$1.17	$5.02								
$3.25	$5.25								
$2.62	$5.12								
$4.29	$10.04								
$1.57	$10.07								
$1.84	$10.09								

Unit 3 Discount Buying

Section 1 Computing the Discount
Part 1

Let us turn our attention now to purchases other than those made at a supermarket. Almost daily we see advertisements similar to the one below.

25% Discount
Everything in Stock

Had we been planning to purchase an $89.95 bicycle from that store, our immediate thought would be, How much has the price been reduced?

To answer this question we must realize that the term 25% Discount is just a short way for writing:

The discount is 25% of the original price.

However,

1. The word "is" implies the equality symbol (=).

and 2. The word "of" implies multiplication (×).

This sentence can be translated as follows:

The discount is 25% of the original price.

↓ ↓ ↓ ↓ ↓

Discount = 25% × $89.95

At this point 25% is renamed as .25 and the discount is computed:

$$\begin{aligned} \text{Discount} &= 25\% \times \$89.95 \\ &= .25 \times \$89.95 \\ &= \$22.4875 \text{ or } \$22.49 \end{aligned}$$

Example 1 Phyllis Samuels was given a 15% discount on an FM radio that was originally priced at $80. Determine the discount Phyllis received.

Solution

$$\begin{aligned} \text{Discount} &= 15\% \text{ of the original price} \\ &= 15\% \times \$80 \\ &= .15 \times \$80 \\ &= \$12 \end{aligned}$$

In the example above we were asked to determine the amount the price had been reduced. Perhaps even more important to us is the answer to the question,

"How much will I have to pay for the article?"

This amount can easily be found by simply subtracting the discount from the original price.

Unit 3

Sale price = Original price − Discount

Example 2 George Erwin was able to purchase a \$21.99 hammock at a discount of 40%. How much did he pay for the hammock?

Solution

$$\begin{aligned} \text{Discount} &= 40\% \text{ of the original price} \\ &= 40\% \times \$21.99 \\ &= .40 \times \$21.99 \\ &= \$8.796 \text{ or } \$8.80 \end{aligned}$$

$$\begin{aligned} \text{Sale price} &= \text{Original price} - \text{Discount} \\ &= \$21.99 - \$8.80 \\ &= \$13.19 \end{aligned}$$

Exercises A

Do you recall how to do the computation in the following exercises? This computation is the same as you will be doing when finding the discount on a purchase.

1. Rename each percent numeral as a decimal.

a.	15%	________	d.	20%	________
b.	28%	________	e.	6%	________
c.	50%	________	f.	9%	________

2. Determine each product to the nearest cent.

a. 20% of $50	$________	e. 10% of $19.95	$________
b. 30% of $75	________	f. 40% of $37.50	________
c. 25% of $84	________	g. 35% of $3.99	________
d. 18% of $69	________	h. 27% of $119.75	________

B

1. Determine each discount.

	Original Price	Discount Rate	Discount
a.	$30	20%	$________
b.	50	40%	________
c.	75	10%	________
d.	129	35%	________
e.	169	18%	________
f.	227	6%	________

2. Determine each discount to the nearest cent.

	Original Price	Discount Rate	Discount
a.	$22.50	15%	$________
b.	8.75	12%	________
c.	19.25	25%	________
d.	109.95	37%	________
e.	235.90	43%	________
f.	359.95	56%	________

3. Determine each sale price.

	Original Price	Discount Rate	Discount	Sale Price
a.	$40	30%	$________	$________
b.	20	10%	________	________
c.	65	60%	________	________
d.	130	45%	________	________
e.	145	31%	________	________

4. Determine each sale price to the nearest cent.

	Original Price	Discount Rate	Discount	Sale Price
a.	$7.50	14%	$________	$________
b.	34.50	38%	________	________
c.	156.85	55%	________	________
d.	249.50	16%	________	________
e.	189.95	25%	________	________

C

1. A garden umbrella that usually sells for $39.95 is placed on sale at a discount of 35%.

a. To the nearest cent, what discount has the purchaser been offered? ________

b. How much will the purchaser have to pay for the umbrella? ________

2. Men's nylon stretch socks that regularly sell for $1.86 a pair can be purchased during a sale at a discount of 28%.

a. What is the normal price for three pairs of these socks? ________

b. How much can a person save by buying three pairs of these socks during the sale? ________

c. What is the cost of three pairs of these socks during the sale? ________

Part 2

The discount rates we have examined so far have been whole numbers such as

15%, 25%, 34%.

When discount rates are quoted, the fraction part of those numbers is usually either

$\frac{1}{4}$, $\frac{1}{2}$, or $\frac{3}{4}$.

For rates involving these fractions, it would be best to rename the fractions as decimals.

$$\tfrac{1}{4} = .25 \qquad \tfrac{1}{2} = .50 \qquad \tfrac{3}{4} = .75$$

Thus, a discount rate quoted as $16\frac{1}{4}\%$,
should be renamed as 16.25%.
Doing this will greatly simplify the computation.

Example 1 Permanent-press dress slacks that are normally \$23.95, are sold during a sale at a discount of 18 1/2%. To the nearest cent, what is the sale price of these slacks?

Solution

$$\begin{aligned}\text{Discount} &= 18\tfrac{1}{2}\% \text{ of original price}\\ &= 18.5\% \times \$23.95\\ &= .185 \times \$23.95\\ &= \$4.43075 \text{ or } \$4.43\\ \text{Sale price} &= \text{Original price} - \text{Discount}\\ &= \$23.95 - \$4.43\\ &= \$19.52\end{aligned}$$

In order to make our computation as simple as possible it is best to rename a discount rate as a decimal numeral rather than fraction numeral. However, in the event those rates are either 33 1/3% or 66 2/3%, the computation would be far easier were they renamed as fractions. Thus,

$$\boxed{33\tfrac{1}{3}\%} = \frac{33\frac{1}{3}}{100} = 33\tfrac{1}{3} \div 100 = \frac{100}{3} \div 100 = \frac{1}{3}$$

Similarly, it can be shown that 66 2/3% = 2/3.

Example 2 Emily Parker waited for a sale to buy her vacuum cleaner. The regular price of the cleaner is \$84.95. When the sale finally did arrive, the discount rate was 33 1/3%. To the nearest cent, how much did Emily have to pay for the cleaner?

Solution

$$\begin{aligned}\text{Discount} &= 33\tfrac{1}{3}\% \text{ of the original price}\\ &= \tfrac{1}{3} \times \$84.95\\ &= \frac{\$84.95}{3}\\ &= \$28.32\end{aligned}$$

$$\begin{aligned}\text{Sale price} &= \text{Original price} - \text{Discount}\\ &= \$84.95 - \$28.32\\ &= \$56.63\end{aligned}$$

Exercises

A

Do you recall how to do the computation in the following exercises? This computation is the same as you will be doing when finding the discount on a purchase.

1. Rename each percent numeral as a decimal.

a. 12.5% ________ d. 4.25% ________
b. 17.5% ________ e. 36.75% ________
c. 23.25% ________ f. 9.75% ________

2. Rename each percent numeral as a decimal.

a. $14\frac{1}{2}\%$ ________ d. $7\frac{1}{4}\%$ ________
b. $5\frac{1}{2}\%$ ________ e. $36\frac{3}{4}\%$ ________
c. $27\frac{1}{4}\%$ ________ f. $4\frac{3}{4}\%$ ________

3. Find the products.

a. 16.5% × 40 ________ d. 6.25% × 45 ________
b. 20.5% × 56 ________ e. 4.75% × .86 ________
c. 9.25% × 30 ________ f. 27.75% × 3.25 ________

4. Find the products to the nearest cent.

a. $12\frac{1}{2}\%$ of \$70 \$________ d. $2\frac{1}{4}\%$ of \$42.50 \$________
b. $6\frac{1}{2}\%$ of \$1.34 ________ e. $27\frac{3}{4}\%$ of \$5.95 ________
c. $15\frac{1}{4}\%$ of \$16 ________ f. $4\frac{3}{4}\%$ of \$37.75 ________

5. Find the products to the nearest cent.

a. $\frac{1}{3}$ × \$28 \$________ d. $\frac{1}{3}$ × \$14.50 \$________
b. $\frac{2}{3}$ × \$47 ________ e. $\frac{2}{3}$ × \$32.75 ________
c. $\frac{2}{3}$ × \$124 ________ f. $\frac{2}{3}$ × \$56.90 ________

B

1. Determine each discount to the nearest cent.

	Original Price	Discount Rate	Discount
a.	\$55	$12\frac{1}{2}\%$	\$________
b.	90	$17\frac{1}{2}\%$	________
c.	70	$5\frac{1}{4}\%$	________
d.	39	$16\frac{1}{4}\%$	________
e.	140	$24\frac{3}{4}\%$	________
f.	175	$36\frac{3}{4}\%$	________

2. Determine each sale price to the nearest cent.

	Original Price	Discount Rate	Discount	Sale Price
a.	$ 8.25	$14\frac{1}{2}$%	$______	$______
b.	23.90	$6\frac{1}{2}$%	______	______
c.	35.75	$8\frac{1}{4}$%	______	______
d.	129.50	$20\frac{1}{4}$%	______	______
e.	229.80	$12\frac{3}{4}$%	______	______

3. Determine the discounts.

	Original Price	Discount Rate	Discount
a.	$ 78	$33\frac{1}{3}$%	$______
b.	96	$33\frac{1}{3}$%	______
c.	135	$66\frac{2}{3}$%	______
d.	189	$66\frac{2}{3}$%	______
e.	258	$66\frac{2}{3}$%	______

4. Determine each sale price to the nearest cent.

	Original Price	Discount Rate	Discount	Sale Price
a.	$10.95	$33\frac{1}{3}$%	$______	$______
b.	37.75	$33\frac{1}{3}$%	______	______
c.	8.97	$66\frac{2}{3}$%	______	______
d.	47.59	$66\frac{2}{3}$%	______	______
e.	98.95	$66\frac{2}{3}$%	______	______

C

1. A ball microphone is placed on sale at a discount rate of 33 1/3%. The normal price of the microphone is $23.95.

a. How much will a person save by purchasing the microphone during the sale? ______

b. What is the sale price of the microphone? ______

2. The Radio Electronics Shop offers its customers a professional model 4-track stereo tape deck at $285 less a discount of 33 1/3%. The Stereo Shop sells this same deck at $299.50 less a discount of 37 1/2%.

a. What is the selling price of the stereo deck at the Radio Electronics shop? ______

b. What is the selling price of the stereo deck at the Stereo Shop? ______

c. How much can be saved by purchasing this deck at the smaller of these two prices? ______

Section 2 Computing the Discount Rate on a Purchase

Consider the advertisement below.

$5 Off on the purchase
of any shirt
Original value $17

A reaction we often have after reading the above is, Just what discount *rate* have we been offered?

The answer to this question lies in the relation that exists between the operations of multiplication and division. For example,

$$30 = 6 \times 5.$$

To check the multiplication above we perform the operation of division.

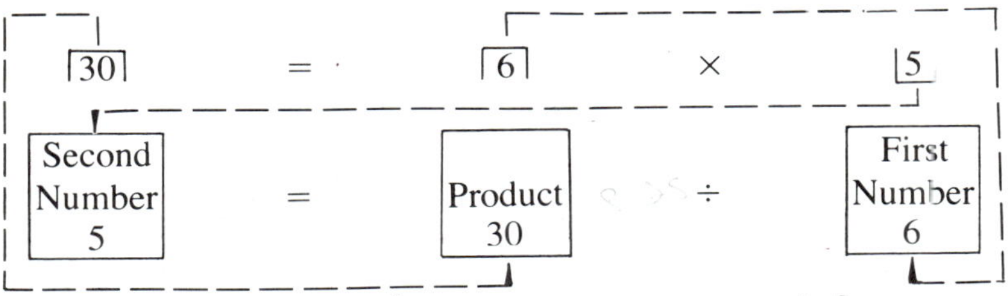

A similar situation exists when finding the discount rate on a purchase. In the preceding unit we learned that the discount is determined as follows:

Discount = Original price × Discount rate

So, it should be true that,

Discount rate = Discount ÷ Original price

Second Number (Discount rate) — **Product** (Discount) — **First Number** (Original price)

Example 1 An automobile tire that normally sells for $40.20 is reduced by $10.50 during a sale. What is the discount rate offered during this sale? Express your answer to the nearest whole percent number.

Solution

Discount rate = Discount ÷ Original price
= $10.50 ÷ $40.20
= .261 (carry division to 3 places)
= 26.1% or 26%

In the solution above it was necessary to rename a decimal as a percent. Prior to now when renaming a percent as a decimal we moved the decimal point two places to the left.

Percent	to	Decimal
73.4%	⟶	.734

To the left

To rename a decimal as a percent, we reverse the direction and move the decimal point two places to the right.

Decimal	to	Percent
.734	⟶	73.4 %

To the right.

Example 2 A suit that regularly sells for $119 is priced during a sale at $89.95. To the whole percent number, what is the discount rate being offered the purchaser?

Explanation The formula for finding the discount rate requires that we know the discount. This can be found by subtracting the selling price from the original value of the suit.

Solution

$$\begin{aligned}\text{Discount} &= \text{Original price} - \text{Selling price}\\ &= \$119.00 - \$89.95\\ &= \$29.05\end{aligned}$$

$$\begin{aligned}\text{Discount rate} &= \text{Discount} \div \text{Original price}\\ &= \$29.05 \div \$119\\ &= .244\\ &= 24.4\% \text{ or } 24\%\end{aligned}$$

Articles are often priced with numbers such as $6.95 or $6.98 rather than $7.00. This is done for a price of $6.98 gives an impression of being a great deal less than $7.00. Rounding these prices to the nearest dollar has little effect upon the discount rate. However, computation with the rounded numbers requires less effort.

Unit 3

Example 3 A can opener that originally sold for $12.99 is priced at $9.98 during a sale. What is the discount rate on this article?

Solution

Original price of $12.99 is rounded to $13.
Discount price of $9.98 is rounded to S10.

$$\begin{aligned}\text{Discount} &= \text{Original price} - \text{Discount price}\\ &= \$13 - \$10\\ &= \$3\end{aligned}$$

$$\begin{aligned}\text{Discount rate} &= \text{Discount} \div \text{Original price}\\ &= \$3 \div \$13\\ &= .231\\ &= 23.1\% \text{ or } 23\%\end{aligned}$$

Exercises

A

Do you recall how to do the computation in the following exercises? This computation is the same as you will be doing when finding the discount rate on a purchase.

1. Rename each decimal as a percent.

a.	.35	________%	d.	.04	________%
b.	.62	________%	e.	.07	________%
c.	.41	________%	f.	.01	________%

2. Rename each decimal as a percent. Round each answer to the nearest whole percent number.

a.	.362	________%	d.	.078	________%
b.	.546	________%	e.	.013	________%
c.	.042	________%	f.	.029	________%

3. Round each to the nearest dollar.

a.	$5.95	$________	c.	$19.96	$________
b.	$14.98	________	d.	$39.99	________

B

1. Determine the discount rates. Round each answer to the nearest whole percent number.

	Original Price	Discount	Discount Rate
a.	$ 5	$ 2	________%
b.	40	25	________%
c.	9	4	________%
d.	29	4	________%
e.	135	16	________%
f.	375	124	________%
g.	527	227	________%

2. Determine each discount rate to the nearest whole number percent numeral.

Item	Original Price	Sale Price	Discount	Discount Rate
a. Camera	$ 48	$ 39	$________	________%
b. Portable radio	27	20	________	________%
c. AM/FM recorder	67	49	________	________%
d. Zoom Super 8	260	195	________	________%
e. 19-inch Color TV	375	298	________	________%

3. Before determining the discount rate on each of the following purchases, round both the original and discount prices to the nearest dollar. Round each answer to the nearest whole percent number.

	Original Price	Sale Price	Approximate Discount	Discount Rate
a. LED watch	$28.95	$17.92	$_____	_____%
b. 4-pc suit	95.98	69.95	_____	_____%
c. Blouse	14.98	8.95	_____	_____%
d. Calculator	125.85	89.92	_____	_____%
e. Exercise bicycle	169.95	109.99	_____	_____%
f. Steel cookware	130.88	63.88	_____	_____%

C

Unit 3

1. Girls' cotton knit polo shirts that regularly sell from $2.50 to $4.00 are placed on sale at $1.50 each.

a. What discount rate would a purchaser receive if she bought a $2.50 polo shirt at the time of the sale? _____

b. What discount rate would a purchaser receive if she bought a $4.00 polo shirt at the time of the sale? _____

2. A national magazine offered to sell 30 issues to subscribers for $3. The regular subscription rate is $5.77. However, if people were to buy the 30 issues at a newsstand, they would have to pay $15.

a. What discount rate would people receive on this special offer over what they would have to pay if they purchased the magazine at a newsstand? _____

b. What discount rate would people receive on this special offer over what they would have to pay if they purchased the magazine at the regular subscription rate? _____

3. Before finding the discount rate on an article, it usually saves a great deal of time if the regular price and the sale price are rounded either to the nearest dime, or the nearest quarter. To what amount would you round each of the following prices?

a.	$1.49	$_____	e.	39¢	_____¢	i.	$16.48	$_____
b.	$1.74	_____	f.	69¢	_____¢	j.	$94.73	_____
c.	$2.24	_____	g.	88¢	_____¢	k.	$49.49	_____
d.	$1.58	_____	h.	19¢	_____¢	l.	$72.29	_____

4. No-iron dungarees that regularly sell for $7.19 can be bought during a sale at $3.74. What discount rate will Ann Perkins receive if she purchases a pair of these dungarees during the sale? Before computing your answer, round the original selling price and the sale price respectively to the nearest dime and nearest quarter. ________

5. The Bradford Shop regularly sells men's socks for 85¢ a pair. During their 50th anniversary sale, they placed these socks on sale at 3 pairs for $1.75.

a. How much would Tom Baxter have to pay for three pairs of these socks if he bought them when they were not on sale? ________

b. What discount rate would Tom receive if he purchased these socks at the time of the sale? Find your answer to the nearest whole percent number. ________

Unit 4 Installment Purchasing

Quite often we do not have the cash to pay for the purchase of items such as a TV set, refrigerator, furniture and the like. Many merchants are more than willing to allow us to buy these articles on the installment plan. That is, we make a small down payment followed by monthly payments until the entire debt is paid.

Under the installment sale, the merchants profit in two ways:

1. They will make the normal profit on the sale of the article.
2. They will make an additional profit because the total of the monthly payments and the down payment is greater than the normal selling price of the article.

The additional charge paid under an installment plan purchase is called by a variety of names. The most common are:

1. Installment Charge
2. Carrying Charge
3. Interest Charge

In the examples that follow we will examine both the down payment and the monthly payments on an installment purchase.

Example 1 A $425 refrigerator can be purchased by making a down payment of 15%. How large is this payment?

Solution

$$\text{Down payment} = \underset{\downarrow}{15\%}\ \underset{\downarrow}{\text{of}}\ \underset{\downarrow}{\text{cash price}}$$
$$= .15 \times \$425$$
$$= \$63.75$$

Example 2 A TV set can be purchased by making monthly payments of $31.25 over a period of 24 months. What is the total cost of this set?

Solution

$$\text{Total cost} = \text{Number of months} \times \text{Monthly payment}$$
$$= 24 \times \$31.25$$
$$= \$750$$

Example 3 The cash price of a washing machine is $269.95. This washer can be purchased by making a 20% down payment and 18 monthly payments of $14.50 each. What is the installment price of this article?

Solution

$$\text{Down payment} = 20\% \text{ of Cash price}$$
$$= .20 \times \$269.95$$
$$= \$53.99$$

$$\text{Total of monthly payments} = \text{Number of months} \times \text{Monthly payment}$$
$$= 18 \times \$14.50$$
$$= \$261$$

$$\text{Installment price} = \text{Down payment} + \text{Total of monthly payments}$$
$$= \$53.99 + \$261$$
$$= \$314.99$$

Unit 4

Example 4 A 10-speed bicycle can be purchased for $139.95. Arrangements can be made to purchase this same bicycle by making a down payment of 15% and 9 monthly payments of $14.95 each. How much can be saved by buying this bicycle for cash?

Solution

$$\begin{aligned}\text{Down payment} &= 15\% \text{ of cash price}\\ &= .15 \times \$139.95\\ &= \$20.99\end{aligned}$$

$$\begin{aligned}\text{Total of monthly payments} &= \text{Number of months} \times \text{Monthly payment}\\ &= 9 \times \$14.95\\ &= \$134.55\end{aligned}$$

$$\begin{aligned}\text{Installment price} &= \text{Down payment} + \text{Total of monthly payments}\\ &= \$20.99 + \$134.55\\ &= \$155.54\end{aligned}$$

$$\begin{aligned}\text{Installment charge} &= \text{Installment price} - \text{Cash price}\\ &= \$155.54 - \$139.95\\ &= \$15.59\end{aligned}$$

Exercises

1. Determine the down payment for each installment purchase.

	Cash Price	Percent Down	Down Payment
a.	$250	10%	$______
b.	375	20%	______
c.	450	25%	______
d.	745	35%	______
e.	2,250	$33\frac{1}{3}\%$	______

2. Determine the down payment to the nearest cent for each purchase.

	Cash Price	Percent Down	Down Payment
a.	$72.50	10%	$______
b.	99.75	20%	______
c.	125.95	5%	______
d.	55.95	30%	______
e.	249.95	$33\frac{1}{3}\%$	______

3. Determine the total amount paid in monthly installments for each purchase.

	Number of Months	Monthly Payment	Total of Monthly Payments
a.	8	$23	$________
b.	12	36	________
c.	16	49	________
d.	9	12.50	________
e.	18	8.75	________
f.	24	54.25	________

4. Determine the installment price for each installment purchase.

	Down Payment	Number of Months	Monthly Payment	Total of Monthly Payments	Installment Price
a.	$ 50	6	$20	$________	$________
b.	75	9	15	________	________
c.	140	12	40	________	________
d.	69	15	18	________	________
e.	18	36	29	________	________

D.P + T.M.P

Unit 4

5. Determine the installment price for each installment purchase.

	Down Payment	Number of Months	Monthly Payment	Total of Monthly Payments	Installment Price
a.	$ 57.50	8	$15.50	$________	$________
b.	49.95	12	10.25	________	________
c.	134.25	36	35.50	________	________
d.	129.95	18	34.50	________	________

6. Determine the installment charge for each installment purchase.

	Down Payment	Number of Months	Monthly Payment	Installment Price	Cash Price	Installment Charge
a.	$ 14.50	12	$ 4.50	$________	$ 54.50	$________
b.	67.95	24	10.25	________	267.95	________
c.	89.75	30	15.45	________	429.75	________
d.	105.98	18	36.75	________	705.98	________

7. Determine the installment charge for each installment purchase.

	Cash Price	Percent Down	Down Payment	Number of Months	Monthly Payment	Installment Price	Installment Charge
a.	$150	20%	$____	7	$22	$____	$____
b.	180	25%	____	12	14	____	____
c.	175	30%	____	18	9	____	____
d.	450	15%	____	24	18	____	____
e.	675	40%	____	36	14	____	____

B

1. Anna Fernicola could purchase a top-loading washer for $278 in cash. Unfortunately, she did not have the money to do this, so she made arrangements to put down a deposit of $35 and pay off the balance in 15 monthly payments of $20 each. How much extra did Anna have to pay by buying the washer on the installment plan? ________

2. The first thing Joe Brooks did on reaching 18 was to buy a used motorcycle for $478.85. Since Joe didn't have the cash to pay, the dealer permitted him to make a down payment of $53.85 and pay off the rest in 30 monthly installments of $19.27 each. How much money could Joe have saved had he had the cash to pay for the motorcycle? ________

3. Joyce Field bought her vacuum cleaner from a door-to-door salesperson by making no down payment, but agreeing to pay $4.75 each week for a year. Had she taken the time to visit the discount house in her neighborhood, she would have found that this same machine could be bought for $139.95 cash. How much money could Joyce have saved had she had the cash and bought the vacuum cleaner at a discount house? ________

4. After long hours of pushing his lawn mower, Bob Cramer went out and bought a deluxe riding mower that could double as a snowplow during the winter season. If he had had the cash, the mower would have cost him $729. But he did not, so he arranged to make a 10% down payment and the rest he agreed to pay off monthly over a two-year period. Each payment was $35.15. How much greater is the installment price than the cash price?

____

5. A 120-watt matched solid-state stereo system can be bought for $319.95 cash. If the buyers don't have cash, the dealer is willing to arrange an easy-pay plan. This consists of making a down payment of 5% and weekly payments of $4.65 each over a two-year period. How much could people save when buying this system if they did not take advantage of the easy-pay plan?

____

The Charge Account Unit 5

Section 1 Minimum Payment

Unit 5

Many department stores have developed deferred-payment plans that are much the same as the arrangements under an installment purchase. Although these plans may vary from store to store, basically they follow either one or the other of the patterns described below.

1. **The Percent Plan** – the minimum payment for any month is a specific percent or fraction of the new balance.
2. **The Table Plan** – the minimum payment for any month is the amount listed in a table such as the one here.

If New Balance Is:	The Minimum Payment Will Be:
$ .01 to $10	Total balance
10.01 to 200	$10
200.01 to 250	$20
250.01 to 300	$25
300.01 to 350	$30
350.01 to 400	$35
Over $400	10% of Balance

Example 1 The new balance on a charge account is $156.24. The payment each month must be at least 1/6 of the monthly balance. What is the smallest payment that can be made on this debt?

Solution

$$\text{Minimum payment} = \tfrac{1}{6} \text{ of new balance}$$
$$= \tfrac{1}{6} \times \$156.24$$
$$= \frac{\$156.24}{6}$$
$$= \$26.04$$

Example 2 The new balance listed on a charge account statement is $247.95. If the Table Plan shown on page 135 is used, what minimum payment must be made?

Explanation

If New Balance Is: → The Minimum Payment Will Be:

Read Down → 200.01 to 250 ——————→ $20

Read across — Minimum payment

Solution

Minimum payment for $247.95 = $20

Example 3 The new balance is $634.75. If the Table Plan is used, what is the minimum payment?

Explanation

If New Balance Is: → The Minimum Payment Will Be:

Read Down → Over $400 ——————→ 10% of Balance

Read Across — Minimum Payment

Solution

$$\text{Minimum payment} = 10\% \text{ of Balance}$$
$$= .10 \times \$634.75$$
$$= \$63.475 \text{ or } \$63.48$$

Exercises

1. Determine the minimum payment on each new balance if the Table Plan is used to determine the payment.

	New Balance	Payment		New Balance	Payment
a.	$ 75.40	$______	d.	$672.50	$______
b.	329.60	______	e.	595.86	______
c.	8.24	______	f.	429.95	______

2. Determine the minimum payment on each new balance if the Percent Plan is used to determine the payment.

	New Balance	Payment	Minimum Payment
a.	$125	10%	$______
b.	176	12%	______
c.	258.25	15%	______
d.	346.83	16%	______
e.	595.97	8%	______

Unit 5

3. Determine the minimum payment on each new balance if the Percent Plan is used to determine the payment.

	New Balance	Fraction	Minimum Payment
a.	$165	$\frac{1}{5}$	$______
b.	234	$\frac{1}{6}$	______
c.	258.45	$\frac{1}{5}$	______
d.	429.50	$\frac{1}{6}$	______
e.	574.82	$\frac{1}{8}$	______
f.	695.73	$\frac{1}{8}$	______

Section 2 Finance Charge

In Section 1 we examined how to determine the minimum payment on a charge account. At this time we want to take a look at the numbers that frequently appear on the bottom line of the monthly statement.

Previous Balance	Plus Finance Charge	Plus Purchases	New Balance	Minimum Payment Due
347.28	5.21	37.59	390.08	35.00

This is the new balance we used in Part 1

This is the minimum payment we found in the table in Part 1

Our concern is with the first two items listed on the bottom line.

Previous Balance	Plus Finance Charge
347.28	5.21

This is the amount still owed after making last month's payment.

This is the charge for not paying the $347.28 last month.

The charge for failure to pay our entire debt varies from store to store. However, the table below is typical of the manner in which these finance rates are expressed.

Monthly Finance Charge	
Previous Balance	Charge
$500 or less	$1\frac{1}{2}$% of balance
Excess over $500	1% of excess

Example 1 Determine the finance charge on a previous balance of $259.95. Use the finance charge table above.

Solution

$$\text{Finance charge} = 1\tfrac{1}{2}\% \text{ of previous balance}$$
$$= 1.5\% \times \$259.95$$
$$= .015 \times \$259.95$$
$$= \$3.89925 \text{ or } \$3.90$$

Example 2 Use the finance charge table above to determine the finance charge on a previous balance of $742.69.

Explanation The finance charge consists of two parts:

1. The charge on the first $500
2. The charge on the excess over $500

Unit 5

Solution

$$\text{Charge on first } \$500 = 1\tfrac{1}{2}\% \text{ of } \$500$$
$$= 1.5\% \times \$500$$
$$= .015 \times \$500$$
$$= \$7.50$$

$$\text{Excess over } \$500 = \$742.69 - \$500$$
$$= \$242.69$$

$$\text{Charge on excess} = 1\% \text{ of excess}$$
$$= .01 \times \$242.69$$
$$= \$2.4269 \text{ or } \$2.43$$

$$\text{Finance charge} = \text{Charge on } \$500 + \text{Charge on excess}$$
$$= \$7.50 + \$2.43$$
$$= \$9.93$$

Example 3 The credit agreement of a department store calls for the following:

1. The minimum payment is 1/8 of the new balance.
2. The finance charge is 1.65% of the previous balance.

With this information, complete the bottom line of the statement below.

Previous Balance	Plus Finance Charge	Plus Purchases	New Balance	Minimum Payment Due
536.29		96.75		

Solution

$$\text{Finance charge} = 1.65\% \text{ of previous balance}$$
$$= .0165 \times \$536.29$$
$$= \$8.85$$

$$\text{New balance} = \text{Previous balance} + \text{Finance charge} + \text{Purchases}$$
$$= \$536.29 + \$8.85 + \$96.75$$
$$= \$641.89$$

$$\text{Minimum payment} = \tfrac{1}{8} \text{ of new balance}$$
$$= \tfrac{1}{8} \times \$641.89$$
$$= \$80.24$$

Completed Bottom Line

Previous Balance	Plus Finance Charge	Plus Purchases	New Balance	Minimum Payment Due
536.29	8.85	96.75	641.89	80.24

Exercises A

Do you recall how to do the computation in the following exercises? This computation is the same as you will be doing when finding finance charges.

1. Rename each percent numeral as a decimal.

a.	1.35%	______	d.	.2%	______	g.	$1\frac{1}{2}$%	______
b.	1.65%	______	e.	.3%	______	h.	$1\frac{1}{4}$%	______
c.	1.85%	______	f.	.1%	______	i.	$1\frac{3}{4}$%	______

2. Determine each product to the nearest cent.

a.	1.45% of $68	$______
b.	1.32% of $75	______
c.	$1\frac{1}{2}$% of $59.50	______
d.	$1\frac{1}{4}$% of $237.50	______
e.	$1\frac{3}{4}$% of $169.75	______

Unit 5

B

1. Use the monthly finance charge table on page 138 to determine the finance charge.

	Previous Balance	Finance Charge		Previous Balance	Finance Charge
a.	$ 84	$______	d.	$136.50	$______
b.	165	______	e.	349.75	______
c.	279	______	f.	467.98	______

2. Use the monthly finance charge table on page 138 to determine the finance charge.

	Previous Balance	Charge On First $500	Excess Over $500	Charge On Excess	Finance Charge
a.	$650	$______	$______	$______	$______
b.	725	______	______	______	______
c.	840.50	______	______	______	______
d.	928.35	______	______	______	______
e.	789.95	______	______	______	______

3. The terms of a department store's Flexible Charge Account are as follows:

Previous Balance		Finance Charge
$700 or less	....	1.65% of balance
Excess over $700	....	1% of excess

Under these conditions, determine the finance charge.

	Previous Balance	Charge On $700 or less	Excess Over $700	Charge On Excess	Finance Charge
a.	$240	$______	$______	$______	$______
b.	595	______	______	______	______
c.	627	______	______	______	______
d.	760	______	______	______	______
e.	854	______	______	______	______
f.	987	______	______	______	______

4. The terms of a department store's credit agreement are as follows:

1. The finance charge is in accordance with the table on page 138.
2. The minimum payment due is in accordance with the table on page 135.

Under these conditions, complete the bottom line of each of these monthly statements.

	Previous Balance	Plus Finance Charge	Plus Purchases	New Balance	Minimum Payment Due
a.	$265	$______	$34	$______	$______
b.	370	______	87	______	______
c.	439	______	29	______	______
d.	584	______	12	______	______
e.	675	______	56	______	______

5. The terms of a department store's credit agreement are as follows:

1. The finance charge is in accordance with the table in exercise 3 above.
2. The minimum payment due is 1/6 of the new balance.

Under these conditions, complete the bottom line of each of these monthly statements.

	Previous Balance	Plus Finance Charge	Plus Purchases	New Balance	Minimum Payment Due
a.	$140	$______	$ 30	$______	$______
b.	260	______	75	______	______
c.	685	______	130	______	______
d.	754	______	125	______	______
e.	937	______	184	______	______

C

1. The previous balance on a charge account is $860.

a. If the credit agreement is in accordance with the table on page 138, what will the finance charge be? ________

b. If the credit agreement is in accordance with the table in exercise B3 above, what will the finance charge be? ________

c. How much can be saved if the first agreement is used? ________

2. We learned earlier that credit life insurance can be purchased when borrowing money on a car loan. The same can be done on a charge account. The cost is .1% of the new balance. If the department store uses the credit agreement in the table on page 138, determine the missing fillins below.

Previous Balance	Plus Finance Charge	Plus Purchases	New Balance	Insurance Premium
$620	$________	$80	$________	$________

Review

Review of Chapter 3

1. Determine the unit price for each purchase. Round each answer to the nearest tenth of a cent.

	Total Cost	Number of Units	Unit Price
a.	57¢	6 ounces	________¢ per ounce
b.	89¢	4 quarts	________¢ per quart
c.	$1.85	2 pounds	________¢ per pound
d.	$3.46	18 ounces	________¢ per ounce

2. To the nearest cent, determine the cost per liter for each purchase.

	Total Cost	Number of Milliliters	Number of Liters	Cost Per Liter
a.	67¢	850	________	________¢
b.	59¢	725	________	________¢
c.	$1.26	1540	________	________¢

3. To the nearest tenth of a cent, determine the cost per kilogram for each purchase.

	Total Cost	Number of Grams	Number of Kilograms	Cost Per Kilogram
a.	62¢	740	________	________¢
b.	$1.75	2,500	________	________¢
c.	$3.62	4,750	________	________¢

4. In the following exercises, which of the two purchases is the better buy?

a. 320 grams for 28¢
675 grams for 51¢ ________

b. 580 ml for $1.65
725 ml for $1.92 ________

5. Determine the amount of money that can be saved by purchasing the less expensive of the two items below.

Two 175 ml cans of frozen lemonade for 75¢
One 525 ml can of frozen lemonade for $1.29 ________

6. Mike Jamieson purchased $7.58 worth of food at the supermarket. He gave the checkout clerk a $10 bill in payment. Indicate the change that Mike received. Use a table similar to the one below.

1¢	5¢	10¢	25¢	50¢	$1	$5

7. A television set that regularly sells for $184 was placed on sale at a 25% discount rate.

a. How much discount did a person receive during the sale? ________

b. How much did a person have to pay for the TV set if he purchased it during the sale? ________

8. During its storewide furniture sale, the Camden Company offered a set of porch furniture at 33 1/3% off. If this porch furniture regularly sells for $237, how much will it cost during the sale? ________

9. An article that regularly sells for $20 can be purchased during a sale at a discount of $5. What discount rate is a purchaser receiving? ________

10. A $119.95 electric guitar can be purchased during a sale for $79.95. By rounding off each of these prices to the nearest dollar find the discount rate to the nearest whole percent number.

____

11. Bruce Devlin purchased a typewriter that regularly sells for $137. Since he did not have the cash, he was permitted to buy it on the following terms: $15 down and $8 per month for 18 months. How much could Bruce have saved had he had the cash to pay for the typewriter? ________

Review

12. The terms of a charge account are as follows:

Previous Balance	Charge
$600 or less	$1\frac{1}{2}\%$ of balance
Excess over $600	1% of excess

Determine the finance charge on a previous balance of $865.

____

13. The terms of a department store charge account are the same as those in exercise 12. The minimum monthly payment must be 1/6 of the new balance. With this information, complete the bottom line of the statement below.

Previous Balance	Plus Finance Charge	Plus Purchases	New Balance	Minimum Payment Due
$462.50	$________	$31.10	$________	$________

Chapter 4

Personal Income

Most of us look forward to the day when we will earn the money to buy the things we want. It is very likely that when that day does arrive our earnings will be computed by either one or a combination of the following ways:

1. Hourly wages
2. Piece-rate wages
3. Commission earnings

In this chapter we will examine the computation of earnings under each of these arrangements. In addition, we will take a look at some of the more common payroll deductions taken from our salaries.

The Hourly-Wage System Unit 1

Section 1 The Regular Hourly Wage

Many of us work under an arrangement in which we earn a specific amount of money for each hour of work. In view of this, if our hourly wage is \$5.84, then for 1 hour's work we will receive

$$1 \times \$5.84 \text{ or } \$5.84.$$

Similarly, for 2 hour's work we will receive

$$2 \times \$5.84 \text{ or } \$11.68.$$

And, in general, our weekly earnings are computed as follows:

$$\text{Weekly earnings} = \text{Number of hours} \times \text{Hourly wage}$$

Example 1 Determine the weekly earnings of Carlos Juarez who works 38 hours each week at an hourly wage of \$6.29.

Solution

$$\begin{aligned} \text{Weekly earnings} &= \text{Number of hours} \times \text{Hourly wage} \\ &= 38 \times \$6.29 \\ &= \$239.02 \end{aligned}$$

Example 2 Sally Clemes works a 37 1/2 hour week at \$4.96 per hour. How much does Sally earn each week?

Explanation As we discovered earlier, it would be best to rename the mixed number 37 1/2 as 37.5 in order to make the computation as simple as possible.

Solution

$$\begin{aligned} \text{Weekly earnings} &= \text{Number of hours} \times \text{Hourly wage} \\ &= 37\tfrac{1}{2} \times \$4.96 \\ &= 37.5 \times \$4.96 \\ &= \$186.00 \end{aligned}$$

Within recent years almost all wage contracts have been designed so that the new hourly raise becomes a fixed percent of the old hourly wage. These contracts will usually contain a statement similar to the following:

The hourly raise is 7% of the previous hourly wage.

The examples below will show what this will mean in terms of a person's earnings.

Example 3 As a result of wage negotiations (bargaining), the hourly wage was increased by 7%. If the hourly wage had been \$7.56, how large is the raise?

Solution

$$\text{Raise} = 7\% \text{ of previous hourly wage}$$
$$= .07 \times \$7.56$$
$$= \$.5292 \text{ or } \$.53 \text{ an hour}$$

Example 4 A new contract calls for wages to increase 19 1/2% over a three-year period. If the hourly wage under the old contract was $6.74, what will the hourly wage be at the end of the three-year period?

Solution

$$\text{Raise} = 19\tfrac{1}{2}\% \text{ of previous hourly wage}$$
$$= 19.5\% \times \$6.74$$
$$= .195 \times \$6.74$$
$$= \$1.3143 \text{ or } \$1.31 \text{ an hour}$$

$$\text{New Hourly Wage} = \text{Previous hourly wage} + \text{Raise}$$
$$= \$6.74 + \$1.31$$
$$= \$8.05$$

Unit 1

Exercises A

1. Determine the weekly earnings for each employee.

	Hourly Wage	Hours Worked	Earnings
a.	$ 4.25	38	$________
b.	5.73	37	________
c.	6.95	40	________
d.	8.51	39	________
e.	10.64	36	________

2. Determine the weekly earnings for each employee.

	Hourly Wage	Hours Worked	Earnings
a.	$3.86	$37\frac{1}{2}$	$________
b.	5.64	$38\frac{1}{2}$	________
c.	7.41	$39\frac{1}{2}$	________
d.	8.53	$40\frac{1}{2}$	________

3. Determine the weekly earnings for each employee.

	Mon	Tues	Wed	Thurs	Fri	Total Hours	Hourly Wage	Earnings
a.	7	7	7	7	7	______	$5.42	$______
b.	8	8	8	8	8	______	6.71	______
c.	8	7	7	8	8	______	3.95	______
d.	6	8	7	7	6	______	7.53	______
e.	$6\frac{1}{2}$	$6\frac{1}{2}$	7	7	7	______	4.86	______
f.	$7\frac{1}{2}$	$7\frac{1}{2}$	$7\frac{1}{2}$	$7\frac{1}{2}$	$7\frac{1}{2}$	______	8.44	______

B

1. Determine the hourly raise as a result of the following wage negotiations.

	Previous Hourly Wage	Percent Increase	Hourly Raise
a.	$ 4.92	9%	$______
b.	10.56	9%	______
c.	7.69	8%	______
d.	6.84	6%	______
e.	9.51	17%	______
f.	5.42	$6\frac{1}{2}$%	______
g.	6.07	$5\frac{1}{2}$%	______
h.	8.79	$14\frac{1}{2}$%	______
i.	4.81	6.8%	______
j.	8.36	7.4%	______

2. Determine the new hourly wage that resulted from the following wage negotiations.

	Previous Hourly Wage	Percent Increase	Hourly Raise	New Hourly Wage
a.	$6.12	7%	$______	$______
b.	5.68	12%	______	______
c.	7.74	18%	______	______
d.	9.17	$6\frac{1}{2}$%	______	______
e.	8.45	$7\frac{1}{2}$%	______	______

3. Determine the weekly earnings after the following raises became effective.

	Previous Hourly Wage	Percent Increase	Hourly Raise	New Hourly Wage	Hours Worked	Earnings
a.	$4.26	8%	$______	$______	40	$______
b.	6.08	9%	______	______	38	______
c.	7.15	11%	______	______	36	______
d.	6.84	5.4%	______	______	37	______
e.	8.21	7.3%	______	______	39	______

C

1. Paul Kramer earns $7.63 an hour for a 38 hour week. He recently received a 6 1/2% raise. How much will his weekly earnings be as a result of this raise? ________

2. As a result of wage negotiations Julia Maello received a raise of 7.8%. Her hourly wage before negotiations was $8.47 and she worked a 37 1/2 hour week.

a. What were Julia's weekly earnings before receiving the raise? ________

b. What were Julia's weekly earnings after receiving the raise? ________

c. How much more will her weekly earnings be after receiving the raise? ________

Section 2 Time-and-a-Half and Double Time Pay

Wage contracts contain information showing the *regular hourly wage* an employee will receive. In addition, the contract also contains a statement pertaining to *overtime pay*. Overtime pay is the money we receive when we work more than:

1. A specified number of hours in one day. or
2. A specified number of hours in one week.

> The overtime hourly wage is usually 1 1/2 times the regular hourly wage.

Overtime hourly wage is sometimes called time-and-a-half pay.

Example 1 If the regular hourly wage is $6.73, what will the overtime hourly wage be?

Solution

$$\begin{aligned} \text{Overtime hourly wage} &= 1\tfrac{1}{2} \times \text{Regular hourly wage} \\ &= 1.5 \times \$6.73 \\ &= \$10.095 \end{aligned}$$

Explanation Notice that $10.095 is not rounded to the nearest cent. No rounding is done until the overtime earnings are computed.

Example 2 Frank Tortorello earns time-and-a-half for all work above 7 hours per day. His regular wage is $7.85. How much will he earn during a day in which he works 9 1/2 hours?

Solution

$$\begin{aligned} \text{Overtime hourly wage} &= 1\tfrac{1}{2} \times \text{Regular hourly wage} \\ &= 1.5 \times \$7.85 \\ &= \$11.775 \end{aligned}$$

$$\begin{aligned} \text{Overtime} &= 9\tfrac{1}{2} - 7 \\ &= 2\tfrac{1}{2} \text{ hours} \end{aligned}$$

$$\begin{aligned} \text{Regular earnings} &= 7 \times \text{Regular hourly wage} \\ &= 7 \times \$7.85 \\ &= \$54.95 \end{aligned}$$

$$\begin{aligned} \text{Overtime earnings} &= 2\tfrac{1}{2} \times \text{Overtime hourly wage} \\ &= 2.5 \times \$11.775 \\ &= \$29.4375 \text{ or } \$29.44 \end{aligned}$$

$$\begin{aligned} \text{Total earnings} &= \text{Regular earnings} + \text{Overtime earnings} \\ &= \$54.95 + \$29.44 \\ &= \$84.39 \end{aligned}$$

Example 3 Mary Sullivan is employed by a company that pays time-and-a-half for all work above 38 hours per week. Her regular hourly wage is $6.47. What will her earnings be for a week in which she was asked to work 45 hours?

Solution

$$\begin{aligned}\text{Overtime hourly wage} &= 1\tfrac{1}{2} \times \text{Regular hourly wage}\\ &= 1.5 \times \$6.47\\ &= \$9.705\end{aligned}$$

$$\begin{aligned}\text{Overtime} &= 45 - 38\\ &= 7 \text{ hours}\end{aligned}$$

$$\begin{aligned}\text{Regular earnings} &= 38 \times \text{Regular hourly wage}\\ &= 38 \times \$6.47\\ &= \$245.86\end{aligned}$$

$$\begin{aligned}\text{Overtime earnings} &= 7 \times \text{Overtime hourly wage}\\ &= 7 \times \$9.705\\ &= \$67.935 \text{ or } \$67.94\end{aligned}$$

$$\begin{aligned}\text{Total earnings} &= \text{Regular earnings} + \text{Overtime earnings}\\ &= \$245.86 + \$67.94\\ &= \$313.80\end{aligned}$$

Unit 1

Many wage contracts also call for bonus pay – extra money – if we are asked to work either on Sundays or holidays.

> Holiday or Sunday hourly wage is usually two times the regular hourly wage.

People will often say that they are earning *double-time* for holiday or Sunday work.

Example 4 A person who normally works for \$5.36 an hour is asked to work 8 1/2 hours on Sunday. If paid at the double-time rate, how much will the employee earn for this day's work?

Solution

$$\begin{aligned}\text{Double-time hourly wage} &= 2 \times \text{Regular hourly wage}\\ &= 2 \times \$5.36\\ &= \$10.72\end{aligned}$$

$$\begin{aligned}\text{Sunday earnings} &= 8\tfrac{1}{2} \times \text{Double-time hourly wage}\\ &= 8.5 \times \$10.72\\ &= \$91.12\end{aligned}$$

Exercises

1. Determine the double-time hourly wage for each regular hourly wage.

	Regular Hourly Wage	Double-time Hourly Wage		Regular Hourly Wage	Double-time Hourly Wage
a.	$4.75	$________	c.	$6.29	$________
b.	3.87	________	d.	8.56	________

2. Determine the time-and-a-half hourly wage for each regular hourly wage.

	Regular Hourly Wage	Time-and-a-half Hourly Wage		Regular Hourly Wage	Time-and-a-half Hourly Wage
a.	$4.00	$________	d.	$3.96	$________
b.	5.60	________	e.	8.75	________
c.	7.50	________	f.	9.53	________

3. The hours shown below were worked at the double-time hourly wage. How much did each employee earn?

	Regular Hourly Wage	Double-time Hourly Wage	Holiday Hours	Holiday Earnings
a.	$4.67	$________	4	$________
b.	5.92	________	7	________
c.	7.83	________	6	________
d.	9.56	________	8	________

4. The hours shown below were worked at the time-and-a-half hourly wage. How much did each employee earn?

	Regular Hourly Wage	Time-and-a-half Hourly Wage	Overtime Hours	Overtime Earnings
a.	$4.20	$________	3	$________
b.	6.30	________	2	________
c.	7.67	________	4	________
d.	5.29	________	5	________
e.	8.19	________	$3\frac{1}{2}$	________
f.	7.53	________	$4\frac{1}{2}$	________

5. The maximum hours per week that an employee works at the regular rate is given in the table below. For overtime hours an employee receives the time-and-a-half rate. Determine the total earnings for each employee.

	Regular Rate	Overtime Rate	Hours Worked	Maximum Hours at Reg. Rate	Regular Earnings	Overtime Hours	Overtime Earnings	Total Earnings
a.	$4.80	$_____	42	40	$_____	_____	$_____	$_____
b.	6.50	_____	45	40	_____	_____	_____	_____
c.	6.09	_____	43	40	_____	_____	_____	_____
d.	7.21	_____	42	38	_____	_____	_____	_____
e.	7.50	_____	$43\frac{1}{2}$	40	_____	_____	_____	_____
f.	8.46	_____	$39\frac{1}{2}$	38	_____	_____	_____	_____
g.	9.18	_____	$40\frac{1}{2}$	37	_____	_____	_____	_____

6. The employees below receive overtime pay if they work more than 7 hours during any one day. How many regular hours and how many overtime hours did each work?

	Mon	Tues	Wed	Thurs	Fri	Regular Hours	Overtime Hours
a.	9	6	5	8	7	_____	_____
b.	7	10	9	7	8	_____	_____
c.	8	5	4	9	6	_____	_____
d.	7	8	$7\frac{1}{2}$	7	$7\frac{1}{2}$	_____	_____
e.	7	$7\frac{1}{2}$	$7\frac{1}{2}$	$7\frac{1}{2}$	$7\frac{1}{2}$	_____	_____

Unit 1

7. The employees below receive overtime pay if they work more than 7 hours during any one day. How much will each earn?

	M	T	W	Th	F	Reg. Hrs.	Over. Hrs.	Reg. Rate	Regular Earnings	Overtime Rate	Overtime Earnings	Total Earnings
a.	7	7	7	7	8	_____	_____	$5.20	$_____	$_____	$_____	$_____
b.	9	7	7	9	7	_____	_____	6.70	_____	_____	_____	_____
c.	8	8	8	8	8	_____	_____	7.52	_____	_____	_____	_____
d.	9	9	7	7	8	_____	_____	8.17	_____	_____	_____	_____

Section 3 Premium Pay

Many companies operate their plants 24 hours a day. Under this arrangement the day is divided into three separate work shifts.

8:00 AM to 4:00 PM	Day Shift
4:00 PM to 12:00 Midnight	Evening Shift
12:00 Midnight to 8:00 AM	Night Shift

People usually prefer to work during the day shift. To induce people to work during either the evening shift or night shift a company will offer them premium pay. The hourly wage for working these shifts is greater than the hourly wage paid to the day-shift employee.

Example 1 The evening shift hourly wage at Lauter, Inc. is 8% more than the regular hourly wage. If the regular hourly wage is \$5.27, how much is the evening shift hourly wage?

Solution

$$\begin{aligned}\text{Premium pay} &= 8\% \text{ of Regular hourly wage}\\ &= .08 \times \$5.27\\ &= \$.4216 \text{ or } \$.42\end{aligned}$$

$$\begin{aligned}\text{Evening shift hourly wage} &= \text{Regular hourly wage} + \text{Premium pay}\\ &= \$5.27 + \$.42\\ &= \$5.69\end{aligned}$$

Example 2 The Ramapo Corporation pays its night shift employees 15% more than its day shift employees. Its regular hourly wage is \$6.24. How much is earned by an employee who works a 38-hour week on the night shift?

Solution

$$\begin{aligned}\text{Premium pay} &= 15\% \text{ of Regular hourly wage}\\ &= .15 \times \$6.24\\ &= \$.94\end{aligned}$$

$$\begin{aligned}\text{Night shift hourly wage} &= \text{Regular hourly wage} + \text{Premium pay}\\ &= \$6.24 + \$.94\\ &= \$7.18\end{aligned}$$

$$\begin{aligned}\text{Weekly earnings} &= \text{Number of hours} \times \text{Night shift hourly wage}\\ &= 38 \times \$7.18\\ &= \$272.84\end{aligned}$$

Exercises A

1. Determine the evening shift hourly wage for each employee.

	Regular Hourly Wage	Premium Rate	Premium Pay	Evening Shift Hourly Wage
a.	$4.50	7%	$______	$______
b.	5.70	7%	______	______
c.	4.85	8%	______	______
d.	6.92	8%	______	______
e.	7.59	10%	______	______
f.	8.67	11%	______	______

2. Determine the night shift hourly wage for each employee.

	Regular Hourly Wage	Premium Rate	Premium Pay	Night Shift Hourly Wage
a.	$4.80	16%	$______	$______
b.	6.50	18%	______	______
c.	9.30	18%	______	______
d.	10.15	20%	______	______
e.	9.85	19%	______	______

Unit 1

3. The information below pertains to different people working for different companies on the evening shift. How much will each earn?

	Regular Hourly Wage	Premium Rate	Premium Pay	Evening Shift Hourly Wage	Hours Worked	Earnings
a.	$6.12	6%	$______	$______	36	$______
b.	7.25	8%	______	______	37	______
c.	7.61	9%	______	______	36	______
d.	5.87	8%	______	______	38	______
e.	9.43	10%	______	______	41	______

4. The information below pertains to different people working for different companies on the night shift. How much will each earn?

	Regular Hourly Wage	Premium Rate	Premium Pay	Night Shift Hourly Wage	Hours Worked	Earnings
a.	$ 4.78	15%	$______	$______	40	$______
b.	6.32	15%	______	______	35	______
c.	9.06	18%	______	______	37	______
d.	10.24	20%	______	______	42	______
e.	8.75	21%	______	______	39	______

B

1. Evelyn Murphy works the 8-hour day shift earning $5.47 per hour.

a. What are her earnings each day? ________

b. If she worked the evening shift she would receive a premium of 9%. What would her hourly wage be on the evening shift? ________

c. How much would she earn for 8 hours work on the evening shift? ________

d. How much more would she earn each day by working the evening shift rather than the day shift? ________

2. The day shift hourly wage for a production worker at the Thompson Tool Company is $7.84. The evening shift worker receives a premium of 9%, while the night shift worker receives a premium of 16 1/2%. The work-week at this company is 36 hours.

a. What is the hourly wage of a production worker on the evening shift? ________

b. How much will a production worker on the evening shift earn each week? ________

c. What is the hourly wage of a production worker on the night shift? ________

d. How much will a production worker on the night shift earn each week? ________

e. How much more does the night shift employee earn each week than the evening shift employee? ________

Unit 2 The Piecework Wage System

Although there are still many people whose earnings depend upon the number of articles they produce, this wage system is rapidly disappearing. People who receive a specified amount of money for each item they make are called *piecework* employees.

Under the piecework wage system the amount an employee receives is usually computed by referring to a wage scale such as the one below.

Weekly Wage Scale

For each of:	Amount Paid Per Article
the first 50	\$2.10
the next 30	\$2.65
the excess above 80	\$2.95

Example 1 Using the wage scale above, determine the earnings of a person who makes 37 articles in one week.

Solution

Earnings = Number of articles × Amount per article
= 37 × \$2.10
= \$77.70

This number is less than 50

Unit 2

Example 2 Using the wage scale above, determine the earnings of a person who makes 106 articles in one week.

Solution

50 at \$2.10

Earnings for first 50 = 50 × \$2.10
= \$105.00

Number of articles remaining = 106 − 50
= 56

30 at \$2.65

Earnings for next 30 = 30 × \$2.65
= \$79.50

Number of articles remaining = 56 − 30
= 26

Excess above 80 at \$2.95

Earnings for remaining 26 = 26 × \$2.95
= \$76.70

Total earnings = \$105.00 + \$79.50 + \$76.70
= \$261.20

Many companies pay their employees on an arrangement that combines the hourly wage system with the piecework system. By doing this, the employees are not only guaranteed a fixed wage but, by producing more, they earn more. The example below is an illustration of this arrangement.

Example 3 The wage agreement at the Lacombe Corporation calls for the following:

1. The work-week is 38 hours at \$5.76 per hour.
2. For each article produced in excess of 75, the employee receives an additional \$1.05.

How much is earned by an employee who produces 109 articles during a week?

Solution

Hourly earnings = 38 × \$5.76
= \$218.88

Excess above 75 articles = 109 − 75
= 34

Piecework earnings = 34 × \$1.05
= \$35.70

Total earnings = Hourly earnings + Piecework earnings
= \$218.88 + \$35.70
= \$254.58

Exercises A

1. Determine the total earnings for the following employees who are paid on a straight piecework basis.

	Number of Items Produced	Payment Per Item	Earnings
a.	87	\$2.09	\$________
b.	110	1.67	________
c.	175	1.38	________
d.	205	.96	________
e.	169	1.35	________

2. Each of the following employees works for the company whose "Graduated Wage Scale" is shown on page 159. How much did each earn? The first exercise is completed for you.

	Number of Items Produced	Number at $2.10	Amount For $2.10 Items	Number at $2.65	Amount For $2.65 Items	Number at $2.95	Amount For $2.95 Items	Total Earnings
a.	74	____	$____	____	$____	____	$____	$____
b.	32	____	____	____	____	____	____	____
c.	65	____	____	____	____	____	____	____
d.	76	____	____	____	____	____	____	____
e.	93	____	____	____	____	____	____	____
f.	117	____	____	____	____	____	____	____

B

Unit 2

1. The following employees work on a combination hourly wage system and a piecework wage system. For each article produced in excess of the fixed number shown, they earn a bonus. Determine each bonus.

	Fixed Number	Number Produced	Excess Articles	Bonus Per Excess Article	Bonus
a.	100	96	____	58¢	$____
b.	100	117	____	89¢	____
c.	70	94	____	$1.10	____
d.	85	116	____	1.24	____
e.	65	103	____	1.37	____
f.	105	162	____	1.09	____

2. The following employees work on a combination hourly wage system and piecework wage system. For each article produced in excess of the fixed number shown, they earn a bonus. How much did each earn?

	Hourly Wage	Hours Worked	Hourly Earnings	Fixed Number	Number Produced	Excess Articles	Bonus Per Excess Article	Bonus	Total Earnings
a.	$4.70	40	$____	50	62	____	$2.45	$____	$____
b.	5.10	38	____	75	95	____	2.25	____	____
c.	5.42	40	____	80	103	____	2.60	____	____
d.	6.51	41	____	70	108	____	.84	____	____
e.	6.09	37	____	110	153	____	.76	____	____

C

1. During the week of June 5, Barbara Faber produced the following number of articles each day.

Monday	Tuesday	Wednesday	Thursday	Friday
85	93	81	86	95

If she receives 63¢ for each article on a straight piecework basis, what were her earnings for the week? ________

2. Hillcrest Electronics pays its employees on a piecework basis of 58¢ per article. To discourage careless work, the company deducts 75¢ from the earnings of an employee for each article that is rejected during inspection. How much did Leo Metz receive during a week in which he produced 485 articles, but had 23 of these articles rejected? ________

3. Jerome Hayes works for a firm where he is paid $5.76 per hour. In addition, he is granted a bonus of 34¢ for each article that he produces over 60 each day. What would be his earnings during a week in which he worked 37 1/2 hours and produced the following number of articles daily?

Monday	Tuesday	Wednesday	Thursday	Friday
63	59	57	68	72

____

Unit 3 The Commission Wage System

People employed in sales work are often paid in terms of the value of the articles they sell. Among these people are the door-to-door salespersons, real estate agents, automobile salespersons, telephone solicitors, and many others. Their earnings are computed as *a fixed percent of their total sales*. People who are paid under this arrangement are said to be working on a *commission*.

Example 1 Harvey Quimby is a salesperson for the Atlantic Boat Corporation. He works on a commission rate of 4%. How much will he earn on the sale of a boat worth $5,460?

Solution

$$\text{Commission} = 4\% \text{ of total sales}$$
$$= .04 \times \$5{,}460$$
$$= \$218.40$$

The commission wage system is often combined with the hourly wage system as a method for paying employees. Under this arrangement our weekly earnings would consist of two parts.

1. The amount we receive based on an hourly wage.
2. The amount we receive as commission for sales.

Unit 3

Example 2 Jane Merrill is a salesperson for the South Shore Department Store. She works a 42-hour week at $4.37 per hour. In addition she receives a commission of 1 1/2% on all sales. What are her total earnings during a week in which she sells $2,300 worth of merchandise?

Solution

$$\text{Hourly earnings} = 42 \times \$4.37$$
$$= \$183.54$$

$$\text{Commission on sales} = 1\tfrac{1}{2}\% \text{ of total sales}$$
$$= 1.5\% \times \$2{,}300$$
$$= .015 \times \$2{,}300$$
$$= \$34.50$$

$$\text{Total earnings} = \text{Hourly earnings} + \text{Commission}$$
$$= \$183.54 + \$34.50$$
$$= \$218.04$$

Exercises

A

1. The following employees work on a straight commission that is based on their total sales. How much will each earn?

	Sales	Commission Rate	Commission
a.	$950	20%	$________
b.	675	22%	________
c.	756	18%	________
d.	1,450	$6\frac{1}{2}$%	________
e.	2,375	$14\frac{1}{2}$%	________
f.	57,200	$5\frac{1}{2}$%	________

2. In the exercises below each employee's commission is computed on the excess sales above the number shown in the second column. Determine each commission.

	Total Sales For the Week	Commission On Sales in Excess of:	Commission Sales	Commission Rate	Commission
a.	$ 965	$ 400	$________	5%	$________
b.	1,340	750	________	6%	________
c.	3,525	1,500	________	$2\frac{1}{2}$%	________
d.	4,637	2,000	________	$3\frac{1}{2}$%	________
e.	5,946	2,500	________	$3\frac{1}{4}$%	________

3. The following employees work on a combination hourly wage system and commission. Determine the earnings for each employee.

	Number of Hours	Hourly Wage	Hourly Earnings	Total Sales	Commission Rate	Commission	Total Earnings
a.	40	$3.90	$________	$ 860	2%	$________	$________
b.	38	4.53	________	1,020	1%	________	________
c.	37	5.06	________	2,340	$1\frac{1}{2}$%	________	________
d.	36	4.95	________	2,007	$2\frac{1}{2}$%	________	________
e.	42	6.72	________	3,430	$2\frac{1}{4}$%	________	________

B

1. Doris Somers, a real estate broker, works on a straight commission of 6 1/2% of the selling price of property. She recently sold Kenneth Glynn's house for $63,750.

a. What was the commission that Doris Somers received? ________

b. How much of the $63,750 was turned over to Kenneth Glynn? ________

2. Patrick Medina sells both life insurance and automobile insurance. His commission on each is as follows:

> 55% of the first year's premium on the life insurance.
> 18% of the annual premium on automobile insurance.

During the period of the one week he sold:

1. Automobile insurance in which the annual premiums amounted to $962.
2. Life insurance in which the first year's premiums amounted to $764.

a. What commission did he receive on the automobile insurance? ________

b. What commission did he receive on the life insurance? ________

c. What was his total commission for the week? ________

3. One of the authors of a textbook receives 3 1/4% of the total sales of the book. During one year 37,464 copies of the book were sold at $7.24 each.

a. What was the total selling price of all the books? ________

b. What commission did the author receive? ________

Payroll Deductions Unit 4

Section 1 Social Security Deductions

Part 1

At the beginning of this chapter reference was made to the fact that the amount we earn is not the amount we take home Our employer is required to withhold money from our salary for a number of reasons. Among these is the payment to the government for the social security insurance we receive. A few of the benefits of this insurance are:

1. An income when we retire
2. An income if we are disabled
3. An income to our family in the event of our death

The cost of this insurance is 6.13% of our earnings. This rate has been changed quite often and may be somewhat different at present. Also, the premiums at this time are based on the first $25,900 of our annual earnings.

Example 1 Ralph Hopkins works a 39-hour week at $6.37 per hour. How large will his social security deduction—payment—be each week?

Solution

$$\text{Weekly earnings} = 39 \times \$6.37$$
$$= \$248.43$$
$$\text{Social security deduction} = 6.13\% \text{ of Earnings}$$
$$= .0613 \times \$248.43$$
$$= \$15.23$$

If our earnings are more than $25,900 during the year, then at some point during the year our payments for social security insurance will cease. Usually, we are interested in knowing which week of the year that will be. To determine this, let us consider a much simpler situation in which we will assume that social security payments are made on only the first $600 of our earnings rather than on $25,900. So, if we earn $600 each week, then payments will be made for 1 week only:

$$1 = \$600 \div \$600$$

Number of weeks (1) — **Maximum amount** ($600) — **Weekly earnings** ($600)

Similarly, if we earn $300 each week, then payments will be made for 2 weeks only:

$$2 = \$600 \div \$300$$

Number of weeks (2) — **Maximum amount** ($600) — **Weekly earnings** ($300)

Judging from the above, we can say that the number of weeks during which payments are made can be found as follows:

Number of weeks = Maximum amount ÷ Weekly earnings

Since the maximum amount on which social security payments are now made is $25,900, this formula becomes,

Number of weeks = \$25,900 ÷ Weekly earnings

This formula may lead to the number of weeks being

43.68.

This means that the entire deduction will be made for the first 43 weeks. The final deduction during the 44th week will be made on but .68 of the weekly earnings.

Example 2 Joan Mason earns $680 each week. During which week of the year will the final deduction be made for social security?

Solution

Number of weeks = \$25,900 ÷ Weekly earnings
= \$25,900 ÷ \$680
= 38.09

Final deduction will be made during the 39th week.

Unit 4

Exercises A

1. Determine the social security deduction for each employee.

	Weekly Earnings	Social Security Tax		Weekly Earnings	Social Security Tax
a.	$153	$_____	e.	$171.50	$_____
b.	246	_____	f.	206.75	_____
c.	307	_____	g.	314.83	_____
d.	328	_____	h.	335.97	_____

2. Determine the weekly social security deduction for each employee.

	Hourly Wage	Hours Worked	Weekly Earnings	Social Security Tax
a.	$5.20	40	$_____	$_____
b.	4.90	38	_____	_____
c.	6.85	39	_____	_____
d.	7.40	37½	_____	_____
e.	6.82	38½	_____	_____
f.	7.96	40½	_____	_____

3. During which week of the year will the final deduction be made for social security insurance?

	Weekly Earnings	Final Week of Deductions		Weekly Earnings	Final Week of Deductions
a.	$600	________	e.	$2,000	________
b.	800	________	f.	3,000	________
c.	920	________	g.	1,200	________
d.	750	________	h.	1,450	________

B

1.

a. What are the yearly maximum earnings from which social security deductions are made? ________

b. What is the largest amount that can be deducted in social security payments during one year? ________

2. Andrew Curran has a monthly salary of $1,275.

a. What are his yearly earnings? ________

b. What will his social security deductions be for the year? ________

3. Roberta Loesh has a biweekly (once every two weeks) salary of $645.

a. What are her annual earnings? ________

b. What will her social security deductions be for the year? ________

Part 2

Rather than compute the social security payments as we had been doing, we can obtain from the federal government a table showing the tax. A small section of this table appears on page 169.

When using the social security table it is best to ignore the columns headed by the words **At least.**

Social Security Employee Tax Table
6.13 percent employee tax deductions

Wages At least	Wages But less than	Tax to be withheld	Wages At least	Wages But less than	Tax to be withheld	Wages At least	Wages But less than	Tax to be withheld	Wages At least	Wages But less than	Tax to be withheld
65.34	65.50	4.01	76.11	76.27	4.67	86.87	87.04	5.33	97.64	97.80	5.99
65.50	65.67	4.02	76.27	76.43	4.68	87.04	87.20	5.34	97.80	97.97	6.00
65.67	65.83	4.03	76.43	76.60	4.69	87.20	87.36	5.35	97.97	98.13	6.01
65.83	66.00	4.04	76.60	76.76	4.70	87.36	87.53	5.36	98.13	98.29	6.02
66.00	66.16	4.05	76.76	76.92	4.71	87.53	87.69	5.37	98.29	98.46	6.03
66.16	66.32	4.06	76.92	77.08	4.72	87.69	87.85	5.38	98.46	98.62	6.04
66.32	66.48	4.07	77.08	77.25	4.73	87.85	88.01	5.39	98.62	98.78	6.05
66.48	66.64	4.08	77.25	77.41	4.74	88.01	88.18	5.40	98.78	98.94	6.06
66.64	66.81	4.09	77.41	77.57	4.75	88.18	88.34	5.41	98.94	99.11	6.07
66.81	66.97	4.10	77.57	77.74	4.76	88.34	88.50	5.42	99.11	99.27	6.08
66.97	67.13	4.11	77.74	77.90	4.77	88.50	88.67	5.43	99.27	99.43	6.09
67.13	67.30	4.12	77.90	78.06	4.78	88.67	88.83	5.44	99.43	99.60	6.10
67.30	67.46	4.13	78.06	78.23	4.79	88.83	88.99	5.45	99.60	99.76	6.11
67.46	67.62	4.14	78.23	78.39	4.80	88.99	89.16	5.46	99.76	99.92	6.12
67.62	67.79	4.15	78.39	78.55	4.81	89.16	89.32	5.47	99.92	100.09	6.13
67.79	67.95	4.16	78.55	78.72	4.82	89.32	89.48	5.48			
67.95	68.11	4.17	78.72	78.88	4.83	89.48	89.65	5.49			
68.11	68.28	4.18	78.88	79.04	4.84	89.65	89.81	5.50			
68.28	68.44	4.19	79.04	79.21	4.85	89.81	89.97	5.51			
68.44	68.60	4.20	79.21	79.37	4.86	89.97	90.14	5.52			
68.60	68.77	4.21	79.37	79.53	4.87	90.14	90.30	5.53			
68.77	68.93	4.22	79.53	79.70	4.88	90.30	90.46	5.54			
68.93	69.09	4.23	79.70	79.86	4.89	90.46	90.62	5.55			
69.09	69.25	4.24	79.86	80.02	4.90	90.62	90.79	5.56			
69.25	69.42	4.25	80.02	80.18	4.91	90.79	90.95	5.57			
69.42	69.58	4.26	80.18	80.35	4.92	90.95	91.11	5.58			
69.58	69.74	4.27	80.35	80.51	4.93	91.11	91.28	5.59			
69.74	69.91	4.28	80.51	80.67	4.94	91.28	91.44	5.60			
69.91	70.07	4.29	80.67	80.84	4.95	91.44	91.60	5.61			
70.07	70.23	4.30	80.48	81.00	4.96	91.60	91.77	5.62			

The multiples of the withholding for FICA on $100 are

Wage	Tax to be withheld
$100	$6.13
200	12.26
300	18.39
400	24.52
500	30.65
600	36.78
700	42.91
800	49.04
900	55.17
1,000	61.30

Unit 4

The numbered steps in the drawing below show how the weekly payment is found for a salary of $77.46.

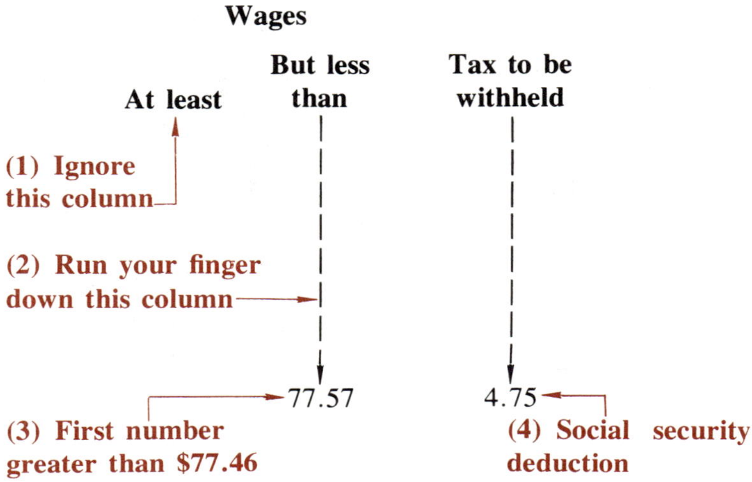

Example 1 Determine the social security deduction on earnings of $89.55.

Solution

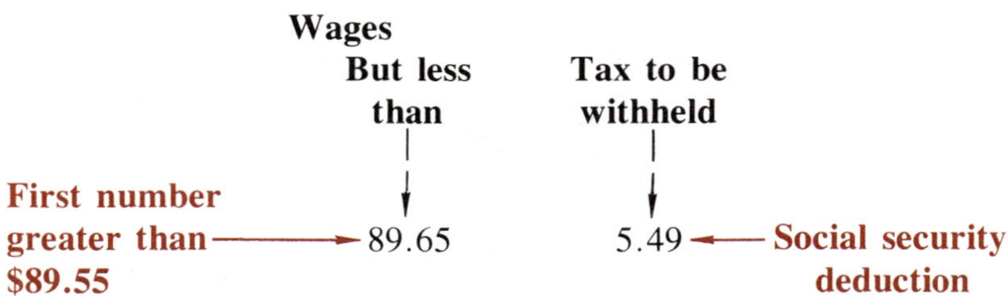

Social security tax on $89.55 = $5.49

Notice that in the lower right corner of the social security tax table another small table is inserted. This table must be used when earnings are greater than $100. The letters FICA in the title of the small table are the abbreviation for Federal Insurance Contributions Act. This is simply another name for the social security insurance.

Example 2 Determine the social security deduction on earnings of $289.25.

Explanation The method for finding the tax is separated into three parts:

(1) Find the tax in the small table on $200: $12.26
(2) Find the tax in the large table of $89.25: $5.47
(3) The sum of the above two is the social security deduction.

$$\$12.26 + \$5.47 = \$17.73$$

Solution

$$\text{Tax on } \$200 = \$12.26$$
$$\text{Tax on } \$89.25 = \$5.47$$
$$\text{Social security tax} = \$12.26 + \$5.47 = \$17.73$$

Unit 4

Exercises A

Use the Social Security Employee Tax Table to compute the deductions in this set of exercises.

1. Determine the social security deduction for each employee.

	Weekly Earnings	Social Security Tax		Weekly Earnings	Social Security Tax
a.	$67.50	$______	f.	$76.29	$______
b.	77.17	______	g.	88.02	______
c.	80.25	______	h.	90.30	______
d.	88.30	______	i.	80.18	______
e.	99.05	______	j.	88.00	______

2. Determine the social security deduction for each employee.

	Weekly Earnings	Tax on $100 Amounts	Tax on Amount Less than $100	Total Tax
a.	$168.50	$_____	$_____	$_____
b.	177.80	_____	_____	_____
c.	288.25	_____	_____	_____
d.	399.30	_____	_____	_____
e.	576.83	_____	_____	_____
f.	787.56	_____	_____	_____

3. Determine the weekly social security deduction for each employee.

	Hourly Wage	Hours Worked	Weekly Earnings	Tax on $100 Amounts	Tax on Amount Less than $100	Total Tax
a.	$ 3.45	23	$_____	$_____	$_____	$_____
b.	5.48	18	_____	_____	_____	_____
c.	4.23	40	_____	_____	_____	_____
d.	7.34	38	_____	_____	_____	_____
e.	9.96	39	_____	_____	_____	_____
f.	12.81	$37\frac{1}{2}$	_____	_____	_____	_____

B

1. Herbert Lowe earns $277.50 each week.

a. What will the weekly deduction for social security be? ________

b. How much FICA tax will be deducted from his salary during a period of one year? ________

2. Hazel Shelton is paid $489.50 biweekly.

a. What will the biweekly deduction for social security be? ________

b. How much FICA tax will be deducted from her salary during a period of one year? ________

3. Erna Frey earns a salary of $1,280 monthly.

a. What will the monthly deduction for social security be? ________

b. How much FICA tax will be deducted from her salary during a period of one year? ________

Section 2 Withholding Tax Deductions

Part 1

For most of us the very largest deduction from our earnings will be made for federal income tax—frequently called *federal withholding tax*. The amount of money deducted depends on two factors:

1. The amount of money we earn
2. The number of people who are dependent upon us for their livelihood—for each of these people we are granted 1 *withholding allowance*.

If we are paid weekly, then *1 withholding allowance is $14.40.* This simply means that before our tax is computed, $14.40 must be subtracted from our earnings for each allowance we claim. As an example, if we claim five allowances, then

$$\begin{aligned}\text{Total withholding allowance} &= 5 \times \$14.40\\ &= \$72.00\end{aligned}$$

If our weekly earnings happen to be $216.50, then the tax will be computed on

$$\$216.50 - \$72.00$$

Weekly earnings (↑ $216.50) **Total withholding allowance** (↑ $72.00)

Therefore, the part of our earnings on which we will have to pay a tax—*taxable wages*—is $144.50.

To find the tax on $144.50 we must refer to the table on page 174. There are two sections to this table—the upper section to be used for a Single Person and the lower for a Married Person.

The numbered steps in the diagram below show how the tax is found for a single person whose taxable income is $144.50.

Federal Withholding Tax

Weekly Payroll Period
Single Person

If amount of taxable wages is: Over	But Not Over	The amount of income tax to be withheld is:
$0	$33	$0
$33	$76	16% of excess over $33
$76	$143	$ 6.88 + 18% of excess over $76
$143	$182	$18.94 + 22% of excess over $143
$182	$220	$27.52 + 24% of excess over $182
$220	$297	$36.64 + 28% of excess over $220
$297	$355	$58.20 + 32% of excess over $297
$355		$76.76 + 36% of excess over $355

Married Person

Over	But Not Over	The amount of income tax to be withheld is:
$0	$61	$0
$61	$105	15% of excess over $61
$105	$223	$ 6.60 + 18% of excess over $105
$223	$278	$27.84 + 22% of excess over $223
$278	$355	$39.94 + 25% of excess over $278
$355	$432	$59.19 + 28% of excess over $355
$432	$509	$80.75 + 32% of excess over $432
$509		$105.39 + 36% of excess over $509

Single Person

If the amount of taxable wages is: Over	But Not Over	The amount of income tax to be withheld is:
(1) Ignore this column	(2) Run your finger down this column	
(3) First number greater than $144.50	$182	$18.94 + 22% of excess over $143

(4) Base Tax: $18.94

(5) Tax on excess: 22% of excess over $143

(6) Tax is the sum of these two

The tax is computed as follows:

$$\text{Base tax} = \$18.94$$

$$\begin{aligned}\text{Excess over } \$143 &= \$144.50 - \$143.00\\ &= \$1.50\end{aligned}$$

$$\begin{aligned}\text{Tax on excess} &= 22\% \text{ of excess over } \$143\\ &= .22 \times \$1.50\\ &= \$.33\end{aligned}$$

$$\begin{aligned}\text{Federal withholding tax} &= \text{Base tax} + \text{Tax on excess}\\ &= \$18.94 + \$.33\\ &= \$19.27\end{aligned}$$

Example 1 Determine the base tax for a married person whose weekly taxable income is $426.

Explanation

But not over

First number greater than $426 → $432 $59.19 + 28% of excess over $355

Base tax → $59.19

Solution Base tax on $426 is $59.19.

Example 2 Determine the tax on excess for a single person whose weekly taxable wages are $324.

Explanation

But not over

First number greater than $324 → $355 $58.20 + <u>32% of excess over $297</u>

Tax on excess → 32% of excess over $297

Solution

$$\begin{aligned}\text{Excess over } \$297 &= \$324 - \$297\\ &= \$27\\ \text{Tax on excess} &= 32\% \text{ of excess}\\ &= .32 \times \$27\\ &= \$8.64\end{aligned}$$

Example 3 Determine the weekly withholding tax for a married person who claims 4 allowances and earns $372 each week.

Solution

$$\text{Total withholding allowance} = 4 \times \$14.40 = \$57.60$$

$$\text{Taxable wages} = \text{Weekly earnings} - \text{Total withholding allowance}$$
$$= \$372 - \$57.60$$
$$= \$314.40$$

$$\text{Base tax on } \$314.40 = \$39.94$$

$$\text{Excess over } \$278 = \$314.40 - \$278 = \$36.40$$

$$\text{Tax on excess} = 25\% \text{ of excess} = .25 \times \$36.40 = \$9.10$$

$$\text{Federal withholding tax} = \text{Base tax} + \text{Tax on excess} = \$39.94 + \$9.10 = \$49.04$$

Most states in the country also raise money through an income tax. The design of the state withholding tax is patterned after that of the federal government. The table below is typical of the form used by many states. Notice that this particular table happens to be used both for single and married persons.

State Withholding Tax
Weekly Payroll Period
1 withholding allowance is $19.80

If amount of taxable wage is: Over	**But Not Over**	**The amount of income tax to be withheld is:**
$0	$384	2.2% of excess over $0
$384		$8.45 + 2.5% of excess over $384

Exercises A

1. Determine the federal base tax for each weekly taxable wage.

	Taxable Wages		Base Tax
a.	$ 98.50	Single	$______
b.	186.20	Single	______
c.	21.40	Single	______
d.	264.57	Single	______
e.	175.94	Married	______
f.	329.75	Married	______

2. Determine the federal tax on excess for each weekly taxable wage.

	Taxable Wages		Excess	Tax on Excess
a.	$195	Single	$______	$______
b.	250	Single	______	______
c.	250	Married	______	______
d.	384	Married	______	______
e.	725	Single	______	______

Unit 4

3. Determine the federal withholding tax for each taxable wage.

	Taxable Wages		Base Tax	Tax On Excess	Withholding Tax
a.	$150	Single	$______	$______	$______
b.	210	Single	______	______	______
c.	320	Married	______	______	______
d.	350	Single	______	______	______
e.	350	Married	______	______	______
f.	675	Married	______	______	______

4. Use the table on page 176 to determine the state withholding tax for each weekly taxable wage.

	Taxable Wages	Base Tax	Tax On Excess	Withholding Tax
a.	$160	$______	$______	$______
b.	275	______	______	______
c.	430	______	______	______
d.	695	______	______	______

B

1. Determine the total federal weekly withholding allowance for each tax payer.

	Number of Allowances	Total With. Allowance		Number of Allowances	Total With. Allowance
a.	2	$	c.	0	$
b.	1		d.	6	

2. Determine the federal taxable wages for each taxpayer.

	Weekly Wages	Number of Allowances	Total With. Allowance	Taxable Wages
a.	$235	2	$	$
b.	287	4		
c.	346	0		
d.	459	6		

3. Determine the federal withholding tax for each taxpayer.

	Weekly Wages	Number Of Allowances	Total With. Allowance	Taxable Wages		Withholding Tax
a.	$230	1	$	$	Single	$
b.	280	3			Single	
c.	304	0			Single	
d.	215	2			Married	
e.	382	4			Married	

4. For the state whose tax table appears on page 176, determine the withholding tax for each taxpayer.

	Weekly Wages	Number Of Allowances	Total With. Allowance	Taxable Wages	Withholding Tax
a.	$260	2	$	$	$
b.	347	1			
c.	440	3			
d.	625	5			

5. Determine the federal withholding tax for each taxpayer.

	Hourly Wage	Number Of Hours	Wages	Number Of Allowances	Taxable Wages		Withholding Tax
a.	$6.31	40	$	3	$	Married	$
b.	7.52	41		2		Married	
c.	8.63	38		1		Single	
d.	9.05	42		4		Single	

C

1. Fred Jackson earns \$6.54 per hour. During the week of July 17 he worked the following number of hours:

Monday	Tuesday	Wednesday	Thursday	Friday
7	8	8	$7\frac{1}{2}$	8

He claims one allowance each for himself, his wife, and his child.

a. What were his total wages for the week? ________

b. What was his total federal withholding allowance? ________

c. What was his federal taxable wage? ________

d. What was his federal withholding tax? ________

e. What was his total state withholding allowance? ________

f. What was his state taxable wage? ________

g. What was his state withholding tax? ________

h. What was the total federal and state tax he had to pay? ________

i. How much money remained after these taxes were deducted? ________

Part 2

Tables exist whereby both the federal and state withholding taxes can be read directly. For most wages, these tables require no computation. A small section of each of these tables appears on page 180.

Example 1 Determine the weekly federal withholding tax for a single person who claims 3 allowances and earns \$263 each week.

Federal Weekly Withholding Tax (Single Persons)

And the wages are		And the number of withholding allowances claimed is										
At least	But less than	0	1	2	3	4	5	6	7	8	9	10 or more
		The amount of income tax to be withheld shall be										
$170	$180	$26.00	$22.80	$19.70	$17.00	$14.40	$11.80	$9.20	$6.60	$4.30	$2.00	$0
180	190	28.30	25.00	21.90	18.80	16.20	13.60	11.00	8.40	5.90	3.60	1.30
190	200	30.70	27.20	24.10	20.90	18.00	15.40	12.80	10.20	7.60	5.20	2.90
200	210	33.10	29.60	26.30	23.10	19.90	17.20	14.60	12.00	9.40	6.80	4.50
210	220	35.50	32.00	28.60	25.30	22.10	19.00	16.40	13.80	11.20	8.60	6.10
220	230	38.10	34.40	31.00	27.50	24.30	21.20	18.20	15.60	13.00	10.40	7.80
230	240	40.90	36.80	33.40	29.90	26.50	23.40	20.20	17.40	14.80	12.20	9.60
240	250	43.70	39.60	35.80	32.30	28.80	25.60	22.40	19.20	16.60	14.00	11.40
250	260	46.50	42.40	38.40	34.70	31.20	27.80	24.60	21.40	18.40	15.80	13.20
260	270	49.30	45.20	41.20	37.20	33.60	30.20	26.80	23.60	20.40	17.60	15.00
270	280	52.10	48.00	44.00	40.00	36.00	32.60	29.10	25.80	22.60	19.50	16.80
280	290	54.90	50.80	46.80	42.80	38.70	35.00	31.50	28.10	24.80	21.70	18.60
290	300	57.70	53.60	49.60	45.60	41.50	37.50	33.90	30.50	27.00	23.90	20.70
300	310	60.80	56.40	52.40	48.40	44.30	40.30	36.30	32.90	29.40	26.10	22.90
310	320	64.00	59.40	55.20	51.20	47.10	43.10	39.00	35.30	31.80	28.30	25.10
$720 and over		208.20	203.00	197.80	192.60	187.40	182.20	177.00	171.90	166.70	161.50	156.30
		Plus 36% of excess over $720										

State Weekly Withholding Tax

Wages		Withholding Allowances										
At Least	Less than	0	1	2	3	4	5	6	7	8	9	10 or more
		Tax to be withheld										
150	153	3.40	3.00	2.50	2.10	1.70	1.30	.80	.40	.00	.00	.00
153	156	3.40	3.00	2.60	2.20	1.70	1.30	.90	.50	.10	.00	.00
156	159	3.50	3.10	2.70	2.20	1.80	1.40	1.00	.50	.10	.00	.00
159	162	3.60	3.10	2.70	2.30	1.90	1.50	1.00	.60	.20	.00	.00
162	165	3.60	3.20	2.80	2.40	1.90	1.50	1.10	.70	.30	.00	.00
165	168	3.70	3.30	2.90	2.40	2.00	1.60	1.20	.70	.30	.00	
168	171	3.80	3.30	2.90	2.50	2.10	1.70	1.20	.80	.40	.00	.00
171	174	3.80	3.40	3.00	2.60	2.10	1.70	1.30	.90	.50	.00	.00
174	177	3.90	3.50	3.10	2.60	2.20	1.80	1.40	.90	.50	.10	.00
177	180	4.00	3.50	3.10	2.70	2.30	1.90	1.40	1.00	.60	.20	.00
180	183	4.00	3.60	3.20	2.80	2.30	1.90	1.50	1.10	.70	.20	.00
183	186	4.10	3.70	3.30	2.80	2.40	2.00	1.60	1.10	.70	.30	.00
186	189	4.20	3.70	3.30	2.90	2.50	2.10	1.60	1.20	.80	.40	.00
189	192	4.20	3.80	3.40	3.00	2.50	2.10	1.70	1.30	.90	.40	.00
192	195	4.30	3.90	3.50	3.00	2.60	2.20	1.80	1.30	.90	.50	.10
195	198	4.40	3.90	3.50	3.10	2.70	2.30	1.80	1.40	1.00	.60	.10
198	201	4.40	4.00	3.60	3.20	2.70	2.30	1.90	1.50	1.00	.60	.20
201	204	4.50	4.10	3.70	3.20	2.80	2.40	2.00	1.50	1.10	.70	.30
204	207	4.60	4.10	3.70	3.30	2.90	2.40	2.00	1.60	1.20	.80	.30
207	210	4.60	4.20	3.80	3.40	2.90	2.50	2.10	1.70	1.20	.80	.40
210	213	4.70	4.30	3.80	3.40	3.00	2.60	2.20	1.70	1.30	.90	.50
213	216	4.80	4.30	3.90	3.50	3.10	2.60	2.20	1.80	1.40	1.00	.50
216	219	4.80	4.40	4.00	3.60	3.10	2.70	2.30	1.90	1.40	1.00	.60
219	222	4.90	4.50	4.00	3.60	3.20	2.80	2.40	1.90	1.50	1.10	.70
222	225	5.00	4.50	4.10	3.70	3.30	2.80	2.40	2.00	1.60	1.20	.70
575& Over		13.30	12.80	12.30	11.80	11.30	10.90	10.40	9.90	9.40	8.90	8.50
		Plus 2.5% of excess over 575.00										

Explanation

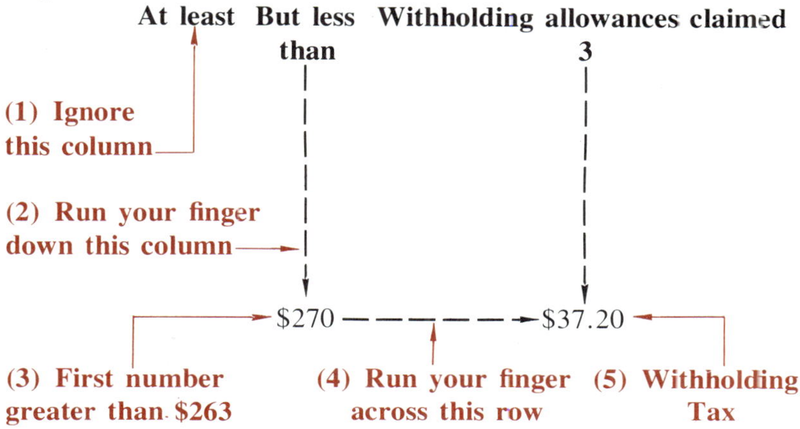

Solution

Federal withholding tax on $263 =$37.20

Example 2 Determine the state withholding tax for a person who earns $634 per week and claims 2 exemptions.

Explanation As in Part 1, when the earnings are larger than the last number in the *Less Than* column, the tax consists of two parts:

(1) a base tax (2) a tax on excess

Solution

Base tax on $634 for 2 exemptions = $12.30

Excess over $575 = $634 − $575
= $59

Tax on excess = 2.5% of excess
= .025 × $59
= $1.48

State withholding tax = Base tax + Tax on excess
= $59 + $1.48
= $60.48

Exercises A

Use the federal and state withholding tax tables on page 180 to compute the deductions for this set of exercises.

1. Find the federal weekly withholding tax for these single persons.

	Earnings	Withholding Allowances	Withholding Tax
a.	$185.40	3	$______
b.	237.50	6	______
c.	294.68	2	______
d.	195.00	0	______
e.	220.00	4	______
f.	290.00	3	______

2. Find the state weekly withholding tax for these taxpayers.

	Earnings	Withholding Allowances	Withholding Tax
a.	$187.50	2	$______
b.	212.36	0	______
c.	219.84	1	______
d.	207.00	4	______
e.	192.00	7	______

3. Determine the state weekly withholding tax for these taxpayers.

	Earnings	Withholding Allowances	Base Tax	Tax On Excess	Withholding Tax
a.	$600	2	$______	$______	$______
b.	640	4	______	______	______
c.	735	3	______	______	______
d.	794	1	______	______	______

4. Determine the federal weekly withholding tax for these single persons.

	Earnings	Withholding Allowances	Base Tax	Tax On Excess	Withholding Tax
a.	$750	3	$______	$______	$______
b.	810	0	______	______	______
c.	845	5	______	______	______
d.	907	8	______	______	______

5. Determine the federal and state withholding taxes and the social security tax on each of the following weekly earnings of a single person. Also determine the total of these three deductions.

	Earnings	Withholding Allowances	Federal Tax	State Tax	Soc. Sec. Tax	Total Tax
a.	$186.50	2	$______	$______	$______	$______
b.	189.20	1	______	______	______	______
c.	198.50	0	______	______	______	______
d.	199.40	3	______	______	______	______
e.	187.56	5	______	______	______	______

B

1. Evelyn Reilly earns $6.40 per hour for a 40-hour week. She is single and claims one withholding allowance.

a. What are her weekly earnings? ______

b. How much is deducted weekly for the federal withholding tax? ______

c. How much is withheld from her salary over a period of one year for federal withholding tax? ______

2. Sam Baker earns $5.23 per hour for a 38-hour week. He is single and claims no withholding allowances.

a. What are his weekly earnings? ______

b. How much is deducted weekly for the federal withholding tax? ______

c. How much is deducted weekly for the state withholding tax? ______

d. How much is deducted weekly for the social security tax? ______

e. What is the total weekly deduction for these three taxes? ______

f. What is his take-home pay, that is, what amount remains after these three deductions are made? ______

3. In addition to the federal and state withholding taxes and social security tax, each of the following single employees has deductions for union dues, hospitalization, and retirement benefits. Determine the take-home pay for each of them. The computation in this exercise will require the use of the federal and state withholding tables on pages 174 and 176 and the social security rate of 6.13%.

	Earnings	Withholding Allowances	Union Dues	Hospitalization	Retirement Benefits	Federal Tax	State Tax	Social Security	Take-Home Pay
a.	$258	2	$ 3.48	$2.69	$ 4.26	$____	$____	$____	$____
b.	317	3	4.29	3.82	6.10	____	____	____	____
c.	431	1	6.54	2.07	14.53	____	____	____	____
d.	594	6	12.50	8.46	23.12	____	____	____	____

Review of Chapter 4

1. Determine the weekly wages of the following hourly employees.

	Hourly Wage	Hours Worked	Earnings
a.	$6.73	41	$______
b.	5.84	$37\frac{1}{2}$	______
c.	7.35	$38\frac{1}{2}$	______

2. Determine the raise in the hourly wage as a result of the following wage negotiations.

	Previous Hourly Wage	Percent Increase	Hourly Raise
a.	$5.04	7%	$______
b.	8.75	$8\frac{1}{2}$%	______

3. Employees of the Standard Elements Corporation received an 8% increase in their hourly wage. The hourly wage was $7.25 before the raise.

a. What new hourly wage do they now receive? ______

b. If they work a 38-hour week, how much more do they now earn per week than before the raise? ______

4. The regular hourly wage at the Precision Motor Company is $6.34.

a. How much will an employee receive who works 5 1/2 hours on Sunday at the double-time rate? ______

b. How much will an employee receive who works 2 1/2 hours overtime at the time-and-a-half rate? ______

5. The employee below receives time-and-a-half pay for working more than 7 hours during any day. How many hours of overtime will this person receive during this week?

Mon	Tues	Wed	Thurs	Fri
9	7	8	$7\frac{1}{2}$	$7\frac{1}{2}$

______

6. Employees of Automatic Dial Corporation earn $6.74 per hour on the day shift.

a. The evening shift employees earn a premium of 12%. What is their hourly wage? ________

b. The night shift employees earn a premium of 18%. What is their hourly wage? ________

c. What are the weekly earnings of a night shift employee for a 36-hour week? ________

7. Silvio Marino works under the following piecework wage scale.

Articles Per Week	Amount Paid Per Article
For each of the first 60	$2.85
For each article above 60	$3.15

How much will he receive during a week in which he produced 84 articles? ________

Review

8. Julia Deering works a 41-hour week at $4.27 per hour. In addition she receives a commission of 2 1/2% on her sales in excess of $1,250. How much will she earn during the week in which she sold $1,958 worth of merchandise? ________

9. Determine the social security deduction on a weekly salary of $261.50 ________

10. Determine the federal withholding tax for a married person having a weekly taxable wage of $327. Use the table on page 174. ________

11. Determine the state withholding tax on a weekly salary of $637 on which two withholding allowances are claimed. Use the table on page 176. ________

12. Joshua Levy earns $198.50 each week. He is single and claims two withholding exemptions.

a. What is his social security tax? ________

b. What is his federal withholding tax? Use the table on page 180. ________

c. What is his state withholding tax? Use the table on page 180. ________

d. What is his take-home pay? ________

1040

Chapter 5

Income Taxes

In the previous chapter we learned that both federal and state withholding taxes are deducted from our earnings each week. However, the total of these deductions is but an approximation of the amount we owe. Therefore, once each year prior to April 15 we must fill out a form – file a return – stating exactly the amount of money we should have paid. In this chapter we will examine just how this form is completed.

Federal Tax – Form W-2 and Form 1040A Unit 1

Part 1

Shortly after January 1 of each year our employer is required to give us the form shown on page 188. The information on this form, called Form W-2, is needed when filling out our income tax return. Of the various items found on Form W-2, only two are important to most taxpayers.

11. Federal income tax withheld

$2,146

This amount is the total federal withholding tax for the year.

12. Wages, tips, etc.

$16,275

This amount is our total earnings for — the year.

Unless our income is fairly high, the entire return that we must file can be completed on either Form 1040A or Form 1040. Single people with an income of more than $20,000, or married people with an income of more than $40,000 must use Form 1040. Since the income for most people is less than these amounts, we will examine Form 1040A only. A copy of this form appears on pages 190 and 191.

1 Control number 222 | 2 Employer's State number

3 Employer's name, address, and ZIP code
General Electronics, Inc.
Highway 240
Glenwood, Oregon 34567

4 Sub-total ☐ Cor-rection ☐ Void ☐

7 Employer's identification number

Make No Entry Here

10 Employee's social security number	11 Federal income tax withheld	12 Wages, tips, other compensation	13 FICA tax withheld	14 Total FICA wages
462-15-7583	\$2,146	\$16,275	\$984,64	\$16,275

15 Employee's name (first, middle, last)
Arthur T. Webb

15 Pension plan coverage? Yes/No

17

18 FICA tips

547 Norman Road
Glenwood, Oregon 34567
19 Employee's address and ZIP code

Wage and Tax Statement

Copy D For employer

Form **W–2**

Department of the Treasury—Internal Revenue Service

A number of the more important fillins for the upper portion of Form 1040A are explained below.

Filing Status

A single person will check here → 1 ☐ Single

A married couple will usually file one form and check here → 2 ☐ Married filing joint return

3 ☐ Married filing separately

4 ☐ Unmarried Head of Household

These people were called withholding allowances in Chapter 4.

Exemptions

In this section for each box we check and for each name we list we receive one exemption.

7 Wages, etc. 7 ________

Here we fill in our year's earnings

8 Interest income 8 ________

Enter interest here. If the amount is more than $400, we cannot use Form 1040A.

9a Dividends ________ **9b Less exclusion** ________ **Balance 9c** ________

Enter income from stocks here. If amount is more than $400, we cannot use Form 1040A.

If dividends are more than $100, write $100 here. If less than $100, write amount of dividends.

10 Adjusted gross income 10 ________

This number is the sum of the numbers on lines 7, 8, and 9c.

The number written on line 10 – *adjusted gross income* – is the amount upon which our tax is computed. If we prefer, we can stop at this point and the Internal Revenue Service (IRS) will determine that tax for us.

Form **1040A** Department of the Treasury—Internal Revenue Service
U.S. Individual Income Tax Return

Use IRS label. Otherwise, please print or type.	Your first name and initial (if joint return, also give spouse's name and initial)	Last name	Your social security number
	Present home address (Number and street, including apartment number, or rural route)		Spouse's social security no.
	City, town or post office, State and ZIP code		Your occupation

▶ Do you want $1 to go to the Presidential Election Campaign Fund? ☐ Yes ☐ No
If joint return, does your spouse want $1 to go to this fund? . . ☐ Yes ☐ No
Note: *Checking Yes will not increase your tax or reduce your refund.*
Spouse's occupation

Filing Status — Check Only One Box

For Privacy Act Notice, see page 5 of Instructions

1 ☐ Single
2 ☐ Married filing joint return (even if only one had income)
3 ☐ Married filing separate return. If spouse is also filing, give spouse's social security number in the space above and enter full name here ▶ ....
4 ☐ Unmarried head of household. Enter qualifying name ▶ .... See page 11 of Instructions.

Exemptions

Always check the box labeled Yourself. Check other boxes if they apply.

5a ☐ Yourself ☐ 65 or over ☐ Blind
b ☐ Spouse ☐ 65 or over ☐ Blind
} Enter number of boxes checked on 5a and b ▶ ☐

c First names of your dependent children who lived with you ▶ ....
} Enter number of children listed ▶ ☐

d Other dependents:

(1) Name	(2) Relationship	(3) Number of months lived in your home.	(4) Did dependent have income of $750 or more?	(5) Did you provide more than one-half of dependent's support?

Enter number of other dependents ▶ ☐

6 Total number of exemptions claimed . . . . Add numbers entered in boxes above ▶ ☐

B of Forms W–2 Here

7 Wages, salaries, tips, and other employee compensation. *(Attach Forms W–2. If you do not have a W–2, see page 7 of Instructions)* 7

8 Interest income *(see page 4 of Instructions)* 8

Please Attach Copy

Please Attach Check or Money Order Here

9a Dividends .... **9b** Exclusion .... Subtract line 9b from 9a ▶ **9c**

(See pages 4 and 8 of Instructions)

10 Adjusted gross income *(add lines 7, 8, and 9c). If under $8,000, see page 2 of Instructions* on *"Earned Income Credit."* **If eligible, enter child's name** ▶ **10**

11a Credit for contributions to candidates for public office. *Enter one-half of amount paid but do not enter more than $25 ($50 if joint return). (See page 8 of Instructions)* . . . **11a**

IF YOU WANT IRS TO FIGURE YOUR TAX, PLEASE STOP HERE AND SIGN BELOW.

b Total Federal income tax withheld *(if line 7 is larger than $17,700, see page 8 of Instructions)* **11b**

c Earned income credit *(from page 2 of Instructions)* **11c**

12 Total *(add lines 11a, b, and c)* **12**

13 Tax on the amount on line 10. *(See Instructions for line 13 on page 9, then find your tax in the Tax Tables on pages 14–25.)* **13**

14 If line 12 is larger than line 13, enter amount to be **REFUNDED TO YOU** . . . . ▶ **14**

15 If line 13 is larger than line 12, enter **BALANCE DUE.** Attach check or money order for full amount payable to "Internal Revenue Service." Write social security number on check or money order . . ▶ **15**

Please Sign Here

Under penalties of perjury, I declare that I have examined this return, including accompanying schedules and statements, and to the best of my knowledge and belief, it is true, correct, and complete. Declaration of preparer (other than taxpayer) is based on all information of which preparer has any knowledge.

▶ Your signature Date | ▶ Spouse's signature (if filing jointly, BOTH must sign even if only one had income)

Paid Preparer's Information	Preparer's signature ▶		Preparer's social security no.	Check if self-employed ▶ ☐
	Firm's name (or yours, if self-employed), address and ZIP code ▶	E.I. No. ▶		
		Date ▶		

☆ U.S. GOVERNMENT PRINTING OFFICE: 1978—O-263-332 13-2687299

Form **1040A**

Unit 1

Exercises

Use the information below to determine the fillins for lines 1 through 10 of Form 1040A.

	Dividends	Wages	Interest	
a.	$ 0	$ 8,600	$ 50	Single - 1 exemption
b.	0	9,840	146	Single - 1 exemption
c.	130	12,428	269	Single - 1 exemption
d.	350	14,950 & 0	178	Joint - husband, wife, 1 child
e.	40	15,143 & 0	264	Joint - husband, wife, 1 child
f.	85	12,651 & 14,329	348	Joint - husband, wife, 2 children

Part 2

Rather than have the IRS compute our tax, we can do this ourselves. To determine the tax, we must refer to a rather lengthy table. A small section of that table appears on page 194. Although the table printed here is designed for use by an unmarried person, there are other tables for:

1. Married Persons Filing Jointly
2. Married Persons Filing Separately
3. Persons Who are Heads of Households

Example 1 Determine the income tax for a single person claiming 2 exemptions who had an adjusted gross income of $11,640.

Explanation This table is read in the same manner as the earlier tax tables we examined. The numbered steps below lead us to the income tax.

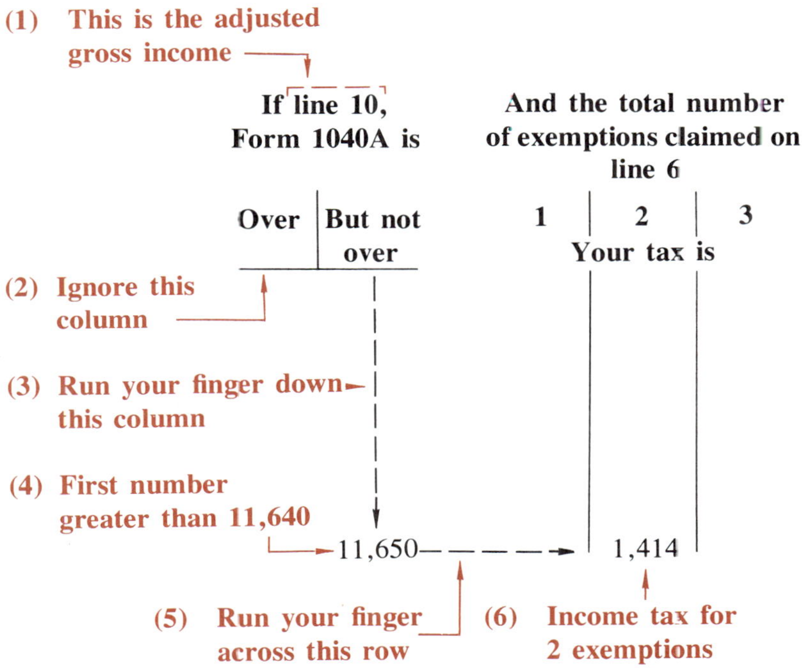

Solution

Income tax = $1,414

We are now prepared to complete Form 1040A. The example below will show how this is done.

Example 2 Determine the fillins for lines 10 through 15 for a single person under the following conditions.

Adjusted gross income: $11,640
Withholding tax: $1,364
Exemptions claimed: 2
Contribution to a political party: $10

1978 Tax Table A—SINGLE (Filing Status Box 1) *(Continued)*

If Form 1040A, line 10, is— Over	But not over	And the total number of exemptions claimed on line 6 is— 1	2	3
		Your tax is—		
11,000	11,050	1,447	1,282	1,117
11,050	11,100	1,459	1,293	1,128
11,100	11,150	1,470	1,304	1,139
11,150	11,200	1,482	1,315	1,150
11,200	11,250	1,493	1,326	1,161
11,250	11,300	1,505	1,337	1,172
11,300	11,350	1,516	1,348	1,183
11,350	11,400	1,528	1,359	1,194
11,400	11,450	1,539	1,370	1,205
11,450	11,500	1,551	1,381	1,216
11,500	11,550	1,562	1,392	1,227
11,550	11,600	1,574	1,403	1,238
11,600	11,650	1,585	1,414	1,249
11,650	11,700	1,597	1,425	1,260
11,700	11,750	1,608	1,436	1,271
11,750	11,800	1,620	1,447	1,282
11,800	11,850	1,631	1,459	1,293
11,850	11,900	1,643	1,470	1,304
11,900	11,950	1,654	1,482	1,315
11,950	12,000	1,666	1,493	1,326
12,000	12,050	1,679	1,505	1,337
12,050	12,100	1,691	1,516	1,348
12,100	12,150	1,704	1,528	1,359
12,150	12,200	1,716	1,539	1,370
12,200	12,250	1,729	1,551	1,381
12,250	12,300	1,741	1,562	1,392
12,300	12,350	1,754	1,574	1,403
12,350	12,400	1,766	1,585	1,414

If Form 1040A, line 10, is— Over	But not over	And the total number of exemptions claimed on line 6 is— 1	2	3
		Your tax is—		
14,000	14,050	2,200	1,998	1,804
14,050	14,100	2,214	2,011	1,816
14,100	14,150	2,227	2,025	1,829
14,150	14,200	2,241	2,038	1,841
14,200	14,250	2,254	2,052	1,854
14,250	14,300	2,268	2,065	1,866
14,300	14,350	2,281	2,079	1,879
14,350	14,400	2,295	2,092	1,891
14,400	14,450	2,308	2,106	1,904
14,450	14,500	2,322	2,119	1,917
14,500	14,550	2,335	2,133	1,930
14,550	14,600	2,349	2,146	1,944
14,600	14,650	2,362	2,160	1,957
14,650	14,700	2,376	2,173	1,971
14,700	14,750	2,389	2,187	1,984
14,750	14,800	2,403	2,200	1,998
14,800	14,850	2,416	2,214	2,011
14,850	14,900	2,430	2,227	2,025
14,900	14,950	2,443	2,241	2,038
14,950	15,000	2,457	2,254	2,052
15,000	15,050	2,472	2,268	2,065
15,050	15,100	2,486	2,281	2,079
15,100	15,150	2,501	2,295	2,092
15,150	15,200	2,515	2,308	2,106
15,200	15,250	2,530	2,322	2,119
15,250	15,300	2,544	2,335	2,133
15,300	15,350	2,559	2,349	2,146
15,350	15,400	2,573	2,362	2,160

If Form 1040A, line 10, is— Over	But not over	And the total number of exemptions claimed on line 6 is— 1	2	3
		Your tax is—		
17,000	17,050	3,053	2,834	2,617
17,050	17,100	3,069	2,849	2,631
17,100	17,150	3,084	2,863	2,646
17,150	17,200	3,100	2,878	2,660
17,200	17,250	3,115	2,892	2,675
17,250	17,300	3,131	2,907	2,689
17,300	17,350	3,146	2,921	2,704
17,350	17,400	3,162	2,936	2,718
17,400	17,450	3,177	2,950	2,733
17,450	17,500	3,193	2,965	2,747
17,500	17,550	3,208	2,979	2,762
17,550	17,600	3,224	2,994	2,776
17,600	17,650	3,239	3,008	2,791
17,650	17,700	3,255	3,023	2,805
17,700	17,750	3,270	3,038	2,820
17,750	17,800	3,286	3,053	2,834
17,800	17,850	3,301	3,069	2,849
17,850	17,900	3,317	3,084	2,863
17,900	17,950	3,332	3,100	2,878
17,950	18,000	3,348	3,115	2,892
18,000	18,050	3,363	3,131	2,907
18,050	18,100	3,379	3,146	2,921
18,100	18,150	3,394	3,162	2,936
18,150	18,200	3,410	3,177	2,950
18,200	18,250	3,425	3,193	2,965
18,250	18,300	3,441	3,208	2,979
18,300	18,350	3,456	3,224	2,994
18,350	18,400	3,472	3,239	3,008

Solution

10. Adjusted gross income $11,640 | 00

11a. Credit for contribution to a candidate 5 | 00

We are permitted to deduct only one-half of the amount contributed.

11b. Total Federal income tax withheld 1,364 | 00

This is the withholding tax listed on Form W-2.

11c. Earned income credit 0 | 00

This credit applies only if our income is less than $8,000.

12. Total (add lines 11a, 11b, and 11c) 1,369 | 00

13. Tax on the amount on line 10 1,414 | 00

We learned how to find this tax in example 1.

14. If line 12 is larger than line 13, enter Refund |

In this example line 12 is not larger than line 13. Hence, we receive no refund.

15. If line 13 is larger than line 12, enter Balance Due 45 | 00

Since the total tax is $1,414, while the amount we paid is $1,369, we still owe $45.

In the example above, had the number on line 12 been $1,523, then the fillins for lines 12, 13, and 14 would have been as follows:

12. Total (add lines 11a, 11b, and 11c) 1,523 | 00

Unit 1

13.	Tax on the amount on line 10	1,414 | 00
14.	If line 12 is larger than line 13, enter Refund	109 | 00

In this situation, the total tax is $1,414 but the amount we paid is $1,523. Hence, the government owes us $109.

Exercises

1. Determine the federal income tax each of the single taxpayers will have to pay.

	Adjusted Gross Income	Number of Exemptions	Tax
a.	$11,347	1	$______
b.	12,186	2	______
c.	14,275	1	______
d.	15,192	3	______
e.	17,341	1	______
f.	12,203	1	______
g.	18,005	3	______

2. Lines 12 and 13 are shown below. Determine the fillin for either line 14 or line 15.

Line	a	b	c	d	e	f
12. Total Paid	1,470	2,180	3,127	1,869	4,716	3,765
13. Tax from Table	1,280	2,068	3,410	1,960	4,124	3,981
14. Refund To You	____	____	____	____	____	____
15. Balance Due	____	____	____	____	____	____

3. Lines 10, 11a and 11b are shown below. Determine the fill-ins for lines 12, 13, 14, and 15. Each taxpayer is single and claims 2 exemptions.

Line	a	b	c	d	e
10	11,726	12,204	14,132	14,569	17,573
11a	0	0	10	15	20
11b	1,476	1,643	2,158	2,035	2,569
12	____	____	____	____	____
13	____	____	____	____	____
14	____	____	____	____	____
15	____	____	____	____	____

4. Determine the fillins for lines 6 through 15 for the following single taxpayers.

	Exemptions	Wages	Interest	Dividends*	Withholding Tax	Line 13	Line 14 or 15
a.	2	$11,327	$ 0	$ 0	$1,565	$______	$______
b.	1	11,702	264	0	1,520	______	______
c.	3	14,224	167	75	1,634	______	______
d.	3	17,182	279	134	2,358	______	______
e.	1	14,823	310	217	2,627	______	______

*Remember to subtract the exclusion.

State Income Tax Unit 2

Most states in the country raise money through an income tax in the same manner as the federal government. Basically, the pattern followed by states for computing the tax include these steps:

1. Compute the adjusted gross income
2. Subtract the total exemptions
3. Subtract the total deductions
4. The amount remaining is the taxable income

Unit 2

As we learned earlier, the term exemptions refers to the allowance we receive for each person dependent upon us for their livelihood. A deduction is simply an expense we are permitted to subtract from our income before computing the tax. Some of these expenses are:

1. interest on loans
2. interest on mortgages
3. gifts to charities or churches.

Notice that the state income tax tables on page 198 are much the same as the Federal Withholding Tax Table we used in chapter 4.

Example 1 How much state tax will a resident of Idaho have to pay on an adjusted gross income of $14,672? She is single, claims 1 dependent, and her deductions total $1,900.

Solution

ARKANSAS

If the taxable income is:
Not over $3,000 . . . .2% of income

Over	But not Over		Of excess Over
$3,000	$6,000	$60 + 2.5%	$3,000
$6,000	$9,000	$135 + 3.5%	$6,000
$9,000	$15,000	$240 + 4.5%	$9,000
$15,000	$25,000	$510 + 6%	$15,000
$25,000	. . . .	$1,110 + 7%	$25,000

Reduction in tax as follows.
Single: $17.50; Married: $35
Head of household: $35; Each dependent: $6

IDAHO

If the taxable income is:
Not over $1,000 . . . .2% of income

Over	But not Over		Of excess Over
$1,000	$2,000	$20 + 4%	$1,000
$2,000	$3,000	$60 + 4.5%	$2,000
$3,000	$4,000	$105 + 5.5%	$3,000
$4,000	$5,000	$160 + 6.5%	$4,000
$5,000	. . .	$225 + 7.5%	$5,000

Exemptions as follows:
Single: $750; Married: $1,500
Head of household: $1,500; Each dependent: $750

NORTH CAROLINA

If the taxable income is:
Not over $2,000 . . . .3% of income

Over	But not Over		Of excess Over
$2,000	$4,000	$60 + 4%	$2,000
$4,000	$6,000	$140 + 5%	$4,000
$6,000	$10,000	$240 + 6%	$6,000
$10,000	. . .	$480 + 7%	$10,000

Exemptions as follows:
Single: $1,000; Married: $2,000
Head of household: $2,000; Each dependent: $600

Solution

$$\text{Total exemptions} = \text{Exemption for herself} + \text{Exemption for dependent}$$
$$= \$750 + \$750$$
$$= \$1{,}500$$

$$\text{Total deductions} = \$1{,}900$$

$$\text{Taxable income} = \text{Adjusted gross income} - \text{Exemptions} - \text{Deductions}$$
$$= \$14{,}672 - \$1{,}500 - \$1{,}900$$
$$= \$11{,}272$$

Tax Computation (See Table on page 000)

$$\text{Base tax} = \$225$$
$$\text{Excess over } \$5{,}000 = \$11{,}272 - \$5{,}000$$
$$= \$6{,}272$$
$$\text{Tax on excess} = 7.5\% \text{ of excess}$$
$$= .075 \times \$6{,}272$$
$$= \$470.40$$

$$\text{Tax} = \text{Base tax} + \text{Tax on excess}$$
$$= \$225 + \$470.40$$
$$= \$695.40$$

Unit 2

In a few states the tax is computed somewhat differently from that shown in Example 1. In those states, to find the taxable income we subtract *only the deductions* from the adjusted gross income.

$$\text{Taxable income} = \text{Adjusted gross income} - \text{Deductions}$$

The tax is now found in the normal manner. However, after determining the tax, it is then necessary to deduct the total exemptions from that number. This pattern is illustrated in the example below.

Example 2 A married resident of the state of Arkansas has an adjusted gross income of $19,684. He has deductions totaling $2,695 and lists 5 dependents. How much state income tax does he have to pay?

Solution

$$\text{Taxable income} = \text{Adjusted gross income} - \text{Deductions}$$
$$= \$19{,}684 - \$2{,}695$$
$$= \$16{,}989$$

Tax computation

$$\text{Base tax} = \$510$$

$$\begin{aligned}\text{Excess over } \$15{,}000 &= \$16{,}989 - \$15{,}000\\ &= \$1{,}989\end{aligned}$$

$$\begin{aligned}\text{Tax on excess} &= 6\% \text{ of excess}\\ &= .06 \times \$1{,}989\\ &= \$119.34\end{aligned}$$

$$\begin{aligned}\text{Income tax} &= \text{Base tax} + \text{Tax on excess}\\ &= \$510 + \$119.34\\ &= \$629.34\end{aligned}$$

$$\begin{aligned}\text{Total exemptions} &= \text{Exemp. for taxpayer} + \text{Exemps. for dependents}\\ &= \$35 + 5 \times \$6\\ &= \$35 + \$30\\ &= \$65\end{aligned}$$

$$\begin{aligned}\text{Tax} &= \text{Income tax} - \text{Total exemptions}\\ &= \$629.34 - \$65\\ &= \$564.34\end{aligned}$$

Exercises A

1. Determine the North Carolina state income tax for the following taxable incomes.

	Taxable Income	Base Tax	Tax on Excess	Tax
a.	$ 7,200	$______	$______	$______
b.	3,500	______	______	______
c.	1,600	______	______	______
d.	14,900	______	______	______
e.	18,900	______	______	______
f.	31,650	______	______	______

2. Determine the North Carolina state income tax for the following adjusted gross incomes.

	Adjusted Gross Income	Deductions		Taxable Income	Tax
a.	$ 8,700	$3,000	Single	$______	$______
b.	11,600	2,500	Single	______	______
c.	14,900	2,800	Married, 1 dependent	______	______
d.	18,300	3,650	Married, 2 dependents	______	______
e.	24,800	4,175	Married, 3 dependents	______	______

3. Determine the Idaho state income tax for the following taxable incomes.

	Taxable Income	Base Tax	Tax on Excess	Tax
a.	$ 7,400	$______	$______	$______
b.	5,300	______	______	______
c.	14,600	______	______	______
d.	27,100	______	______	______
e.	2,500	______	______	______

4. Determine the Idaho state income tax for the following adjusted gross incomes.

	Adjusted Gross Income	Deductions		Taxable Income	Tax
a.	$ 8,500	$1,700	Single	$______	$______
b.	11,300	1,950	Married, 2 dependents	______	______
c.	4,275	530	Single	______	______
d.	16,420	4,500	Head of house, 1 dep.	______	______
e.	24,950	6,400	Married, 4 dependents	______	______

Unit 2

5. Determine the Arkansas state income tax, before exemptions, for the following taxable incomes.

	Taxable Income	Base Tax	Tax on Excess	Tax
a.	$ 5,700	$______	$______	$______
b.	2,300	______	______	______
c.	8,760	______	______	______
d.	17,340	______	______	______
e.	35,650	______	______	______

6. Determine the Arkansas state income tax, after exemptions, for the following gross incomes.

	Adjusted Gross Income	Deductions		Taxable Income	Tax
a.	$10,400	$2,300	Single	$______	$______
b.	12,800	2,500	Married, 1 dependent	______	______
c.	15,700	2,940	Head of house, 1 dep.	______	______
d.	21,400	4,520	Married, 3 dependents	______	______
e.	36,900	3,480	Single	______	______

B

1.

a. Two men have exactly the same net income and the same number of dependents, but one is married, while the other is classified as the head of a household. Will one pay a smaller state income tax than the other if they live in Arkansas? Justify your answer. ________________

b. Would there be any difference in their taxes if they lived in North Carolina? ________

2. James Fields is unmarried and has no dependents. His net income last year was $12,658.

a. If he lived in Idaho how much state income tax did he have to pay? ________

b. If he lived in North Carolina, how much would his state income tax have been? ________

c. How much more – or less – would he have had to pay in North Carolina than in Idaho? ________

3. In Arkansas, how much more state income tax would a married man with two dependents have to pay than one with four dependents if they both had exactly the same net income? Assume that the net income of each would require a tax of more than $100. ________

4. How much more – or less – would a single woman with no dependents and a net income of $16,000 have to pay in state income tax if she lived in North Carolina rather than in Arkansas? ________

Review of Chapter 5

1. Determine the federal income tax for the following single taxpayers. Use the table on page 194.

	Adjusted Gross Income	Number of Exemptions	Tax
a.	$11,375	1	$________
b.	14,539	2	________
c.	17,031	3	________

2. Lines 12 and 13 of Form 1040A are shown below. Determine the fillin for either line 14 OR line 15.

Line		a	b
12	Total Paid	2,675	3,856
13	Tax from Table	2,904	3,527
14	Refund to You	________	________
15	Balance Due	________	________

3. Lines 10, 11a, and 11b of Form 1040A are shown below. Determine the fillins for lines 12, 13, and 14 or 15. Each taxpayer is single and claims 1 exemption. Use Tax Table, page 194.

Line		a	b
10	Adjusted gross income	14,912	17,623
11a	Credit for contribution	10	15
11b	Withholding tax	2,695	2,843
12	Total Paid	________	________
13	Tax from Table	________	________
14	Refund to You	________	________
15	Balance Due	________	________

4. Determine the North Carolina state income tax for the following taxable incomes.

	Taxable Income	Base Tax	Tax on Excess	Tax
a.	$ 8,000	$________	$________	$________
b.	4,600	________	________	________
c.	16,500	________	________	________

Review

5. Determine the Arkansas state income tax, before exemptions, for the following taxable incomes.

	Taxable Income	Base Tax	Tax on Excess	Tax
a.	$ 7,000	$________	$________	$________
b.	12,800	________	________	________
c.	16,250	________	________	________

6. Determine the Idaho state income tax for the following adjusted gross incomes.

	Adjusted Gross Income	Deductions		Taxable Income	Tax
a.	$ 9,600	$2,300	Single	$________	$________
b.	11,200	3,100	Married, 3 dependents	________	________
c.	14,800	5,200	Married, 1 dependent	________	________

7. The state income tax return of a married man living in Arkansas shows that he has deductions totaling $3,872 and 4 dependents. If his adjusted gross income is $23,615, how much is his state income tax? ________

Chapter 6

Banking — The Savings Account

In a previous chapter we learned how to compute earnings. In this chapter we will examine one of the ways in which these earnings can be put to work to earn more money for us.

The Deposit Slip Unit 1

Most people realize that keeping all of their money at home is simply an invitation for theft. For this reason, they have turned to the banks where not only is that money protected against theft, but they also receive periodic payments from their bank for the use of that money.

Each time money is deposited at a bank it is necessary to complete a deposit slip similar to the one below.

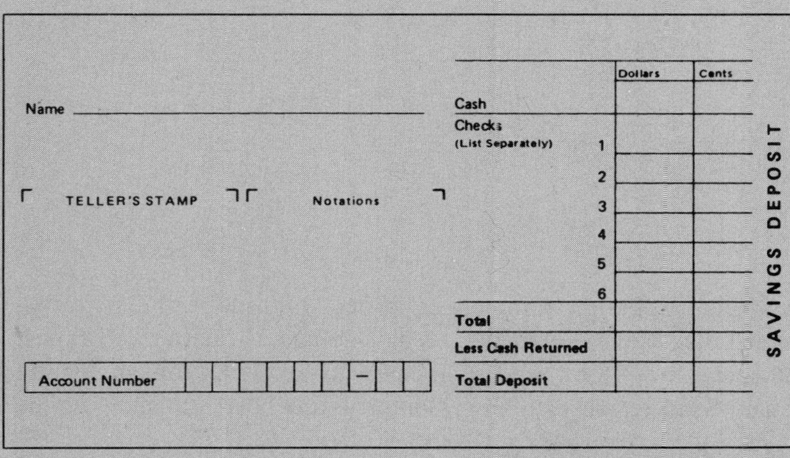
Name

TELLER'S STAMP

Notations

Account Number

		Dollars	Cents
Cash			
Checks (List Separately)	1		
	2		
	3		
	4		
	5		
	6		
Total			
Less Cash Returned			
Total Deposit			

SAVINGS DEPOSIT

Frequently we fail to align the decimal points when adding numbers. Therefore, the line in the deposit slip that separates the Dollars from the Cents is used as a replacement for the decimal point. This line forces us to arrange the numbers carefully.

Each of the terms of the deposit slip are explained in the figure below.

(1) This fill-in is the total of all bills and coins deposited. → Cash

(2) Each check deposited is listed separately as the fill-ins for these lines. → Checks 1–6

(3) This fill-in is the total of the cash and all the checks. → Total

(4) If any money is to be returned to us, that amount is listed here. → Less Cash Returned

(5) To determine this fill-in we subtract the amount we want returned from the total we give the bank teller. → Total Deposit

		Dollars	Cents
Cash			
Checks (List Separately)	1		
	2		
Endorse all checks	3		
	4		
	5		
	6		
Total			
Less Cash Returned			
Total Deposit			

Example Philip Demarest deposited 4 ten-dollar bills, 6 five-dollar bills, 8 one-dollar bills, 2 half-dollars, 3 quarters, 7 dimes, one check for $18.62, and another for $146.17. Of this amount he wanted $125 returned to him. Complete a deposit slip showing the fill-ins for this deposit.

Solution

(1) Total of bills and coins. → Cash

(2) Each check listed separately. → Checks (List Separately)

(3) Total of cash and all checks. → Total

(4) Amount returned to Philip Demarest. → Less Cash Returned

(5) Total deposited: $245.24 − $125.00 = $120.24 → Total Deposit

		Dollars	Cents
Cash		80	45
Checks (List Separately)	1	18	62
	2	146	17
Endorse all checks	3		
	4		
	5		
	6		
Total		245	24
Less Cash Returned		125	00
Total Deposit		120	24

Exercises A

Unit 1

1. Complete each deposit slip.

a.

		Dollars	Cents
Cash		26	23
Checks			
(List Separately)	1	54	00
	2	19	00
Total			
Less Cash Returned			
Total Deposit			

b.

		Dollars	Cents
Cash		19	56
Checks			
(List Separately)	1	135	00
	2	67	40
Total			
Less Cash Returned			
Total Deposit			

c.

		Dollars	Cents
Cash		12	67
Checks			
(List Separately)	1	47	00
	2	83	50
	3	135	65
Total			
Less Cash Returned			
Total Deposit			

d.

		Dollars	Cents
Cash		39	58
Checks			
(List Separately)	1	68	37
	2	107	26
	3	354	92
Total			
Less Cash Returned			
Total Deposit			

2. Complete each deposit slip.

a.

	Dollars	Cents
Cash	8	75
Checks		
(List Separately) 1	69	27
2	85	41
Total		
Less Cash Returned	25	00
Total Deposit		

b.

	Dollars	Cents
Cash	16	17
Checks		
(List Separately) 1	56	48
2	134	00
Total		
Less Cash Returned	47	00
Total Deposit		

c.

	Dollars	Cents
Cash	36	92
Checks		
(List Separately) 1	87	25
2	60	35
3	98	49
4	136	57
Total		
Less Cash Returned	65	00
Total Deposit		

d.

	Dollars	Cents
Cash	18	37
Checks		
(List Separately) 1	195	15
2	247	64
3	206	18
4	398	39
Total		
Less Cash Returned	475	00
Total Deposit		

B

Complete a deposit slip for each of the following deposits. Draw the deposit slips similar to those in Part A.

1. Deposit: 6 ten-dollar bills, 8 five-dollar bills, 4 one-dollar bills, 1 half-dollar, 3 quarters, 4 dimes, 1 check for $46.50. ________

2. Deposit: 3 ten-dollar bills, 12 five-dollar bills, 9 one-dollar bills, 2 half-dollars, 2 quarters, 7 dimes, 4 nickles, 1 check for $126.75, 1 check for $37.50. ________

3. Deposit: 2 twenty-dollar bills, 5 ten-dollar bills, 7 one-dollar bills, 8 nickles, 14 pennies, 1 check for $67.30, 1 check for $140. Money to be returned: $75. ________

4. Deposit: 15 one-dollar bills, 12 half-dollars, 20 quarters, 25 dimes, 15 nickles, 40 pennies, 1 check for $217.20, 1 check for $86.30, 1 check for $109.25. Money to be returned: $135. ________

5. Deposit: 27 one-dollar bills, 20 half-dollars, 25 quarters, 50 dimes, 25 nickles, 200 pennies, 1 check for \$150, 1 check for \$53.40, 1 check for \$117.60. Money to be returned: \$175.

____

The Simple Interest Formula Unit 2

Section 1 Annual Periods of Time

Having deposited our money in a bank we now turn to an investigation of just how the bank will pay us for the use of this money.

It is very likely that we have seen signs such as the following displayed in a bank.

Annual Interest Rate 5%

Unit 2

Through this information we are being told that we will receive 5% of the deposit if our money remains in the bank for *one full year*. The payment we receive from the bank is called the *interest*, while the 5% is the *annual interest rate*.

Example 1 How much interest will a deposit of \$6,200 earn in a period of 1 year? The annual interest rate is 7%.

Solution

Annual interest rate (points to 7%)

Interest = 7% of the deposit
= .07 × \$6,200
= \$434

One year's earnings on \$6,200 (points to \$434)

The amount of money deposited is referred to as the *principal.* Based on the solution to Example 1, it would appear that the interest for one year is found by *multiplying the principal by the interest rate.*

Interest = Principal × Interest Rate.

This is usually expressed as the formula

$$I = P \times R.$$

Example 2 How much interest will a deposit of $7,350 earn in a period of one year if the interest rate is 6 1/2%?

Solution

$$\begin{aligned} I &= P \times R \\ &= \$7{,}350 \times 6\tfrac{1}{2}\% \\ &= \$7{,}350 \times 6.5\% \\ &= \$7{,}350 \times .065 \\ &= \$477.75 \end{aligned}$$

Exercises A

Do you recall how to do the computation in the following exercises? This computation is the same as you will be doing when finding the interest on a deposit.

1. Rename each percent numeral as a decimal.

a.	5%	________	f.	$6\frac{1}{2}\%$	________
b.	4%	________	g.	$7\frac{1}{2}\%$	________
c.	6%	________	h.	$5\frac{1}{4}\%$	________
d.	5.4%	________	i.	$6\frac{1}{4}\%$	________
e.	6.7%	________	j.	$7\frac{3}{4}\%$	________

2. Find the products.

a.	2,000 × .05	c.	4,000 × .065	e.	6,150 × .0525
b.	3,600 × .07	d.	3,400 × .065	f.	8,430 × .0675

3. Find the products.

a.	\$1,500 × 6%	\$________	e.	\$7,400 × 6.5%	\$________
b.	\$2,800 × 8%	________	f.	\$8,000 × $4\frac{1}{2}\%$	________
c.	\$4,300 × 3.5%	________	g.	\$6,500 × $6\frac{1}{4}\%$	________

B

1. Determine the annual interest for each deposit.

	Deposit	Annual Interest Rate	Interest
a.	\$8,000	5%	\$________
b.	6,500	7%	________
c.	875	6%	________
d.	3,275	8%	________

2. Determine the annual interest for each deposit.

	Principal	Annual Interest Rate	Interest
a.	\$4,000	6.5%	\$________
b.	2,300	7.5%	________
c.	1,700	5.25%	________
d.	3,500	6.25%	________
e.	4,250	5.75%	________

3. Determine the annual interest for each deposit.

Unit 2

	Principal	Annual Interest Rate	Interest
a.	\$1,400	$5\frac{1}{2}\%$	\$________
b.	5,200	$6\frac{1}{2}\%$	________
c.	650	$5\frac{1}{4}\%$	________
d.	2,340	$6\frac{1}{4}\%$	________
e.	4,265	$6\frac{3}{4}\%$	________

C

1. An investment of \$5,000 is made at an annual interest rate of 6%.

a. How much interest will this investment earn in a period of one year? ________

b. How much interest will this investment earn in a period of two years? ________

c. How much interest will this investment earn in a period of five years? ________

2. An investment of \$6,000 is made at an annual interest rate of 8%.

a. How much interest will this investment earn in a period of one year? ________

b. How much interest will this investment earn in a period of one-half year? ________

c. How much interest will this investment earn in a period of four months? ________

Section 2 Monthly Periods of Time

Banks do not actually wait an entire year before adding interest to our account. The bank that pays interest each half year will simply add one-half of the full year's interest to our account each period.

$$\text{One-half year's interest} = \text{Full year's interest} \times \tfrac{1}{2}$$

Similarly,

$$\text{One-third year's interest} = \text{Full year's interest} \times \tfrac{1}{3}$$
$$\text{One-fourth year's interest} = \text{Full year's interest} \times \tfrac{1}{4}$$

The interest we receive depends upon the fraction of the year for which this interest is paid. Therefore, this will change the interest formula as follows:

$$I = \underbrace{P \times R}_{} \times T$$

This number is the full year's interest. (points to $P \times R$)

This number is the fraction of year for which the interest is paid. (points to T)

Example 1 Determine the semiannual interest payment on \$1,648 if the interest rate is 5 1/2%.

Explanation There are two points that have to be clarified before showing the solution to this problem:

1. A semiannual interest payment is the interest we receive each half-year.
2. It is not necessary to state that the interest rate of 5 1/2% is for a period of one year. Interest rates are always quoted for a one-year period unless otherwise stated.

Solution

$$\begin{aligned} I &= P \times R \times T \\ &= \$1{,}648 \times 5\tfrac{1}{2}\% \times \tfrac{1}{2} \\ &= \$1{,}648 \times .055 \times \tfrac{1}{2} \\ &= \$90.64 \times \tfrac{1}{2} \\ &= \$45.32 \end{aligned}$$

$\frac{1}{2}$ — Fraction of year

\$90.64 — Interest for 1 year

Example 2 A bank pays interest once every 3 months. Determine the interest for that period on \$5,640 at an interest rate of 6 1/4%.

Solution

$$\begin{aligned} I &= P \times R \times T \\ &= \$5{,}640 \times 6\tfrac{1}{4}\% \times 3 \text{ months} \\ &= \$5{,}640 \times .0625 \times \tfrac{3}{12} \\ &= \$5{,}640 \times .0625 \times \tfrac{1}{4} \\ &= \$352.50 \times \tfrac{1}{4} \\ &= \$88.13 \end{aligned}$$

$\frac{3}{12}$ — Fraction of year

$\frac{1}{4}$ — $\frac{3}{12}$ renamed as $\frac{1}{4}$

\$352.50 — Interest for 1 year

Unit 2

Exercises A

Do you recall how to do the computation in the following exercises? This computation is the same as you will be doing when finding the interest for monthly periods.

1. Find the products.

a. $600 × .05 × $\frac{1}{2}$	$______		d. $2,500 × .08 × $\frac{1}{4}$	$______
b. $750 × .07 × $\frac{1}{2}$	______		e. $3,600 × .045 × $\frac{1}{6}$	______
c. $1,400 × .06 × $\frac{1}{3}$	______		f. $4,000 × .065 × $\frac{3}{4}$	______

2. Find the products.

a. $88 × 6% × $\frac{1}{2}$	$______		c. $2,750 × 6$\frac{1}{2}$% × $\frac{1}{6}$	$______
b. $1,260 × 5$\frac{1}{2}$% × $\frac{1}{4}$	______		d. $4,200 × 8% × $\frac{1}{2}$	______

B

1. Determine the semiannual interest payments.

	Principal	Interest Rate	Interest
a.	$6,000	7%	$______
b.	7,320	6%	______
c.	5,000	5.4%	______
d.	8,200	6.2%	______
e.	4,800	5$\frac{1}{2}$%	______
f.	6,500	6$\frac{1}{4}$%	______

2. Determine the quarterly interest payments.

	Principal	Interest Rate	Interest
a.	$2,500	7%	$______
b.	3,700	6.3%	______
c.	8,400	7$\frac{1}{2}$%	______
d.	6,800	6$\frac{3}{4}$%	______

3. Determine the interest payments.

	Principal	Interest Rate	Interest Period	Interest
a.	$9,000	6%	1 month	$______
b.	8,100	8%	1 month	______
c.	6,600	5%	2 months	______
d.	3,900	5.2%	2 months	______
e.	5,700	6.4%	3 months	______
f.	2,420	5$\frac{1}{2}$%	3 months	______

C

1. Eleanor Rossner deposited $4,000 in a bank that paid an interest rate of 5.87%.

a. How much interest will be added to her account at the end of one year if interest is paid once each year? ________

b. Including the interest, how much money will there be in her account at the end of one year? ________

2. The Greenboro Savings Bank pays interest semiannually at an interest rate of 5 1/4%. George Wells deposited $6,000 in this bank.

a. How much interest will he receive at the end of the first interest period? ________

b. Including interest, how much money will there be in his account at the end of that interest period? ________

3. A deposit of S15,000 is made at a bank that pays interest semiannually.

a. How large will the interest payment be at the end of a half year if the interest rate is 7%? ________

b. How large will the interest payment be at the end of a half year if the interest rate is 5%? ________

c. How much more is the payment at the 7% rate than at the 5% rate? ________

Section 3 Daily Periods of Time

Our final step on this unit will be to compute interest when the period of time is measured in days rather than months or years. If interest is paid for 29 days it will be necessary for us to express 29 days as a fraction of a year. Since there are 365 days in one year, then 29 days is 29/365ths of a year.

Many years ago, before the invention of the computor, banks were given permission to use 360 as the number of days in a year. This was done in order to simplify the computation. Hence, when determining bank interest, 29 days will not be the fraction of the year shown above, but rather 29/360ths of a year. Even this change is not very helpful when trying to find the interest on, let us say, \$574 at 7 1/2% for 29 days.

$$
\begin{aligned}
I &= P \times R \times T \\
&= \$574 \times 7\tfrac{1}{2}\% \times 29 \text{ days} \\
&= \$574 \times .075 \times \tfrac{29}{360}
\end{aligned}
$$

Fraction of year (points to $\tfrac{29}{360}$)

$$
= \$43.05 \times \tfrac{29}{360}
$$

Interest for one year (points to \$43.05)

$$
= \$3.47
$$

As we can see from the solution above the computation of interest for daily periods is quite difficult. In view of this, tables such as the one below are used.

Simple Interest on \$1
(360-Day Year)
Rates

Days	5%	5.25%	6.0%	6.5%	7.0%	7.5%
1	.0001389	.0001458	.0001667	.0001806	.0001944	.0002083
2	.0002778	.0002917	.0003333	.0003611	.0003889	.0004167
3	.0004167	.0004375	.0005000	.0005417	.0005833	.0006250
4	.0005556	.0005833	.0006667	.0007222	.0007778	.0008333
5	.0006944	.0007292	.0008333	.0009028	.0009722	.0010417
6	.0008333	.0008750	.0010000	.0010833	.0011667	.0012500
7	.0009722	.0010208	.0011667	.0012639	.0013611	.0014583
8	.0011111	.0011667	.0013333	.0014444	.0015556	.0016667
9	.0012500	.0013125	.0015000	.0016250	.0017500	.0018750
10	.0013889	.0014583	.0016667	.0018056	.0019444	.0020833
11	.0015278	.0016042	.0018333	.0019861	.0021389	.0022917
12	.0016667	.0017500	.0020000	.0021667	.0023333	.0025000
13	.0018056	.0018958	.0021667	.0023472	.0025278	.0027083
14	.0019444	.0020417	.0023333	.0025278	.0027222	.0029167
15	.0020833	.0021875	.0025000	.0027083	.0029167	.0031250
16	.0022222	.0023333	.0026667	.0028889	.0031111	.0033333
17	.0023611	.0024792	.0028333	.0030694	.0033056	.0035417
18	.0025000	.0026250	.0030000	.0032500	.0035000	.0037500
19	.0026389	.0027708	.0031667	.0034306	.0036944	.0039583
20	.0027778	.0029167	.0033333	.0036111	.0038889	.0041667
21	.0029167	.0030625	.0035000	.0037917	.0040833	.0043750
22	.0030556	.0032083	.0036667	.0039722	.0042778	.0045833
23	.0031944	.0033542	.0038333	.0041528	.0044722	.0047917
24	.0033333	.0035000	.0040000	.0043333	.0046667	.0050000
25	.0034722	.0036458	.0041667	.0045139	.0048611	.0052083
26	.0036111	.0037917	.0043333	.0046944	.0050556	.0054167
27	.0037500	.0039375	.0045000	.0048750	.0052500	.0056250
28	.0038889	.0040833	.0046667	.0050556	.0054444	.0058333
29	.0040278	.0042292	.0048333	.0052361	.0056389	.0060417
30	.0041667	.0043750	.0050000	.0054167	.0058333	.0062500
31	.0043056	.0045208	.0051667	.0055972	.0060278	.0064583

Example 1 Determine the interest on $1 for 18 days at an annual interest rate of 6.5%.

Explanation

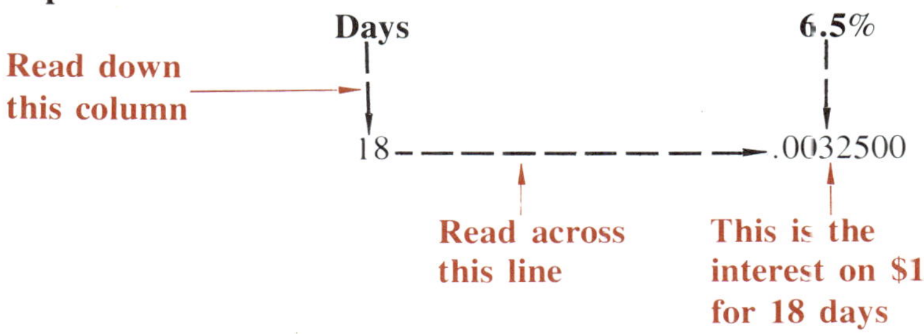

Solution

Interest on $1 at 6.5% for 18 days = $.0032500

Example 2 Determine the interest on $537 for 26 days at an annual interest rate of 5.25%.

Explanation The interest on 537 dollars is 537 times as much as the interest on 1 dollar.

Solution

Interest on $1 for 26 days = $.0037917
Interest of $537 for 26 days = 537 × $.0037917
= $2.04

Example 3 Determine the interest on $2,700 for 42 days at an annual interest rate of 7%.

Explanation By adding the interest for 30 days to the interest for 12 days we can determine the interest for 42 days.

Solution

Interest on $1 for 30 days = $.0058333
\+ Interest on $1 for 12 days = $.0023333

Total Interest on $1 for 42 days = $.0081666
Interest on $2,700 for 42 days = 2,700 × $.0081666
= $22.05

Exercises

1. Determine the interest on $1 for each period. Do not round your answer.

	Number of Days	Interest Rate	Interest
a.	17	5%	$________
b.	24	6%	________
c.	29	7%	________
d.	10	7.5%	________
e.	21	$6\frac{1}{2}$%	________
f.	14	$5\frac{1}{4}$%	________

2. Determine the interest to the nearest cent for each deposit.

	Deposit	Number of Days	Interest Rate	Interest On $1	Interest On Deposit
a.	$2,000	27	5%	$________	$________
b.	3,000	16	5%	________	________
c.	4,000	25	6%	________	________
d.	5,000	12	7%	________	________
e.	900	16	6.5%	________	________
f.	750	21	7.5%	________	________
g.	4,300	14	$5\frac{1}{4}$%	________	________
h.	1,200	29	$6\frac{1}{2}$%	________	________

3. Determine the interest on $1 for each period. Do not round your answer.

	Number Of Days	Interest Rate	Interest
a.	40	6%	$________
b.	50	7%	________
c.	35	5%	________
d.	45	7.5%	________
e.	48	$6\frac{1}{2}$%	________
f.	57	$7\frac{1}{2}$%	________
g.	53	$5\frac{1}{4}$%	________

4. Determine the interest to the nearest cent in the following exercises.

	Deposit	Number Of Days	Interest Rate	Interest On $1	Interest On Deposit
a.	$5,000	38	5%	$________	$________
b.	7,000	44	6%	________	________
c.	4,300	52	7%	________	________
d.	6,500	55	6.5%	________	________
e.	5,600	47	7.5%	________	________

Computing the Interest On a Savings Account Unit 3

Section 1 The Smallest Balance Method

In the previous unit we learned how interest on a deposit can be determined. Banks, however, credit this interest to our account in many different ways. In this section we will take a look at the method that gives the depositor the least amount of interest. It is called the smallest balance method for computing interest.

Most banks credit interest to a depositor's account on the last day of each month. Under the smallest balance method, that interest is computed on the smallest amount of money the person had in the bank at any time during the month. As an example, consider the following information:

Amount on deposit from June 1 to June 28: $10,000
Withdrawal on June 28: $9,000
Balance from June 28 through June 30: $1,000

Although $10,000 was in the account for almost the entire month, the interest for the month will be computed on $1,000 for that is the smallest balance.

In the examples and exercises in this section we will use the smallest balance method for computing interest.

Example 1 A savings account showed the following activity for the month of May.

Balance May 1: $3,400
Deposit May 7: $250
Withdrawal May 12: $500
Deposit May 25: $1,400

Determine the amount of money on which interest was computed this month.

Solution

Balance May 1 = $3,400

Balance May 7 = $3,400 + $250 ← Deposit
= $3,650

Balance May 12 = $3,650 − $500 ← Withdrawal
= $3,150

Balance May 25 = $3,150 + $1,400 ← Deposit
= $4,550

Balance for computation of interest = $3,150 ← Smallest balance

In order to compute the monthly interest, it is necessary to know the number of days in each month. The table below will help to recall this.

Days in Each Month

January	31	July	31
February	28 (29)	August	31
March	31	September	30
April	30	October	31
May	31	November	30
June	30	December	31

Example 2 A savings account showed the following activity during the month of July:

Balance July 1: $2,700
Withdrawal July 3: $900
Deposit July 12: $3,000
Deposit July 27: $850

If the interest rate is 5 1/4%, how much interest was credited to the account on July 31?

Solution

$$\text{Balance July 1} = \$2,700$$

Withdrawal

$$\text{Balance July 3} = \$2,700 - \$900 = \$1,600$$

Deposit

$$\text{Balance July 12} = \$1,600 + \$3,000 = \$4,600$$

Deposit

$$\text{Balance July 27} = \$4,600 + \$850 = \$5,450$$

$$\text{Smallest balance} = \$1,600$$

July has 31 days

$$\text{Interest on \$1 for 31 days} = \$.0045208$$

$$\text{Interest on \$1,600 for 31 days} = 1,600 \times \$.0045208 = \$7.23$$

Exercises A

1. Determine the balance on which the interest will be computed for the following months.

a. Balance April 1: $2,700
Deposit April 7: $460
Deposit April 15: $610
Withdrawal April 24: $1,000

______

b. Balance Oct. 1: $3,250
Withdrawal Oct. 9: $670
Deposit Oct. 16: $785
Withdrawal Oct. 21: $1,200

______

c. Balance Feb. 1: $960
Deposit Feb. 8: $420
Withdrawal Feb. 11: $535
Withdrawal Feb. 18: $275

______

d. Balance June 1: $6,400
Deposit June 5: $820
Withdrawal June 9: $2,000
Deposit June 17: $515

______

e. Balance May 1: $793
Deposit May 6: $186
Deposit May 9: $2,750
Withdrawal May 25: $825

______

f. Balance July 1: $2,672
Deposit July 7: $958
Withdrawal July 10: $596
Withdrawal July 19: $467

______

Unit 3

2. Determine the interest added to these accounts for the month.

a. Interest rate: 5%
Balance June 1: $2,000
Withdrawal June 9: $600

____

b. Interest rate: 7%
Balance March 1: $5,000
Deposit March 8: $340

____

c. Interest rate: 5.25%
Balance September 1: $3,100
Deposit September 6: $250

____

d. Interest rate: 6.5%
Balance October 1: $8,300
Withdrawal October 18: $900

____

3. Determine the interest added to these accounts for the month.

a. Interest rate: 5%
Balance April 1: $900
Deposit April 7: $600
Withdrawal April 20: $300
Withdrawal April 26: $150

____

b. Interest rate: 5.25%
Balance June 1: $6,200
Withdrawal June 10: $700
Deposit June 18: $500
Withdrawal June 24: $900

____

c. Interest rate: 5%
Balance July 1: $5,000
Deposit July 7: $400
Deposit July 12: $800
Withdrawal July 26: $950

____

d. Interest rate: 5.25%
Balance March 1: $7,620
Withdrawal March 12: $3,260
Deposit March 20: $1,640
Withdrawal March 26: $910

____

B

1. Regina Taylor has $6,000 on deposit in a bank that pays an interest rate of 5 1/4%. She is expecting delivery of a car on April 29 and needs the $6,000 to pay for the car. How much interest can she save by paying for the car on May 1 rather than April 29?

____

2. Steve Caputo's savings account showed the following activity for the month of September.

Balance September 1: $5,300
Withdrawal September 6: $2,500
Deposit September 8: $3,700

The bank in which this money is kept pays a 5% interest rate.

a. How much interest was credited to his account for the month of September? ________

b. If the dates of the withdrawal and the deposit could have been interchanged, how much interest would have been credited to his account? ________

c. How much money would Steve have saved had he been able to interchange the dates of the withdrawal and the deposit? ________

Section 2 Day In to Day Out Method

The method that we are about to examine for crediting interest to an account gives the depositor a greater return than any other method used by banks. Under this arrangement the depositor receives interest for each day his money remains in the bank. It is called the *day in to day out method* for computing interest.

Unit 3

In applying this method it is necessary to determine the number of days from a date during a month to the last day of the month. As an example, if a deposit is made on March 15, then during that month the deposit *will earn* interest from March 15 to March 31. To determine the number of days from the first date to the second, we simply subtract 15 from 31.

$$\text{Number of days from March 15 to March 31} = 31 - 15 = 16$$

Similarly, a *withdrawal* on March 7 will *not* earn interest for 24 days of that month.

$$\text{Number of days from March 7 to March 31} = 31 - 7 = 24$$

In the examples and exercises in this section we will use the day in to day out method for computing interest.

Example 1 Determine the interest earned for July on a \$3,500 deposit made on July 12 at a bank that pays a 7 1/2% interest rate.

Solution

$$\begin{aligned}\text{Number of days from July 12 to July 31} &= 31 - 12\\ &= 19\\ \text{Interest on \$1 for 19 days} &= \$.0039583\\ \text{Interest on \$3,500 for 19 days} &= 3{,}500 \times \$.0039583\\ &= \$13.85\end{aligned}$$

Example 2 Determine the interest lost during October on a withdrawal of \$1,970 made on October 23 at a bank that pays an interest rate of 6 1/2%.

Solution

$$\begin{aligned}\text{Number of days from October 23 to October 31} &= 31 - 23\\ &= 8\\ \text{Interest on \$1 for 8 days} &= \$.0014444\\ \text{Interest on \$1,970 for 8 days} &= 1{,}970 \times \$.0014444\\ &= \$2.85\end{aligned}$$

Example 3 An account showed the following activity during the month of June:

Balance June 1: \$4,800
Deposit June 9: \$560
Withdrawal June 16: \$975

If this bank paid an interest rate of 5 1/4%, how much interest was credited to this account June 30?

Solution

Interest on \$1 for 30 days = \$.0043750
Interest on \$4,800 for 30 days = 4,800 × \$.0043750
= \$21.00

Balance June 1 (\$4,800)

Interest on \$1 for 21 days = \$.0030625
Interest on \$560 for 21 days = 560 × \$.0030625
= \$1.72

Deposit June 9 (\$560)

Interest on \$1 for 14 days = \$.0020417
Interest on \$975 for 14 days = 975 × \$.0020417
= \$1.99

Withdrawal June 16 (\$975)

	Balance Interest		**Deposit Interest**		**Withdrawal Interest**
Total interest =	\$21.00	+	\$1.72	−	\$1.99

= \$24.71

Exercises A

1. Determine the number of days for which each deposit will earn interest during the month in which the deposit was made.

	Date of Deposit	Number of days		Date of Deposit	Number of Days
a.	April 7	________	c.	September 3	________
b.	July 16	________	d.	December 22	________

2. Determine the number of days for which each withdrawal will lose interest during the month in which the withdrawal was made.

	Date of Withdrawals	Number of Days		Date of Withdrawal	Number of Days
a.	May 5	________	c.	January 14	________
b.	November 12	________	d.	June 6	________

Unit 3

B

1. Determine the interest earned by each deposit during the month.

	Deposit	Date of Deposit	Interest Rate	Number of Days	Interest
a.	$ 500	6/5	6%	______	$______
b.	2,000	8/16	7%	______	______
c.	3,800	11/3	6.5%	______	______
d.	4,900	3/21	$5\frac{1}{4}$%	______	______

2. Determine the interest lost by each withdrawal during this month.

	Withdrawal	Date of Withdrawal	Interest Rate	Number of Days	Interest
a.	$ 600	4/10	5%	______	$______
b.	2,100	7/21	6%	______	______
c.	3,600	9/3	$7\frac{1}{2}$%	______	______

C

1. Determine the interest credited to each account during this month.

	Balance on 1st Day of Month	Deposit	Date of Deposit	Interest Rate	Interest
a.	$2,000	$600	6/8	5%	$______
b.	3,000	500	5/12	6%	______
c.	5,000	400	7/16	7%	______
d.	4,600	800	4/7	$6\frac{1}{2}$%	______
e.	2,500	350	8/9	$5\frac{1}{4}$%	______

2. Determine the interest credited to each account during this month.

	Balance on 1st Day of Month	Withdrawal	Date of Deposit	Interest Rate	Interest
a.	$1,000	$300	3/8	5%	$______
b.	7,000	200	5/14	6%	______
c.	4,500	750	1/10	7%	______
d.	6,800	810	4/4	$6\frac{1}{2}$%	______

3. Determine the interest credited to each account during this month.

	Balance on 1st Day of Month	Deposit	Withdrawal	Interest Rate	Interest
a.	$4,000	$300 (5/6)	$700 (5/9)	6%	$______
b.	2,000	1,000 (7/3)	400 (7/15)	5%	______
c.	6,000	800 (4/20)	2,000 (4/9)	7%	______
d.	7,000	1,500 (6/22)	3,000 (6/5)	7.5%	______
e.	5,000	750 (11/3)	450 (11/15)	6.5%	______

D

1. On June 1, George White had $4,500 on deposit in his account at the Wayside Trust Company. During the month of June the account showed the following activity:

June 14: Withdrawal of $1,600
June 27: Deposit of $700

The account paid an interest rate of 5%.

a. How much interest was credited to Mr. White's account on June 30? ____________

b. How much money did he have in the account after interest was credited? ________

2. The Mutual Savings Bank of Bayside pays an interest rate of 5 1/4% on certain accounts. One of these accounts showed the following activity for the month of March.

Balance on March 1: $1,400
Withdrawal March 5: $250
Deposit March 10: $620
Withdrawal March 17: $350

a. How much interest was credited to this account on March 31? ________

b. How much money did the account contain after the interest was credited? ________

Compound Interest Unit 4

Section 1 The Compound Interest Formula

Within recent years people have been depositing money in special bank accounts and allowing these accounts to remain dormant, that is, no deposits or withdrawals are made. We are now going to examine how their money grows under this arrangement.

Let us consider the situation in which a deposit of $6,000 is made at a bank that pays an interest rate of 8% once each year. The interest earned on this deposit for 1 year can be computed in the usual manner.

$$\begin{aligned} I &= P \times R \times T \\ &= \$6{,}000 \times .08 \times 1 \\ &= \$480 \end{aligned}$$

The amount of money on deposit at the end of the first year will be the sum of the original $6,000 plus the interest.

$$\begin{aligned} \text{Amount of money} &= \text{Original Principal} + \text{Interest} \\ A &= \$6{,}000 + \$480 \\ &= \$6{,}480 \end{aligned}$$

Rather than having to compute the amount of money in the bank at the end of one year as we did above, it can be found more rapidly by simply multiplying $6,000 by 1.08.

Amount on deposit during 1st year

↓

$$\begin{aligned} \text{First year } A &= \$6{,}000 \times 1.08 \\ &= \$6{,}480 \end{aligned}$$

Similarly, the amount of money in the bank at the end of the second year can be found by multiplying $6,480 by the same 1.08.

Amount on deposit during 2nd year

↓

$$\begin{aligned} \text{Second year } A &= \$6{,}480 \times 1.08 \\ &= \$6{,}998.40 \end{aligned}$$

The amount of money on deposit at the end of two years will be $6,998.40. In arriving at this number we multiplied $6,000 by 1.08 and that product was then multiplied again by 1.08. This can be expressed as follows:

$$\text{Amount after 2 years} = \$6{,}000 \times 1.08 \times 1.08$$

Earlier we learned that a short way for writing 1.08×1.08 is 1.08^2. The number 1.08^2 is read as *1.08 to the second power*. Therefore, the amount of money after two years can be expressed as,

$$\text{Amount after 2 years} = \$6{,}000 \times 1.08^2$$

The amount of money after three years can be expressed as,

$$\text{Amount after 3 years} = \$6{,}000 \times 1.08^3$$

The amount of money after any number of years can be expressed as,

$$\text{Amount after n years} = \$6{,}000 \times 1.08^n$$

In the illustration above the interest for each year was computed not only on the original principal but also on all the interest that had been added to the account. This process of paying interest on both the original principal and the interest earned by that principal is called *compounding the interest*. If interest is added to the account only once a year, we say that the interest is *compounded annually*.

In the examples and exercises in this unit we will consider the account as remaining dormant after the deposit is made.

Unit 4

Example 1 $9,000 is deposited in an account that pays an interest rate of 6% compounded annually. Express the amount of money that will be in this account after 7 years.

Solution

$$\text{Amount after 7 years} = \$9{,}000 \times 1.06^7$$

Original deposit → $9,000

Example 2 $5,000 is deposited in an account that pays an interest rate of 7% compounded annually. Determine the amount of money that will be in this account at the end of 3 years.

Solution

$$
\begin{aligned}
A &= \$5{,}000 \times 1.07^3 \\
&= \$5{,}000 \times [1.07 \times 1.07] \times 1.07 \\
&= \$5{,}000 \times [1.1449 \times 1.07] \\
&= \$5{,}000 \times 1.225043 \\
&= \$6{,}125.22
\end{aligned}
$$

Exercises

A

1. Express the amount of money that will be in the following accounts after the number of years shown. Interest is compounded annually. (See Example 1 above.)

	Original Deposit	Interest Rate	Number of Years	Answer
a.	\$ 3,000	7%	5	\$__________
b.	8,000	6%	9	__________
c.	7,500	5%	12	__________
d.	12,400	8%	25	__________
e.	6,000	$6\frac{1}{2}$%	10	__________
f.	9,000	$7\frac{1}{2}$%	15	__________
g.	15,000	$5\frac{1}{4}$%	8	__________

2. Determine the amount of money that will be in the following accounts after the number of years shown. Interest is compounded annually. (See Example 2 above)

	Original Deposit	Interest Rate	Number of Years	Amount on Deposit at End of Period
a.	\$ 4,000	6%	2	\$__________
b.	5,000	8%	2	__________
c.	9,000	7%	2	__________
d.	12,000	6%	3	__________
e.	14,000	8%	3	__________
f.	10,000	$6\frac{1}{2}$%	2	__________
g.	16,000	$7\frac{1}{2}$%	3	__________

B

1. \$8,000 is deposited in an account that pays a 6% interest rate compounded annually.

a. How much money will be in this account at the end of two years? __________

b. How much interest was earned by the \$8,000 over the two-year period? __________

2. $50,000 was deposited in an account for a period of three years.

a. If interest is compounded annually at an interest rate of 6%, how large will the account be at the end of the three-year period? __________

b. If interest is compounded annually at an interest rate of 8%, how large will the account be at the end of the three-year period? __________

c. How much more money will be earned under the 8% rate than under the 6% rate? __________

Section 2 The Compound Interest Table

The greatest difficulty concerning compound interest is the computation involved in raising a number to a power. In the illustration

$$\text{Amount after 15 years} = \$7{,}000 \times 1.08^{15},$$

if we had no calculator, we would have to spend a great deal of time finding the product of fifteen 1.08's.

In order to eliminate this tedious work, a table was developed that will give us this answer immediately. This table appears on page 232. The example below illustrates how the table is used.

Unit 4

Example 1 Find the value of 1.08^{15}.

Explanation

Number of Periods — **8%**

Read down → 15 – – – – – – – – – – – – – – 3.1722

Read across — **This is 1.08^{15}**

Solution

$$1.08^{15} = 3.1722$$

COMPOUND INTEREST TABLE
Value of $1

Number of Periods	1.5%	2%	2.5%	3%	3.5%	4%	5%	6%	7%	8%	9%	10%
1	1.0150	1.0200	1.0250	1.0300	1.0350	1.0400	1.0500	1.0600	1.0700	1.0800	1.0900	1.1000
2	1.0302	1.0404	1.0506	1.0609	1.0712	1.0816	1.1025	1.1236	1.1449	1.1664	1.1881	1.2100
3	1.0457	1.0612	1.0769	1.0927	1.1087	1.1248	1.1576	1.1910	1.2250	1.2597	1.2950	1.3310
4	1.0614	1.0824	1.1038	1.1255	1.1475	1.1699	1.2155	1.2625	1.3108	1.3605	1.4116	1.4641
5	1.0773	1.1041	1.1314	1.1593	1.1877	1.2167	1.2763	1.3382	1.4026	1.4693	1.5386	1.6105
6	1.0934	1.1262	1.1597	1.1941	1.2293	1.2653	1.3401	1.4186	1.5007	1.5869	1.6771	1.7716
7	1.1098	1.1487	1.1887	1.2299	1.2723	1.3159	1.4071	1.5036	1.6058	1.7138	1.8280	1.9487
8	1.1265	1.1717	1.2184	1.2668	1.3168	1.3686	1.4775	1.5938	1.7182	1.8059	1.9926	2.1436
9	1.1434	1.1951	1.2489	1.3048	1.3629	1.4233	1.5513	1.6895	1.8385	1.9990	2.1719	2.3579
10	1.1605	1.2190	1.2801	1.3439	1.4106	1.4802	1.6289	1.7908	1.9672	2.1589	2.3674	2.5937
11	1.1779	1.2434	1.3121	1.3842	1.4600	1.5395	1.7103	1.8983	2.1049	2.3316	2.5804	2.8531
12	1.1956	1.2682	1.3449	1.4258	1.5111	1.6010	1.7959	2.0122	2.2522	2.5182	2.8127	3.1384
13	1.2136	1.2936	1.3785	1.4685	1.5640	1.6651	1.8856	2.1329	2.4098	2.7196	3.0658	3.4523
14	1.2318	1.3195	1.4130	1.5126	1.6187	1.7317	1.9799	2.2609	2.5785	2.9372	3.3417	3.7975
15	1.2502	1.3459	1.4483	1.5580	1.6753	1.8009	2.0789	2.3966	2.7590	3.1722	3.6425	4.1772
16	1.2690	1.3728	1.4845	1.6047	1.7340	1.8730	2.1829	2.5404	2.9522	3.4259	3.9703	4.5950
17	1.2880	1.4002	1.5216	1.6528	1.7947	1.9479	2.2920	2.6928	3.1588	3.7000	4.3276	5.0545
18	1.3073	1.4282	1.5597	1.7024	1.8575	2.0258	2.4066	2.8543	3.3799	3.9960	4.7171	5.5599
19	1.3270	1.4568	1.5987	1.7535	1.9225	2.1068	2.5270	3.0256	3.6165	4.3157	5.1417	6.1159
20	1.3469	1.4859	1.6386	1.8061	1.9898	2.1911	2.6533	3.2071	3.8697	4.6610	5.6044	6.7275
21	1.3671	1.5157	1.6796	1.8603	2.0594	2.2788	2.7860	3.3996	4.1406	5.0338	6.1088	7.4003
22	1.3876	1.5460	1.7216	1.9161	2.1315	2.3699	2.9253	3.6035	4.4304	5.4365	6.6586	8.1403
23	1.4084	1.5769	1.7646	1.9736	2.2061	2.4647	3.0715	3.8198	4.7405	5.8715	7.2579	8.9543
24	1.4295	1.6084	1.8087	2.0328	2.2833	2.5633	3.2251	4.0489	5.0724	6.3412	7.9111	9.9497
25	1.4509	1.6407	1.8539	2.0938	2.3673	2.6658	3.3864	4.2919	5.4274	6.8485	8.6231	10.8347

This table can be interpreted differently than we did above. Notice that the heading of the table is *Value of $1*. Using Example 1 as an illustration, the number 3.1722 is the amount of money to which $1 will grow in 15 years at an interest rate of 8% compounded annually:

$1 grows to $3.17.

The numbers in the body of the Compound Interest Table are called *compounding factors*.

Example 2 $14,000 is deposited in an account that pays an interest rate of 7% compounded annually. Determine the amount of money that will be in this account at the end of 20 years.

Solution

$$A = \$14{,}000 \times 1.07^{20}$$

$$= \$14{,}000 \times 3.8697$$

Compounding factor (3.8697)

$$= \$54{,}175.80$$

Unit 4

Most banks add interest to our account more often than once each year. As an example, if interest is compounded semiannually, this would mean that interest is added to our account every half year. Under this condition, an annual interest rate of 8% becomes a half year rate of 4%:

$\frac{1}{2}$ of 8% is 4%.

Similarly, if interest is added to our account every quarter of a year, then an annual rate of 8% becomes a quarterly rate of 2%:

$\frac{1}{4}$ of 8% is 2%.

Example 3 A bank pays an annual interest rate of 7% compounded quarterly. What interest rate is paid for the quarterly period?

Solution

$$\text{Quarterly interest rate} = 7\% \div 4 = 1.75\%$$

Example 4 If a bank compounds interest quarterly, how many times will the interest be added to an account over a period of 5 years?

Solution

Number of years → 5; Number of times each year → 4

$$\text{Number of interest periods} = 5 \times 4 = 20$$

Example 5 $12,000 is deposited at a bank that pays an interest rate of 7% compounded semiannually. Determine the amount of money that will be in this account at the end of 9 years.

Solution

$$\text{Semiannual interest rate} = 7\% \div 2 = 3.5\%$$

$$\text{Number of interest periods} = 9 \times 2 = 18$$

$$A = \$12{,}000 \times 1.035^{18} = \$12{,}000 \times 1.8575 = \$22{,}290$$

(1.8575 — Compounding factor)

Exercises A

Use the compound interest table when finding the answers for this set of exercises.

1. Find the value of each quantity.

a. 1.07^{12} ________
b. 1.09^{14} ________
c. 1.02^{21} ________
d. 1.10^{5} ________
e. 1.015^{9} ________
f. 1.025^{17} ________
g. 1.035^{19} ________
h. 1.08^{7} ________

2. If interest is compounded annually, find the amount to which $1 will grow at each interest rate. Round your answer to the nearest cent.

	Interest Rate	Number of Years	Value of $1
a.	5%	9	$________
b.	6%	12	________
c.	7%	15	________
d.	8%	20	________
e.	9%	17	________
f.	2%	25	________
g.	10%	25	________

3. Determine the semiannual interest rate for each annual rate.

	Annual Rate	Semiannual Rate		Annual Rate	Semiannual Rate
a.	8%	________%	c.	7%	________%
b.	12%	________%	d.	9%	________%

4. Determine the quarterly interest rate for each annual rate.

	Annual Rate	Quarterly Rate		Annual Rate	Quarterly Rate
a.	4%	________%	c.	6%	________%
b.	12%	________%	d.	9%	________%

5. Determine the monthly interest rate for each annual rate

	Annual Rate	Monthly Rate		Annual Rate	Monthly Rate
a.	24%	________%	c.	18%	________%
b.	36%	________%	d.	6%	________%

Unit 4

6. If interest is compounded semiannually, determine the number of periods it will be credited to an account during the years shown.

	Years	Periods		Years	Periods
a.	3	________	c.	9	________
b.	5	________	d.	14	________

7. If interest is compounded quarterly, determine the number of periods it will be credited to an account during the years shown.

	Years	Periods		Years	Periods
a.	6	________	c.	14	________
b.	11	________	d.	23	________

8. If interest is compounded monthly, determine the number of periods it will be credited to an account during the years shown.

	Years	Periods		Years	Periods
a.	3	________	c.	9	________
b.	7	________	d.	16	________

B

1. Interest is compounded annually on the following deposits. Determine the amount of money that will be in each account after the number of years shown.

	Original Deposit	Interest Rate	Number of Years	Value of Deposit
a.	$4,000	5%	3	$________
b.	7,000	6%	12	________
c.	5,000	8%	16	________
d.	6,400	9%	18	________
e.	9,500	7%	25	________

2. Interest is compounded semiannually on the following deposits. Determine the amount of money that will be in each account after the number of years shown.

	Original Deposit	Annual Rate	Semiannual Rate	Number of Years	Number of Periods	Value of Deposit
a.	$5,000	6%	________%	4	________	$________
b.	6,000	8%	________%	9	________	________
c.	9,000	10%	________%	12	________	________
d.	4,500	7%	________%	5	________	________
e.	8,300	5%	________%	11	________	________

3. Interest is compounded quarterly on the following investments. Determine the amount of money to which each investment will grow after the number of years shown.

	Investment	Annual Rate	Quarterly Rate	Number of Years	Number of Periods	Value of Investment
a.	$ 2,000	16%	________%	3	________	$________
b.	10,000	8%	________%	5	________	________
c.	6,500	12%	________%	6	________	________
d.	7,600	6%	________%	2	________	________

4. Interest is compounded monthly on each of the following investments. Determine the amount of money to which each investment will grow after the number of years shown.

	Investment	Annual Rate	Monthly Rate	Number of Years	Number of Periods	Value of Investment
a.	$ 4,000	24%	________%	2	________	$________
b.	6,000	36%	________%	1	________	________
c.	15,000	18%	________%	$1\frac{1}{2}$	________	________
d.	20,000	30%	________%	$\frac{1}{2}$	________	________

C

1. If you run your finger down the 4% column on the compound interest table on page 232, you will discover that the first numeral that appears as $2 or over is the numeral $2.0258. This occurs at the 18th period. You can interpret this to mean that $1 becomes approximately $2 at a 4% interest rate compounded annually. Another way of looking at this is to say that money doubles itself in 18 years at an interest rate of 4% compounded annually.

How long will it take money to double itself at each of the following rates if interest is compounded annually?

a. 5% ________ **c.** 4% ________ **e.** 8% ________
b. 6% ________ **d.** 7% ________ **f.** 10% ________

Unit 4

2. How long will it take money to triple itself at each of the following rates if interest is compounded annually?

a. 5% ________ **b.** 7% ________ **c.** 10% ________

3. Joseph Jackson deposited $5,000 at the Industrial Savings Bank, which paid interest annually. The account remained dormant for a period of 15 years, after which he withdrew the money.

a. How much money would Joseph have received had the rate of interest been 5%? ________

b. How much money would Joseph have received had the rate of interest been 8%? ________

c. How much more would he have received at the 8% rate than at the 5% rate? ________

4. The Bank of Dearborn pays an annual interest rate of 6% to its depositors. Carolyn Sloane opened an account there with a deposit of $12,000 and then allowed the account to remain dormant for a period of 5 years.

a. How much money would Carolyn have in the account at the end of the five-year period if interest is compounded semiannually? ________

b. How much money would Carolyn have in the account at the end of the five-year period if interest is compounded quarterly? ________

c. How much more money would Carolyn have in the bank if interest is compounded quarterly rather than semiannually? ________

5. Fifty dollars was deposited at a bank that paid a 5% interest rate, compounded annually. If the account remained dormant for a period of 40 years, how large would the account be at the end of that period? ________

Unit 5 The Time Savings Account

Several years ago the federal government placed limits on the interest rates that can be paid by banks. For commercial banks, the largest rate that can be offered depositors on a regular savings account is 5%. A savings bank is permitted to offer 5 1/4% on these same accounts. The regular savings account is one in which we can deposit or withdraw our money at any time.

There are, though, special accounts called either Time Savings Accounts or Savings Certificates that pay a much higher rate of interest. These accounts have several restrictions that the regular account does not have. Among these are:

1. A certain minimum deposit is required – usually, at least $500 or $1,000
2. Should we withdraw any of the *original* deposit before a specified period of time, we will lose a great deal of the interest that our money earned.

In view of these requirements, people will usually deposit a relatively large amount of money in an account, which will then remain dormant for the period of time specified. However, the interest the deposit earns can be withdrawn without any penalty.

Interest on these savings certificates is usually compounded daily. This requires that the interest rate be determined for a period of one day. As an example, if the annual rate is 6%, then the daily rate will be

$$6\% \div 360 \text{ or } .016667\%$$

Number of days in a bank year. (points to 360)

Example 1 On May 1 a time savings account showed a balance of $5,420. This account pays a 6% annual interest rate that is compounded daily. How large will this account be on May 31?

Explanation Since May has 31 days, the interest will be compounded 31 times during this month.

Unit 5

Solution

$$\text{Daily interest rate} = 6\% \div 360$$
$$= .016667\%$$
$$\text{Number of periods} = 31$$

$$A = \$5{,}420 \times 1.00016667^{31}$$

Number of periods (points to the exponent 31)

.016667% is renamed as .00016667

$$= \$5{,}420 \times 1.005180$$
$$= \$5{,}448.08$$

The most difficult task in the example above is in computing the value of 1.00016667^{31}. Fortunately, there are tables such as the one on page 000 where these values can be found. The number of days listed are for either one month or three months since some banks credit interest to accounts every month while others do it every three months.

As in the earlier Compound Interest Table, the numbers in the body of this table are often called compounding factors. Therefore, in Example 1 above the number 1.005180 is the compounding factor for *a 6% rate on a 31-day month.*

This same number, $1.005180 is the value of *$1 at a 6% annual rate compounded daily for 31 days.*

Compound Interest Table
Value of $1 - Interest Compounded Daily

Number of Days	Annual Rate 5½%	6%	6¾%	7%	7½%	8%
28	1.004287	1.004677	1.005263	1.005459	1.005850	1.006241
30	1.004594	1.005012	1.005640	1.005850	1.006269	1.006688
31	1.004747	1.005180	1.005829	1.006045	1.006479	1.006912
89	1.013689	1.014943	1.016826	1.017454	1.018713	1.019972
90	1.013844	1.015112	1.017017	1.017652	1.018925	1.020199
91	1.013999	1.015281	1.017207	1.017850	1.019137	1.020426
92	1.014154	1.015450	1.017398	1.018048	1.019350	1.020653

Example 2 What is the factor for a 7% rate on a 30-day month?

Explanation

Number of Days — Annual Rate 7%

(1) **Read down** → 30 - - - - - - - - - - - 1.005850

(2) **Read across** (3) **Factor**

Solution

Factor for 7% rate on 30-day month = 1.005850

Example 3 On April 1 a time savings account showed a balance of $7,000. This account pays a 7 1/2% annual rate compounded daily. If interest is added to this account every three months, how large will this account be on June 30?

Solution

$$\text{Number of days in April, May, and June} = 30 + 31 + 30 = 91$$

$$\text{Factor for } 7\tfrac{1}{2}\% \text{ on 91 days} = 1.019137$$

$$\text{Amount after 91 days} = \$7,000 \times \text{Factor} = \$7,000 \times 1.019137 = \$7,133.96$$

Exercises

A

Interest is compounded daily in this set of exercises.

1. Determine the compounding factors.

	Interest Rate	Number of Days	Factor
a.	6%	28	______
b.	7%	89	______
c.	$6\frac{3}{4}$%	30	______
d.	$5\frac{1}{2}$%	91	______
e.	8%	90	______

Unit 5

2. Determine the compounding factors for these months.

	Interest Rate	Month	Number of Days	Factor
a.	$5\frac{1}{2}$%	May	______	______
b.	$7\frac{1}{2}$%	August	______	______
c.	7%	February	______	______
d.	$6\frac{3}{4}$%	November	______	______
e.	6%	July	______	______

3. Determine the compounding factor for the following quarterly periods - every three months. (See table, page 220.)

	Interest Rate	First Date of Quarterly Period	Number of Days	Factor
a.	7%	January 1	______	______
b.	8%	March 1	______	______
c.	6%	September 1	______	______
d.	$5\frac{1}{2}$%	November 1	______	______
e.	$6\frac{3}{4}$%	December 1	______	______

B

1. Determine the amount of money in the following accounts at the end of the period shown.

	Amount at Beginning of Period	Interest Rate	Number of Days	Factor	Amount at End of Period
a.	$ 1,000	$5\frac{1}{2}$%	30	________	$________
b.	4,000	6%	31	________	________
c.	7,000	8%	28	________	________
d.	12,000	$6\frac{3}{4}$%	89	________	________
e.	18,000	7%	91	________	________
f.	26,000	$7\frac{1}{2}$%	90	________	________

2. Determine the amount of money in the following accounts at the end of the month shown.

	Amount at Beginning of Month	Interest Rate	Month	Number of Days	Factor	Amount at End of Month
a.	$ 2,000	7%	June	________	________	$________
b.	3,000	8%	March	________	________	________
c.	11,000	6%	October	________	________	________
d.	24,000	$5\frac{1}{2}$%	April	________	________	________

3. Determine the amount of money in the following accounts at the end of the quarterly period shown.

	Amount at Beginning of Period	Interest Rate	First Date of Quarterly Period	Number of Days	Factor	Amount at End of Period
a.	$ 3,000	6%	May 1	________	________	$________
b.	5,000	7%	April 1	________	________	________
c.	14,000	8%	October 1	________	________	________
d.	20,000	$6\frac{3}{4}$%	February 1	________	________	________

C

1. On March 3 Cathy Jung opened a time savings account with a deposit of $10,000.

a. If the account paid an annual interest rate of 7 1/2%, how much money will the account show on March 31? ________

b. How much interest did her deposit earn during the month of March? ________

2. A time savings account was opened on September 2 with a deposit of $30,000.

a. If the annual rate of interest is 5 1/2%, how large will the account be on September 30? ________

b. If the annual rate of interest is 8%, how large will the account be on September 30? ________

c. How much more will the account earn at the 8% rate than at the 5 1/2% rate? ________

3. A time savings account was opened on May 3 with a deposit of $20,000. The annual interest rate on this account was 7 1/2%.

a. How large will the account be on May 31? ________

b. How large will the account be on June 30? You will have to use your answer to part (a) above. ________

c. How much interest was earned by this account during the two-month period? ________

Growth of Regular Deposits Unit 6

Unit 6

Many people plan ahead to anticipate their need for money some five or ten years in the future. As an example, at the birth of a child the parents will often begin to set aside a fixed amount of money each year to provide for the child's college education.

Let us consider the situation where $2,000 is deposited each year under the following conditions:

(a) The annual interest rate is 7%

(b) Interest is compounded only once each year

Our concern now is to determine the amount of money that will accumulate after a period of three years. To do this, we make use of the compound interest formula and treat each deposit separately.

1. Value of first deposit after 3 years $= \$2{,}000 \times 1.07^3$
$= \$2{,}450.09$

2. Value of second deposit after 2 years $= \$2{,}000 \times 1.07^2$
$= \$2{,}289.80$

3. Value of third deposit after 1 year $= \$2{,}000 \times 1.07^1$
$= \$2{,}140.00$

The sum of these three amounts is the total amount of money that has accumulated over the three-year period.

$$\begin{aligned}\text{Total value}&\\ \text{after three years} &= \$2{,}450.09 + \$2{,}289.80 + \$2{,}140.00\\ &= \$6{,}879.89\end{aligned}$$

Obviously, the work above is very time consuming. So, as usual, a table was developed to aid us. The table below shows how annual deposits of $1 will grow over the years. We can also consider the numbers in this table to be factors in the same manner as we had with the numbers in the compound interest tables.

Growth of $1 Annual Deposits

Years	Rate 5%	6%	7%	8%	9%
1	1.0500	1.0600	1.0700	1.0800	1.0900
2	2.1525	2.1836	2.2149	2.2464	2.2781
3	3.3101	3.3746	3.4399	3.5061	3.5731
4	4.5256	4.6371	4.7507	4.8666	4.9841
5	5.8019	5.9753	6.1533	6.3359	6.5233
6	7.1420	7.3938	7.6540	7.9228	8.2004
7	8.5491	8.8975	9.2598	9.6366	10.0285
8	10.0266	10.4913	10.9780	11.4876	12.0210
9	11.5779	12.1808	12.8164	13.4866	14.1929
10	13.2068	13.9716	14.7836	15.6455	16.5603
11	14.9171	15.8699	16.8884	17.9771	19.1407
12	16.7130	17.8821	19.1406	20.4953	21.9534
13	18.5986	20.0151	21.5505	23.2149	25.0192
14	20.5786	22.2760	24.1290	26.1521	28.3609
15	22.6575	24.6725	26.8881	29.3243	32.0034
16	24.8404	27.2129	29.8402	32.7502	35.9737
17	27.1324	29.9057	32.9990	36.4502	40.3013
18	29.5390	32.7600	36.3790	40.4463	45.0185
19	32.0660	35.7856	39.9955	44.7620	50.1601
20	34.7193	38.9927	43.8642	49.4229	55.7645
21	37.5052	42.3923	48.0057	54.4568	61.8733
22	40.4305	45.9958	52.4361	59.8933	68.5319
23	43.5020	49.8156	57.1767	65.7648	75.7898
24	46.7271	53.8645	62.2490	72.1059	83.7009
25	50.1135	58.1564	67.6765	78.9544	92.3240

Example 1 Deposits of $1 are made annually over a period of 15 years. If interest is compounded annually at 7%, how large will the account be at the end of this period?

Explanation

	Years		Rate
			7%
(1) Read down →	15	- - - - - - - - - - - - -	26.8881 ←
		(2) Read across	**(3) Total value of account— or factor**

Solution

Total value of $1
annual deposits after 15 years = $26.8881
or $26.89

Example 2 Annual deposits of $2,000 are made over a 16-year period. The interest rate is 8% compounded annually. How large is the account at the end of this period?

Solution

Factor on 8% for 16 years = 32.7502
Total value after 16 years = $2,000 × Factor
= $2,000 × 32.7502
= $65,500.40

Exercises A

Interest is compounded annually in this set of exercises.

1. In each exercise deposits of $1 are made annually. How large will each account be at the end of the period shown?

	Interest Rate	Number of Years	Value of Account at Close of Period
a.	6%	8	$______
b.	8%	19	______
c.	9%	23	______
d.	5%	12	______
e.	7%	17	______

2. The following deposits are made annually over the period of years shown. How large will each account be at the end of the period shown?

	Annual Deposits	Interest Rate	Number of Years	Factor	Value of Account at Close of Period
a.	$ 400	5%	12	______	$______
b.	600	6%	18	______	______
c.	1,200	8%	25	______	______
d.	2,300	7%	9	______	______
e.	3,500	9%	16	______	______
f.	4,600	7%	15	______	______

3. The following deposits are made annually over the period of years shown. How much interest will these accounts earn over this period?

	Annual Deposits	Interest Rate	Number of Years	Factor	Value of Account at Close of Period	Interest Earned
a.	$ 800	5%	10	______	$______	$______
b.	800	9%	10	______	______	______
c.	600	7%	6	______	______	______
d.	1,100	8%	18	______	______	______
e.	2,500	6%	22	______	______	______

B

1. On the day George was born, his parents began to put aside $500 each year to pay for George's college education. The money was deposited annually in a bank that paid a 6% interest rate, compounded annually. On George's 19th birthday, his parents made the last deposit of $500 and then turned the account over to George. How much money was in the account at that time? ______

2. On the day that Frank Roberts was 45 years old, he and his wife decided to deposit $2,000 each year on his birthday so that they would have their own retirement fund when Mr. Roberts retired at 65. Their deposits were made at a bank where interest was compounded annually.

a. How much money will be in their account on Mr. Roberts' 65th birthday if the annual rate of interest is 5%? The last deposit was made on that day. ______

b. How much money will be in their account on Mr. Roberts' 65th birthday if the annual rate of interest is 8%? ______

c. How much better off financially would the Roberts have been had the bank paid 8% rather than 5%? ______

3. Each Christmas for 25 years, Margaret Riley and the other employees of the Arco Drill Company received a bonus of $250. Instead of spending the money, she deposited it in a special bank account to which she made no other deposits. Interest was compounded annually at 7%.

a. How large was the account at the end of the 25-year period? ________

b. How much interest had been added to the account over the period of years? ________

A Review of Chapter 6

1. Complete each deposit slip.

a.

		Dollars	Cents
Cash		17	82
Checks			
(List Separately)	1	69	00
	2	53	25
Total			
Less Cash Returned			
Total Deposit			

b.

		Dollars	Cents
Cash		14	96
Checks			
(List Separately)	1	85	65
	2	127	40
	3	93	57
Total			
Less Cash Returned		25	00
Total Deposit			

Review

2. Determine the interest for each account for one period.

	Principal	Interest Rate	Period	Interest
a.	$ 700	5%	Annually	$______
b.	800	6%	Annually	______
c.	1,400	6%	Semiannually	______
d.	2,000	8%	Semiannually	______
e.	2,400	8%	Quarterly	______
f.	1,600	7%	Semiannually	______

3. Use the Simple Interest Table, page 216, to determine the interest for each deposit of $1.

	Number of Days	Interest Rate	Interest
a.	14	6%	$______
b.	27	7.5%	______
c.	45	$5\frac{1}{4}$%	______

4. Use the Simple Interest Table, page 216, to determine the interest to the nearest cent for each deposit.

	Deposit	Days	Number Rate	Interest on $1	Interest
a.	$4,000	20	6%	$______	$______
b.	2,000	26	7%	______	______
c.	6,000	40	6½%	______	______

5. A savings account showed the following activity during the month of June:

Balance June 1: $3,400
Deposit June 7: $1,000
Withdrawal June 16: $800

The *smallest balance method* for computing interest is used on this account. Determine the interest credited to this account on June 30, if the annual rate is 5 1/4%. ______

6. Use the *day in to day out method* for computing interest to determine the interest credited to the account in problem 5.

____

7. Each of the following accounts remained dormant for the period of time shown. How large will the account be at the end of the period if interest is compounded annually? In determining your answer, use the Compound Interest Table on page 232.

	Amount On Deposit At Beginning Of Period	Interest Rate	Period Of Years	Factor	Amount On Deposit At End Of Period
a.	$6,000	5%	24	$______	$______
b.	8,500	7%	15	______	______

8. Each of the following accounts remained dormant for the period of time shown. How large will the account be at the end of the period if interest is compounded periodically as indicated? In determining your answer, use the Compound Interest Table.

	Original Deposit	Interest Compounded	Years	Interest Rate	Factor	Account At End Of Period
a.	$6,000	Semiannually	8	4%	$______	$______
b.	3,000	Semiannually	12	5%	______	______
c.	8,000	Quarterly	5	10%	______	______

9. Interest is compounded daily and credited to the following accounts for the number of days shown. How much money will there be in these accounts at the end of the period? In finding your answer, use the table on page 240.

	Amount at Beginning of Period	Interest Rate	Number of Days	Factor	Amount at End of Period
a.	$ 5,000	6%	30	______	$______
b.	16,000	7%	91	______	______

10. The following deposits are made annually over the period of years shown. If interest is compounded annually, how large will each account be at the close of the period? In finding your answer, use the table on page 244.

	Annual Deposits	Interest Rate	Period Of Years	Factor	Account At End Of Period
a.	$ 200	4%	16	______	$______
b.	800	6%	22	______	______
c.	1,400	9%	18	______	______

B

In finding your answer to each of the following problems, use whatever tables in this chapter you feel are necessary.

1. An $8,000 savings account remained dormant over a period of 10 years in a bank where interest is compounded semiannually at 5%. How much interest was added to this account over the 10-year period? ______

Review

2. Joanne Arben opened a savings account with a $10,000 deposit, and then she permitted the account to remain dormant over a period of 10 years. The bank paid an interest rate of 7%. How much more will Joanne have in her account at the end of the 10 years if interest is compounded semiannually rather than annually? ______

3. At 20, George West decided that when he reached 40, he would take a trip around the world. To prepare for this, he planned to deposit $400 annually at a bank that paid an interest rate of 6% compounded annually.

a. At the end of 20 years, how much money did George have for his world trip? ______

b. How much more would George have had if he had kept his money in a bank that paid 7% compounded semiannually? ______

SOLD
SOLD

Chapter 7

Banking Services

Until now our concern with banking has been only in terms of the money our savings will earn. In this chapter we will take a look at a few of the other services offered by banks.

The Checking Account Unit 1

Section 1 Writing Checks

Many banks offer their depositors a choice among several types of checking accounts. Usually, we will find that with one of these not only is there no charge for the checks we write but also, we are not required to keep a minimum balance in our account.

There are many reasons for which we would want a checking account. Of these, the two below stand out:

1. Bills can be paid by mail without being concerned whether the payment will be lost or stolen.
2. When the cashed check is returned to us by the bank, we have a record of the fact that the bill had been paid.

The sample check on page 252 is typical of most checks. Notice that the amount of the check is written in two ways:

1. Arabic numerals
2. English words.

This is done to prevent a dishonest person from changing the amount. For the same reason, a wavy line is drawn after the word "and" at the time we write the amount of the check in English words.

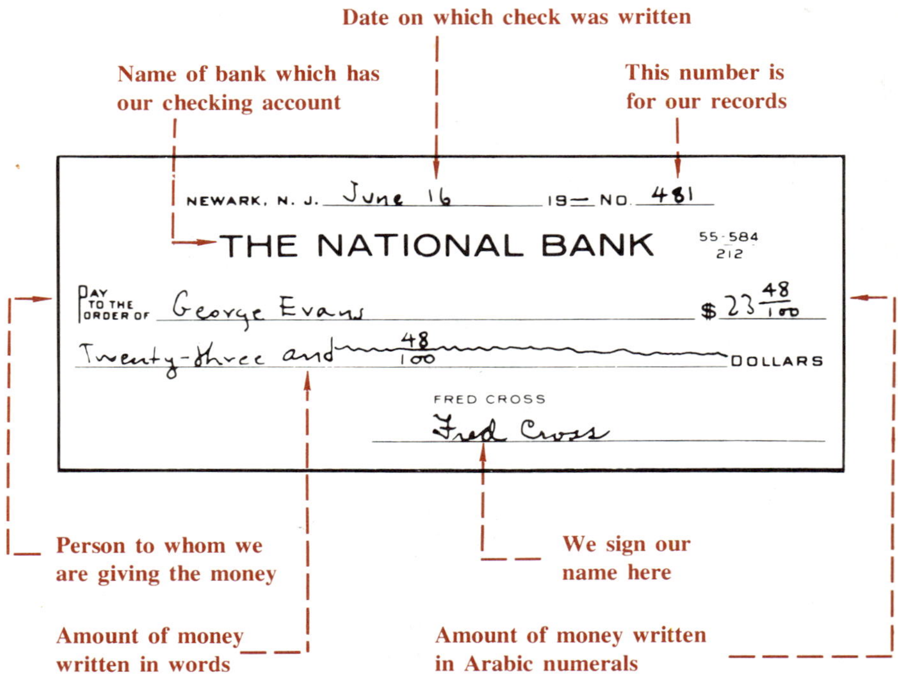

Example 1 A check is written for \$56.73. How will this amount be written in Arabic numerals and English words?

Solution

Arabic numerals: $\$56\frac{73}{100}$

English words: Fifty-six and $\frac{73}{100}$ Dollars

Write "and" after the whole number of dollars

Example 2 A check is written for \$108. How will this amount be written in Arabic numerals and English words?

Solution

Write 00 here

Arabic numerals: $\$108\frac{00}{100}$

English words: One hundred eight and $\frac{00}{100}$ Dollars

Do not write "and" here

Example 3 A check is written for $2,587.06. How will this amount be written in Arabic numerals and English words?

Solution

Write 0 here

Arabic numerals: $\$2{,}587\frac{06}{100}$

English words:

Two thousand five hundred eighty-seven and $\frac{06}{100}$ Dollars

Exercises

1. A check is written for each of the following amounts. How will the Arabic numerals appear on the check?

a.	$27.52	________	**f.**	$39.05	________
b.	$9.16	________	**g.**	$126.07	________
c.	$164.35	________	**h.**	$6	________
d.	$209.28	________	**i.**	$108	________
e.	$3,518.67	________	**j.**	$2,395	________

Unit 1

2. A check is written for each of the following amounts. How will the English words appear on the check?

a. $7.25 ____________________

b. $16.04 ____________________

c. $275.08 ____________________

d. $25 ____________________

e. $150 ____________________

f. $2,346.75 ____________________

g. $2,046.75 ____________________

h. $2,006.75 ____________________

i. $2,000.75 ____________________

j. $2,000.05 ____________________

3. Try to obtain a number of blank checks from your local bank. If not, draw check forms similar to the one on page 252 and then write a check for each of the following. Fill in the present date and your own check numbers.

	Issued to:	Amount	Issued by:
a.	Smith Brothers Oil Company	$87.45	Eugene Farrell
b.	Richard Peterson	17.04	Edward Hawkins
c.	Williams Helms, Inc.	39.07	Claire Dobson
d.	Leonard Motors	57	Bernard Mosby
e.	Henry Schmidt	135	Leslie Meehan
f.	Fireman's Insurance Company	857.19	Joann Williams
g.	May's Transport Company	1,526.50	Alma Jackson

Section 2 The Checkbook Stub

When we write a check we also complete a checkbook stub such as the one below. The stub provides us with a good deal of information. One of the more important pieces of information supplied by the stub is a record of the balance that remains in the account after each check is written.

NO. 481 $ 23.48 ← Amount for which check was written

DATE June 16 19 79

TO George Evans ← Person for whom check was written

FOR Lawn Mower Repair

	DOLLARS	CENTS	
BALANCE BROUGHT FORWARD	472	68	← Balance written at bottom of previous page of checkbook
AMOUNT DEPOSITED	120	50	← Deposit made before writing check
TOTAL	593	18	← Total in checking account after deposit: 472.68 + 120.50 = 593.
AMOUNT THIS CHECK	23	48	← Amount for which check was written
BALANCE BROUGHT FORWARD	569	70	← Balance remaining after writing check: 593.18 − 23.48 = 569.7

Example 1 Complete the following checkbook stub.

	Dollars	Cents
Balance Brought Forward	519	73
Amount Deposited		
Total		
Amount This Check	186	45
Balance Carried Forward		

Solution

	Dollars	Cents	
Balance Brought Forward	519	73	
Amount Deposited			← No deposit
Total	519	73	← 519.73 + 0 = 519.73
Amount This Check	186	45	
Balance Carried Forward	333	28	← 519.73 − 186.45 = 333.28

Unit 1

Example 2 A checkbook stub shows a balance of $254.29. Before writing a check for $57.84 a deposit of $107.15 is made. Complete the checkbook stub with this information.

Solution

	Dollars	Cents	
Balance Brought Forward	254	29	
Amount Deposited	107	15	
Total	361	44	← 254.29 + 107.15 = 361.44
Amount This Check	57	84	
Balance Carried Forward	303	60	← 361.44 − 57.84 = 303.60

Exercises

1. Complete each of the following check stubs.

a.

	Dollars	Cents
Balance Brought Forward	560	75
Amount Deposited		
Total		
Amount This Check	35	40
Balance Carried Forward		

c.

	Dollars	Cents
Balance Brought Forward	604	57
Amount Deposited		
Total		
Amount This Check	87	62
Balance Carried Forward		

b.

	Dollars	Cents
Balance Brought Forward	704	28
Amount Deposited		
Total		
Amount This Check	39	52
Balance Carried Forward		

d.

	Dollars	Cents
Balance Brought Forward	300	02
Amount Deposited		
Total		
Amount This Check	16	75
Balance Carried Forward		

2. Complete each of the following checkbook stubs.

a.

	Dollars	Cents
Balance Brought Forward	314	16
Amount Deposited	60	—
Total		
Amount This Check	75	—
Balance Carried Forward		

e.

	Dollars	Cents
Balance Brought Forward	863	57
Amount Deposited	347	50
Total		
Amount This Check	168	89
Balance Carried Forward		

b.

	Dollars	Cents
Balance Brought Forward	182	71
Amount Deposited	62	85
Total		
Amount This Check	109	50
Balance Carried Forward		

f.

	Dollars	Cents
Balance Brought Forward	903	08
Amount Deposited	69	73
Total		
Amount This Check	784	92
Balance Carried Forward		

c.

	Dollars	Cents
Balance Brought Forward	436	78
Amount Deposited	120	95
Total		
Amount This Check	39	40
Balance Carried Forward		

g.

	Dollars	Cents
Balance Brought Forward	1,486	53
Amount Deposited	195	—
Total		
Amount This Check	575	39
Balance Carried Forward		

d.

	Dollars	Cents
Balance Brought Forward	607	16
Amount Deposited	159	—
Total		
Amount This Check	56	47
Balance Carried Forward		

h.

	Dollars	Cents
Balance Brought Forward	2,504	17
Amount Deposited	596	20
Total		
Amount This Check	795	75
Balance Carried Forward		

Unit 1

3. Draw checkbook stub forms similar to the one on page 254 and complete each of them with the following information.

a. Balance brought forward: $227.89
Check issued: May 9, Service Bureau Corp., $56.20 ________

b. Balance brought forward: $509.18
Check issued: February 14, Ann Nagy, $173.50 ________

c. Balance brought forward: $294.06
Check issued: July 28, Paul Underwood, $62.37
Deposit: July 27, $114.75 ________

d. Balance brought forward: $413.82
Check issued: October 3, Turnpike Sales, $49.95
Deposit: October 1, $316.85 ________

e. Balance brought forward: $712.58
Check issued: April 14, Home Oil Company, $128.42
Deposit: April 12, $106.50 ________

f. Balance brought forward: $1,241.20
Check issued: June 10, Ruth Morrow, $471.88
Deposit: June 7, $247.50 ________

Section 3 The Monthly Statement

Once each month our bank will send us a statement such as the one below. This statement contains the following information:

1. A list of our checks that were cashed during the month.
2. A list of the deposits we made during the month.
3. The bank's record of the amount of money in our account at the close of the month.

Depositor
William Allyn
53 Somerset Road
Westbrook, N.Y.

Account Number
005678-8
Statement Date
6/1

Previous Balance	No. Deposits	Deposits	No. Checks	Checks
856.41	2	424.70	5	433.54

Date	Deposits	Checks	Balance
9/2		83.21	773.20
9/9		41.50	731.70
9/11		273.20	458.50
9/16	320.50		779.00
9/17		9.14	769.86
9/29	104.20		874.06
9/30		26.49	847.57

856.41 + 424.70 − 433.54 = 847.57

Previous Balance — Total Deposits — Total Checks — Present Balance

When we receive this statement we should take the time to compare our checkbook records with the bank's records. There are several reasons why we would want to do this.

1. We may have made an error in computation.
2. A bank employee may have programmed incorrect information into the computer.
3. We may have written checks that we forgot to record on our checkbook stub.
4. People for whom we had written checks may not have cashed them.

The process of comparing our own records with those of the bank is called *reconciling the two sets of records.*

Example Reconcile the checkbook balance with the bank statement balance based on the following information:

Balance recorded in checkbook: \$423.78
Balance recorded on bank statement: \$510.54
Checks written but not cashed: \$14.67, \$27.50, \$5.84
Checks written but not recorded on stub: \$19.75
Deposit made but not recorded on stub: \$56.50

Unit 1

Explanation When reconciling our records with those of the bank our objective is to determine the *actual balance* that both records should show. This means:

For our records

1. If we forgot to subtract any checks we wrote, then we must subtract them.
2. If we forgot to add any deposits we made, then we must add them.

For the bank records

1. If checks have been written but not cashed, they must be subtracted from the bank records. These checks will soon arrive at the bank.

Solution

Our Records		Bank Records		
Checkbook balance:	$423.78	Bank statement balance:		$510.54
Add:		Subtract:		
Deposit not recorded:	+ 56.50	Checks not cashed:	14.67	
	480.28		29.50	
Subtract:			5.84	50.01
Check not recorded:	− 19.75	Actual bank balance:		$460.53
Actual checkbook balance:	$460.53			

Exercises

A

For each of the exercises below, draw a reconciliation statement similar to the one in the example above.

1. Reconcile the bank balance with the checkbook balance for each exercise.

	Bank Balance	Checkbook Balance	Checks Not Cashed	Actual Balance
a.	$247.10	$203.90	$43.20	$______
b.	513.25	446.45	$66.80	______
c.	605.90	588.00	$17.90	______
d.	630.64	555.77	$57.25, $17.62	______
e.	483.77	449.34	$9.48, $24.95	______
f.	943.57	869.50	$16.58, $4.69, $52.80	______

2. Reconcile the bank balance with the checkbook balance for each exercise.

	Bank Balance	Checkbook Balance	Checks Not Recorded	Checks Not Cashed	Actual Balance
a.	$567.20	$611.10	$67.50	$23.60	$______
b.	481.58	453.77	$56.95	$84.76	______
c.	695.81	733.77	$74.26	$14.80, $21.50	______
d.	763.47	698.07	$21.85, $37.50	$124.75	______
e.	907.64	884.34	$81.25, $10.49	$61.24, $53.80	______

B

1. Based on the information below, reconcile the checkbook balance with the bank balance.
Checkbook balance: \$694.23   Bank balance: \$620.63
Check written that has not been cashed: \$12.75
Charge for bank service: \$1.15
Check written but never recorded: \$85.20 ________

2. Based on the information below, reconcile the checkbook balance with the bank balance:
Checkbook balance: \$467.82   Bank balance: \$631.63
Checks written that have not been cashed: \$62.45, \$18.75
Charge for bank service: \$.95
Deposit made but never recorded: \$83.56 ________

3. Based on the information below, reconcile the checkbook balance with the bank balance.
Checkbook balance: \$762.54   Bank balance: \$744.72
Checks written that have not been cashed: \$12.75, \$42.80, \$26.18
Charge for bank service: \$1.40
Check written but never recorded: \$169.40
Deposit made but never recorded: \$71.25 ________

Unit 1

4. When Fred Johnson received his bank statement on May 2 he noticed that the bank balance recorded on it was \$513.78. In reconciling his checkbook balance of \$647.52 with the bank balance he discovered that a deposit of \$84.50 he had mailed to the bank on April 30 had not arrived in time to be added to his account. In addition, his examination revealed the following information:
Checks written but not cashed: \$20.75, \$12.50
Service charge: \$1.25
Check written but never recorded: \$81.24
Reconcile the bank balance with Fred's checkbook balance.

Unit 2 Check Cashing Charges

Because of the great number of bad checks that are written, there are many banks that will not cash checks for people who are not depositors. Therefore, people who have no bank accounts are forced to turn to check cashing firms when cashing their paychecks, social security checks, or any other check.

As might be expected, the cost of cashing checks varies quite widely. There is, however, a similarity between the manner in which the charges are quoted. The two tables below—one from an agency in a large eastern city, the other from an agency in a large central city—are typical of check cashing costs.

Ace Check Cashing Agency

Charges

Amount of check	Cost
Checks written on in-state banks:	
\$0 to \$40.00	25¢
\$40.01 to 49.99	5¢ + $\frac{1}{2}$% of check
\$50.00 and over	10¢ + $\frac{1}{2}$% of check
Checks written on out-of-state banks:	
\$0 to \$20.00	25¢
\$20.01 to \$49.99	5¢ + $1\frac{1}{4}$% of check
\$50.00 and over	10¢ + $1\frac{1}{4}$% of check

Example 1 Determine the charge for cashing a \$154.75 in-state check at the Ace Check Cashing Agency.

Explanation

Checks written on in-state banks ← Use in-state section of table

Amount of check	Cost
50.00 and over	10¢ + $\frac{1}{2}$% of check

\$154.75 is more than \$50 → 50.00

Base charge → 10¢

Additional charge → $\frac{1}{2}$% of check

Solution

$$\text{Base charge} = \$.10$$

$$\text{Additional charge} = \tfrac{1}{2}\% \text{ of check}$$

$$= .5\% \times \$154.75$$

$$= .005 \times \$154.75$$

$$= \$.77$$

$$\text{Total charge} = \text{Base charge} + \text{Additional charge}$$

$$= \$.10 + \$.77$$

$$= \$.87$$

Notice that the rates in the table below are the very same whether the checks are written on in-state banks or out-of-state banks. Notice that the rates in the second table are much higher than those in the first table for checks of small amounts.

Modern Check Cashing Service

Charges

At least	But less than	Cost
$.01	$1.01	$.30
1.01	5.01	.55
5.01	10.01	.60
10.01	15.01	.65
15.01	20.01	.80
20.01	45.01	.90
45.01	100.01	1.00
100.01	200.01	$1.00 + Cost of excess above $100
200.01	300.01	$2.00 + Cost of excess above $200
300.01	....	1% of check

Unit 1

Example 2 Determine the charge for cashing a $27.50 check at the Modern Check Cashing Service.

Solution

	At least	But less than	Cost
Ignore this column → (At least)		45.01 →	.90

First number larger than 27.50 → 45.01

.90 ← Charge for cashing check

Charge = $.90

Example 3 Determine the charge for cashing a \$217.64 check at the Modern Check Cashing Service.

Explanation The amount of the check is separated into two parts:

\$200 + \$17.64

A separate charge is then made on each of these parts:

Charge on \$200	Charge on \$17.64
\$2.00	\$.80
Base charge	**Charge on excess—see table**

Solution

Charge on \$217.64 = Charge on \$200 + Charge on \$17.64
= \$2.00 + \$.80
= \$2.80

Exercises A

Do you recall how to do the computation in the following exercises? This computation is the same as you will be doing when finding the check-cashing charges.

1. Rename each percent numeral as a decimal.

a. 1% ________ **c.** $1\frac{3}{4}$% ________ **e.** $\frac{1}{4}$% ________
b. $1\frac{1}{2}$% ________ **d.** $\frac{1}{2}$% ________ **f.** $\frac{3}{4}$% ________

2. Find the product.

a. 1% of \$374 \$________ **e.** $\frac{3}{4}$% of \$80 \$________
b. 1% of \$567.25 ________ **f.** $\frac{3}{4}$% of \$183.50 ________
c. $\frac{1}{2}$% of \$43.50 ________ **g.** $1\frac{1}{4}$% of \$34 ________
d. $\frac{1}{2}$% of \$165.70 ________ **h.** $1\frac{1}{4}$% of \$106.80 ________

B

1. Determine the charge for cashing each check at the Modern Check Cashing Service.

	Check	Charge		Check	Charge
a.	$23.75	$________	d.	$83.65	$________
b.	7.98	________	e.	2.19	________
c.	46.50	________	f.	15.00	________

2. Determine the total charge for cashing each check at the Modern Check Cashing Service.

	Check	Base Charge	Charge On Excess	Total Charge
a.	$125	$________	$________	$________
b.	117	________	________	________
c.	168	________	________	________
d.	194	________	________	________
e.	103	________	________	________
f.	101.50	________	________	________

3. Determine the charge for cashing each check at the Modern Check Cashing Service.

	Check	Charge		Check	Charge
a.	$460	$________	d.	$314.20	$________
b.	375	________	e.	475.85	________
c.	536	________	f.	568.43	________

Unit 2

4. Find the charge for cashing each check at Ace Check Cashing Agency.

	Check	Bank	Charge
a.	$28.50	In-state	$________
b.	10.75	Out-of-state	________
c.	19.50	Out-of-state	________
d.	4.29	In-state	________

5. Determine the charge for cashing each in-state check at the Ace Check Cashing Agency.

	Check	Base Charge	Additional Charge	Total Charge
a.	$ 45.20	$________	$________	$________
b.	49.50	________	________	________
c.	70.60	________	________	________
d.	146.50	________	________	________
e.	275.80	________	________	________

6. Determine the charge for cashing each out-of-state check at the Ace Check Cashing Agency.

	Check	Base Charge	Additional Charge	Total Charge
a.	$ 25	$______	$______	$______
b.	37	______	______	______
c.	68	______	______	______
d.	146	______	______	______
e.	293	______	______	______

7. The maximum check cashing charges permitted in a large eastern city are shown below.

$\frac{3}{4}$% of the value of the check
30¢ minimum charge

Determine the charge for cashing each check.

	Check	Charge		Check	Charge
a.	$80	$______	d.	$ 16.50	$______
b.	20	______	e.	52.40	______
c.	95	______	f.	247.30	______

C

1. Charles Putnam cashed a check for $158 at the Ace Check Cashing Agency.

a. What will the charge be if this is an in-state check? ______

b. What will the charge be if this is an out-of-state check? ______

c. How much more will he have to pay if the check is from an out-of-state bank? ______

2. Nancy Evans has a weekly paycheck for $186.50. The check is drawn on an in-state bank and is cashed weekly at the Ace Check Cashing Agency.

a. What charge does she pay each week for cashing her paycheck? ______

b. How much does she pay over the period of one year for simply cashing her checks? ______

3. A check for $237.50 is drawn on an out-of-state bank.

a. How much will the charge be if it is cashed at the Ace Check Cashing Agency? ________

b. How much will the charge be if it is cashed at the Modern Check Cashing Service? ________

c. How much more is the charge at the Ace Check Cashing Agency? ________

Bank Loans Unit 3

Section 1 The Installment Loan

Banks earn their money by lending money. The most common of the personal loans is the installment loan. In reality, this loan is no different than the automobile loan that we examined much earlier in our work. However, this type of loan is used by people who need money to repair their home, build an addition to their home, buy a boat, pay for the cost of a vacation, and many other similar reasons.

Advertisements such as the one below frequently appear in newspapers.

National Sussex Bank
You'll Like Our Low Bank Rates (12.5%)

Unpaid Balance	Amount Per Month 12 Months	18 Months	30 Months
$1,500	$133.62	$ 91.82	$58.48
2,000	178.17	122.43	77.97
2,400	213.80	146.92	93.56

Unit 3

The monthly payments in the advertisement above, and in fact for any installment loan, are obtained by consulting a monthly installment factor table similar to the one below.

Monthly Installment Factors

Interest Rate	Number of Months 12	18	24	30
10%	.0879159	.0600571	.0461449	.0378114
10.5%	.0881486	.0602876	.0463760	.0380443
11%	.0883817	.0605185	.0466078	.0382781
11.5%	.0886151	.0607500	.0468403	.0385127
12%	.0888488	.0609821	.0470735	.0387481
12.5%	.0890829	.0612146	.0473073	.0389844

Example 1 Determine the monthly installment factor on an 18 month loan at an annual rate of 11%.

Solution

Interest Rate **18 Months**

Read down

11% — — — — — — — — .0605185

Read across **Factor**

Monthly installment factor = .0605185

The monthly installment factors do have a meaning other than simply being numbers that are found by some mathematical formula. Thus, the answer in Example 1 can be interpreted as follows:

> If \$1 is borrowed at an interest rate of 11% for a period of 18 months, the monthly payments will be \$.0605185 or a little over 6¢.

Example 2 \$3,500 is borrowed from a bank on a 24-month installment loan. If the annual interest rate is 10 1/2%, how large are the monthly payments?

Solution

Monthly installment factor = .0463760

Amount borrowed **Factor**

Monthly payment = \$3,500 × .0463760
= \$162.32

Example 3 \$5,200 is borrowed on a 30-monthly installment loan at a 12 1/2% interest rate. How much money will be repaid over the 30-month period?

Solution

$$\text{Monthly installment factor} = .0389844$$

Amount borrowed → $5,200; Factor → .0389844

$$\text{Monthly payment} = \$5{,}200 \times .0389844$$
$$= \$202.72$$

Number of months → 30; Monthly payment → $202.72

$$\text{Total repaid} = 30 \times \$202.72$$
$$= \$6{,}081.60$$

Exercises

A

1. Determine the monthly installment factor for each exercise.

	Interest Rate	Number of Months	Factor
a.	10%	18	______
b.	11%	12	______
c.	12.5%	30	______
d.	11.5%	24	______
e.	12%	18	______

2. Determine the monthly payment for each monthly installment loan.

	Amount Borrowed	Interest Rate	Number of Months	Factor	Monthly Payment
a.	$3,000	11%	12	______	$______
b.	2,000	12%	24	______	______
c.	2,500	10%	18	______	______
d.	4,600	10.5%	30	______	______
e.	2,700	$11\frac{1}{2}$%	24	______	______

Unit 3

3. Determine the total repaid for each monthly installment loan.

	Amount Borrowed	Interest Rate	Number of Months	Factor	Monthly Payment	Total Repaid
a.	$1,000	10.5%	18	______	$______	$______
b.	600	11%	12	______	______	______
c.	1,400	12%	24	______	______	______
d.	2,200	10%	30	______	______	______
e.	3,700	$11\frac{1}{2}$%	24	______	______	______

4. Determine the interest paid for each monthly installment loan. That is, determine how much more the borrower repaid than he received.

	Amount Borrowed	Interest Rate	Number of Months	Factor	Monthly Payment	Total Repaid	Interest
a.	$4,000	11%	12	______	$______	$______	$______
b.	6,000	10%	30	______	______	______	______
c.	2,800	12%	18	______	______	______	______
d.	5,900	11%	24	______	______	______	______

B

1. Eric Swenson borrowed $3,400 on a monthly installment loan at an interest rate of 10.5%.

a. How much will he repay if the loan is for a 12-month period? ______

b. How much will he repay if the loan is for a 30-month period? ______

c. How much more will the 30-month loan cost him than the 12-month loan? ______

2. Maria Cotugno borrowed $4,700 on a 24-month installment loan.

a. How much will she return if the interest rate is 10%? ______

b. How much will she return if the interest rate is 12.5%? ______

c. How much more will the 12.5% loan cost her than the 10% loan? ______

3. Using the Monthly Installment Factors table on page 267 show how the monthly payment of $133.62 for the $1,500 twelve-month loan was arrived at in the advertisement on page 267.

Section 2 The Passbook Loan

A frequent method used for borrowing money from a bank is the *passbook loan*. Under this arrangement we would leave our passbook (bankbook) at the bank until the debt is paid. The bank is certain its money will be repaid because the amount we are permitted to borrow is less than the amount we have on deposit. In reality, the bank is simply returning to us some of our own money but it is charging us for the use of this money.

Usually, the passbook loan and other *security* loans are *demand loans*. This means that the bank can demand its money be repaid at any time it desires; however, banks very rarely make this demand. On the other hand, we are permitted to pay off the debt at any time we desire.

To compute the cost of the loan, reference is made to a Daily Factor Table such as that on page 272. This table is exactly the same as the Simple Interest Table we used earlier, however the rates are higher. Depending on the bank, the interest is paid either monthly or quarterly—some banks give us the choice. *In each of the examples and exercises in this section, interest will be paid on the last day of each month–except, of course, for the last payment.*

Example 1 $1,250 was borrowed on March 8 at an interest rate of 9 1/2%. How large will the first interest payment be?

Solution

Number of days from March 8 to March 31 = 31 − 8
= 23

Factor for 23 days at $9\frac{1}{2}\%$ = .0060694 ← See table, page 272

Amount borrowed → $1,250; Factor → .0060694

Interest = $1,250 × .0060694
= $7.59

Unit 3

Example 2 $2,675 was repaid on April 17. If the interest rate was 11%, how large was the last interest payment?

Explanation Interest had been paid through the last day of the previous month. Hence, the only interest that remained to be paid was for the 17 days of April.

Solution

Factor for 17 days at 11% = .0051944 ← See table, page 272

Amount borrowed → $2,675; Factor → .0051944

Interest = $2,675 × .0051944
= $13.90

DAILY FACTOR TABLE
or
SIMPLE INTEREST ON $1
(360-Day Year)

Rates

Days	8%	8.5%	9%	9.5%	10%	11%	12%
1	.0002222	.0002361	.0002500	.0002639	.0002778	.0003056	.0003333
2	.0004444	.0004722	.0005000	.0005278	.0005556	.0006111	.0006667
3	.0006667	.0007083	.0007500	.0007917	.0008333	.0009167	.0010000
4	.0008889	.0009444	.0010000	.0010556	.0011111	.0012222	.0013333
5	.0011111	.0011806	.0012500	.0013194	.0013889	.0015278	.0016667
6	.0013333	.0014167	.0015000	.0015833	.0016667	.0018333	.0020000
7	.0015556	.0016528	.0017500	.0018472	.0019444	.0021389	.0023333
8	.0017778	.0018889	.0020000	.0021111	.0022222	.0024444	.0026667
9	.0020000	.0021250	.0022500	.0023750	.0025000	.0027500	.0030000
10	.0022222	.0023611	.0025000	.0026389	.0027778	.0030556	.0033333
11	.0024444	.0025972	.0027500	.0029028	.0030556	.0033611	.0036667
12	.0026667	.0028333	.0030000	.0031667	.0033333	.0036667	.0040000
13	.0028889	.0030694	.0032500	.0034306	.0036111	.0039722	.0043333
14	.0031111	.0033056	.0035000	.0036944	.0038889	.0042778	.0046667
15	.0033333	.0035417	.0037500	.0039583	.0041667	.0045833	.0050000
16	.0035556	.0037778	.0040000	.0042222	.0044444	.0048889	.0053333
17	.0037778	.0040139	.0042500	.0044861	.0047222	.0051944	.0056667
18	.0040000	.0042500	.0045000	.0047500	.0050000	.0055000	.0060000
19	.0042222	.0044861	.0047500	.0050139	.0052778	.0058056	.0063333
20	.0044444	.0047222	.0050000	.0052778	.0055556	.0061111	.0066667
21	.0046667	.0049583	.0052500	.0055417	.0058333	.0064167	.0070000
22	.0048889	.0051944	.0055000	.0058056	.0061111	.0067222	.0073333
23	.0051111	.0054306	.0057500	.0060694	.0063889	.0070278	.0076667
24	.0053333	.0056667	.0060000	.0063333	.0066667	.0073333	.0080000
25	.0055556	.0059028	.0062500	.0065972	.0069444	.0076389	.0083333
26	.0057778	.0061389	.0065000	.0068611	.0072222	.0079444	.0086667
27	.0060000	.0063750	.0067500	.0071250	.0075000	.0082500	.0090000
28	.0062222	.0066111	.0070000	.0073889	.0077778	.0085556	.0093333
29	.0064444	.0068472	.0072500	.0076528	.0080556	.0088611	.0096667
30	.0066667	.0070833	.0075000	.0079167	.0083333	.0091667	.0100000
31	.0068889	.0073194	.0077500	.0081806	.0086111	.0094722	.0103333

Exercises A

1. Determine the simple interest factor.

	Number of Days	Interest Rate	Factor
a.	5	8%	______
b.	18	9%	______
c.	23	11%	______
d.	12	12%	______
e.	29	8.5%	______
f.	4	$9\frac{1}{2}$%	______

2. Determine the interest for each loan.

	Loan	Number of Days	Interest Rate	Factor	Interest
a.	$ 600	18	8%	______	$______
b.	500	9	10%	______	______
c.	1,200	28	8.5%	______	______
d.	2,500	21	11%	______	______
e.	750	14	12%	______	______

3. Determine the interest payment on the day each loan was repaid. (See Example 2)

	Loan	Date Repaid	Number of Days	Interest Rate	Factor	Interest
a.	$ 800	April 12	______	9%	______	$______
b.	1,000	Sept. 23	______	11%	______	______
c.	1,400	Oct. 6	______	12%	______	______
d.	2,300	Feb. 18	______	9.5%	______	______
e.	3,500	Dec. 11	______	8.5%	______	______

4. Determine the first interest payment for each loan.

	Loan	Date Borrowed	Number of Days	Interest Rate	Factor	Interest
a.	$ 700	June 10	______	8%	______	$______
b.	1,600	July 16	______	11%	______	______
c.	2,800	Oct. 14	______	12%	______	______
d.	3,700	Jan. 25	______	8.5%	______	______
e.	4,600	March 8	______	9.5%	______	______

Unit 3

5. Determine the amount of money given to the bank on the date each loan was repaid. Include both the amount of the loan and the last interest payment on the loan.

	Loan	Date Repaid	Number of Days	Interest Rate	Factor	Interest	Total Payment
a.	$2,000	Jan. 16	______	10%	______	$______	$______
b.	1,300	Nov. 27	______	9%	______	______	______
c.	2,700	March 18	______	11%	______	______	______
d.	3,300	June 24	______	8.5%	______	______	______
e.	750	July 9	______	12%	______	______	______

B

1. Linda Tucker borrowed $10,000 from her bank for a period of 30 days.

a. How much interest will she pay if the interest rate is 8%?

______

b. How much interest will she pay if the interest rate is 12%? ________

c. How much will she save if she is able to get the loan at the 8% rate rather than the 12% rate ? ________

2. Patrick Turner borrowed $4,000 from the Wayne State Bank at an interest rate of 10%. He borrowed the money on May 17 and repaid the debt on July 26.

a. How much was the interest payment on the last day of May? ________

b. How much was the interest payment on the last day of June? ________

c. How much was the interest payment on July 26? ________

d. What was the total interest paid on this debt? ________

Section 3 The Discount Bearing Loan

Under this third method for borrowing money from a bank, we must pay in advance for the cost of the loan. Thus, were we to borrow $500 and the cost for borrowing this money was $30, then the bank would give us $470. ← ($500 − $30)

Each of the numbers above has its own special name:

Principal	**Discount**	**Proceeds**
↓	↓	↓
$500	$30	$470

An arrangement in which we pay for the cost of the loan at the time we borrow the money is called *discounting a loan*. In the illustration above, we borrowed $500 but received only $470. However, when the debt is due the amount we must return to the bank is $500.

Everything we learned prior to now concerning interest will also apply to discount. Therefore, the interest formula:

$$I = P \times R \times T$$

This number is a full year's interest

This number is the fraction of a year for which the money was borrowed.

becomes the discount formula:

$$D = P \times R \times T$$

Example 1 Determine the discount on a \$3,000 loan borrowed at a discount rate of 9% for a period of 48 days.

Solution

$$\text{Discount} = P \times R \times T$$

Fraction of year

$$= \$3{,}000 \times 9\% \times \frac{48}{360}$$

$$= \$270 \times \frac{2}{15}$$

One year's discount

$$= \$36$$

Example 2 Determine the proceeds on a \$2,400 loan borrowed at a discount rate of 10 1/2 % for a period of 42 days.

Solution

$$\text{Discount} = P \times R \times T$$

$$= \$2{,}400 \times 10\tfrac{1}{2}\% \times \frac{42}{360}$$

$$= \$\overset{40}{\cancel{2{,}400}} \times .105 \times \frac{7}{\underset{1}{\cancel{60}}}$$

$$= \$29.40$$

$$\text{Proceeds} = \text{Principal} - \text{Discount}$$

$$= \$2{,}400 - \$29.40$$

$$= \$2{,}370.60$$

Unit 3

Exercises

A

Do you recall how to do the computation in the following exercises? This computation is the same as you will be doing when finding the discount on a loan.

1. Reduce each fraction to lowest terms.

a.	$\frac{9}{36}$	________	**f.**	$\frac{60}{360}$	________	**k.**	$\frac{18}{360}$	________
b.	$\frac{90}{360}$	________	**g.**	$\frac{120}{360}$	________	**l.**	$\frac{72}{360}$	________
c.	$\frac{18}{36}$	________	**h.**	$\frac{30}{360}$	________	**m.**	$\frac{24}{360}$	________
d.	$\frac{180}{360}$	________	**i.**	$\frac{80}{360}$	________	**n.**	$\frac{48}{360}$	________
e.	$\frac{40}{360}$	________	**j.**	$\frac{36}{360}$	________	**o.**	$\frac{84}{360}$	________

2. Find the products.

a. $800 \times \frac{90}{360}$ ________ **c.** $1{,}200 \times \frac{72}{360}$ ________

b. $300 \times \frac{30}{360}$ ________ **d.** $3{,}600 \times \frac{48}{360}$ ________

B

1. Determine the discount for each loan

	Principal	Discount Rate	Number of Days	Discount
a.	$ 800	9%	30	$________
b.	700	10%	36	________
c.	1,800	11%	40	________
d.	1,200	12%	60	________
e.	1,500	9%	48	________
f.	2,400	11%	18	________
g.	2,000	$10\frac{1}{2}$%	72	________

2. Determine the proceeds for each loan.

	Principal	Discount Rate	Number of Days	Discount	Proceeds
a.	$ 400	9%	90	$________	$________
b.	2,000	9%	100	________	________
c.	1,500	10%	24	________	________
d.	2,300	11%	72	________	________
e.	3,600	12%	80	________	________
f.	2,700	$10\frac{1}{2}$%	84	________	________

C

1. Howard Fleming borrowed $1,620 from a bank for a period of 80 days at a discount rate of 10%.

a. What was the discount on this loan? ________

b. How much did he receive from the bank? ________

c. How much did he return to the bank? ________

2. On May 23 Mary Cosgrove borrowed $1,800 from a bank for a period of 3 months at a discount rate of 12%.

a. If the bank computes the charge on a monthly basis, how much will the discount be? (What fraction of a year is 3 months?) ________

b. If the bank computes the charge on a daily basis, how much will the discount be? (There are 92 days in this period. What fraction of a bank year is 92 days?) ________

c. How much will Mary save if the discount is computed on a monthly basis rather than a daily basis? ________

3. Use the Daily Factor Table, page 272, to determine the proceeds for each loan.

	Principal	Discount Rate	Number of Days	Factor	Discount	Proceeds
a.	$5,000	10%	15	________	$________	$________
b.	6,000	11%	23	________	________	________
c.	2,500	12%	27	________	________	________

Review

4. A loan of $4,000 was made at a discount rate of 11% for a period of 18 days.

a. If the discount formula is used for computation, what will the discount be? ________

b. If the Daily Factor Table is used for computation, what will the discount be? ________

Review of Chapter 7

1. A check is written for each amount. How will the English words appear on the check?

a. $23.57 ____________________

b. $35 ____________________

c. $182.05 ____________________

2. Draw a check form similar to the one on page 252 and then write a check to Bicycle Repair Service for $12.95.

3. Draw a checkbook stub form similar to the one on page 254 and complete this form based on the information below.
Balance brought forward: $169.75
Check issued: March 17, Susan Riley, $23.60
Deposit: March 15, $84 ________

4. Reconcile the checkbook balance with the bank balance based on the information below.
Checkbook balance: $527.48 Bankbook balance: $781.48
Checks written that have not been cashed: $45.30, $86.50
Check written but never recorded: $17.80
Deposit made but never recorded: $140 ________

5. Determine the charge for cashing each of these checks at the Modern Check Cashing Service whose rates appear on page 263.

	Check	Charge		Check	Charge
a.	$ 17	$________	c.	$247	$________
b.	125	________	d.	384	________

6. Julia Hernandez borrowed $4,000 on a monthly installment loan for a period of 24 months at an interest rate of 12%. (See Table on page 267.)
a. What is the monthly installment factor on this loan? ________
b. What is the monthly payment on this loan? ________
c. How much did Julia return to the bank over the 24-month period? ________
d. How much more did she return to the bank than she received? ________

7. Use the Daily Factor Table on page 272 to determine the interest on a loan of $1,600 for a period of 27 days at an interest rate of 11%. ________

8. A passbook loan for $3,000 was made on April 12 and repaid June 17. The interest rate was 10% and interest had to be paid on the last day of each month. Use the Daily Factor Table to compute the answers below.

a. How much interest was paid on the last day of April? ________

b. How much interest was paid on the last day of May? ________

c. How much interest was paid on the 17th of June? ________

9. Simeon Fletcher borrowed $1,800 at a discount rate of 11% for a period of 90 days.

a. What discount was charged for this period of time? ________

b. What were the proceeds that Simeon received at the time he borrowed the money? ________

c. How much did Simeon return at the time he paid off the debt? ________

Review

PERSONAL LOANS

Chapter 8

Small Loans

There are many people who are unable to borrow money from a bank. These are people who have little or no collateral such as a passbook, stocks, jewelry, art work and the like which banks require from the people to whom they grant a loan. In view of this, the person who is in need of money but can not borrow from a bank turns to other sources. In this chapter we will examine the cost of borrowing money from a few of these lending agencies.

The Small-Loan Agency Unit 1

Each day as we thumb our way through the newspaper we find advertisements such as:

> "**IN DEBT?** We will assist you in getting out of debt. One weekly payment can possibly pay your bills and overdue debts."

> "I'm behind in my house payments . . . bills are coming due . . . the kids need clothes . . . who'll lend a hand? **WE WILL!**"

> "It's time to fix up your home! Get a **FIX-UP** loan."

> "Don't be bothered by bills. Here's the **CASH** you need."

> "Exclusive for Engineers! Dear engineer . . . **Borrow $200 to $1,000** or more in complete confidence by mail! Write us at RST"

The companies that place these advertisements are called *small loan agencies*. These small amounts, though, may be from as little as $50 to as much as $25,000 depending on the laws of the state in which the agency is located.

Each state governs the rate of interest that small loan agencies are permitted to charge. These rates are quoted on either a monthly or yearly basis. And, as the loan increases, the rate decreases. The rate table below shows the legal maximum loan rates in one of the western states.

Maximum Interest Rates on Small Loans

2 1/2% per month on unpaid balance up to $300

2% per month on unpaid balance over $300 but under $500

1 1/2% per month on unpaid balance over $500 but under $1,000

1% per month on unpaid balance over $1,000

Many of the advertisements of small loan agencies call upon us to borrow money under a *consolidation loan*. This is an arrangement whereby we borrow enough money to wipe clean our bills. We would then be left with one monthly payment to the loan agency rather than many small payments on our many small debts.

However, most of our bills—electric, gas, food, dentist, car repair, plumber and the like—require no interest payment. On the other hand, the loan agency's consolidation loan does carry with it a rather substantial charge. Therefore, it may be wise to examine the cost, if any, on our small debts before borrowing money at interest in order to eliminate them.

Example 1 The RST Consumer Loan Company mailed the advertisement on page 283 to people throughout the country.

Facts Chart

Select Your Loan Here . . .

Amount of Loan	Number of Monthly Payments	Amount of Each Monthly Payment	Annual Rate
$1,000	36	$41.34	29.34%
900	36	37.53	30.04%
800	36	33.70	30.82%
700	36	29.85	31.79%
600	30	29.21	33.07%
500	24	28.82	34.61%

Determine the amount a borrower will repay the loan company over the 36-month period on the $800 loan.

Solution

$$\text{Monthly payment} = \$33.70$$

$$\begin{aligned}\text{Total repaid} &= \text{Number of months} \times \text{Monthly payment}\\ &= 36 \times \$33.70\\ &= \$1{,}213.20\end{aligned}$$

It might be well to notice that the interest rates in both the example above and the one that follows are quite a bit above the 5% or 6% rates we receive from a bank for the use of our money.

The lending agency below advertises loans that are to be paid back weekly. However, the period of the loan is recorded in months. This is probably done for two reasons:

1. Twelve months gives an impression of many fewer payments than fifty-two weeks.
2. $22.49 per week seems like a great deal less money than the $98.30 that will have to be paid monthly.

When doing the computation for the loans in this table, it will be necessary to rename months as a fraction of a year. Thus,

6 months is 6/12 of a year.
Similarly, 15 months is 15/12 of a year.

Example 2 The Economy Credit Corporation placed the following advertisement in a newspaper.

WE'LL LEND A HELPING HAND
$25 to $2,500

If You Owe	Pay as Low as	Time Length
$ 500	$20.85 per week	6 months
1,000	22.49 per week	12 months
1,500	28.01 per week	15 months
2,000	32.28 per week	18 months
2,500	32.51 per week	24 months

Annual Rate: 31.65%

How much interest will the borrower have to pay on a $2,000 loan?

Solution

Fraction of a year → $\frac{18}{12}$; Weeks in a year → 52

$$\text{Number of weeks in 18 months} = \frac{18}{12} \times 52 = 78$$

$$\text{Weekly payment} = \$32.28$$

Number of weeks → 78; Amount per week → $32.28

$$\text{Total repaid} = 78 \times \$32.28 = \$2,517.84$$

$$\text{Interest} = \text{Total repaid} - \text{Amount borrowed} = \$2,517.84 - \$2,000 = \$517.84$$

Exercises A

Use the Facts Chart of Example 1 when doing the following exercises.

1. Determine the total repaid for each loan.

	Loan	Monthly Payment	Number of Months	Total Repaid
a.	$1,000	$______	______	$______
b.	700	______	______	______
c.	500	______	______	______

2. Determine the interest paid for each loan.

	Loan	Monthly Payment	Number of Months	Total Repaid	Interest
a.	$600	$________	________	$________	$________
b.	900	________	________	________	________

B

Use the advertisement of Example 2 when doing the following exercises.

1. Determine the amount repaid for each loan.

	Loan	Number of Months	Number of Weeks	Weekly Payment	Total Repaid
a.	$1,000	________	________	$________	$________
b.	2,500	________	________	________	________

2. Determine the interest paid for each loan.

	Loan	Number of Months	Number of Weeks	Weekly Payment	Total Repaid	Interest
a.	$ 500	________	________	$________	$________	$________
b.	1,500	________	________	________	________	________

C

Use the advertisement below when doing the exercises that follow it.

SELECT YOUR LOAN FROM THIS TABLE

(Rates set by law)

24-Month Plan

Amount of Loan	Monthly Payment	Annual Rate
$ 400	$22.09	28.66%
500	27.18	26.96%
800	41.88	22.95%
1,200	61.00	19.87%
1,500	75.23	18.47%

Unit 1

1. Determine the amount repaid for each loan.

	Loan	Monthly Payment	Amount Repaid
a.	$ 400	$________	$________
b.	1,200	________	________

2. Determine the cost for each loan.

	Loan	Monthly Payment	Amount Repaid	Interest
a.	$ 800	$______	$______	$______
b.	1,500	______	______	______

D

Use the advertisement below when doing the exercises that follow it.

TERMS TO SUIT YOUR BUDGET

Cash To You	24 Months	18 Months	12 Months	Annual Rate of Interest
$ 500	$30.18	$ 37.00	$ 50.85	38.46%
1,000	58.14	71.83	99.60	34.28%
1,500	83.98	104.61	146.33	30.14%

1. Determine the amount repaid for each loan.

	Loan	Period of Loan	Monthly Payment	Amount Repaid
a.	$ 500	18 months	$______	$______
b.	1,000	24 months	______	______
c.	1,500	12 months	______	______

2. Determine the cost for each loan.

	Loan	Period of Loan	Monthly Payment	Amount Repaid	Interest
a.	$1,000	12 months	$______	$______	$______
b.	500	12 months	______	______	______
c.	1,500	18 months	______	______	______

E

1. Determine the interest that will have to be paid on the $500 loan for 24-months in the advertisement of Exercise C.

______

2. Determine the interest that will have to be paid on the $500 loan for 24-months in the advertisement of Exercise D.

______

3. How much money can be saved by borrowing the $500 from the loan company of Exercise C than from the one in Exercise D? ________

F

The Confidential Cash Loan Corporation mailed advertising circulars to people urging them to "Consolidate Your Bills." The advertisement read as follows:

CUT YOUR MONTHLY PAYMENTS 1/3 TO 1/2 AND GET EXTRA CASH BESIDES

(Annual rate 29.49%)

Take 24 months to pay

Here's a typical problem that many families face.

Accounts	Amount Owed	Monthly Payments
Clothing	$ 48	$ 6
Appliance	100	10
Hospital	110	10
Doctor	24	12
Dentist	105	15
Car Repair	20	5
Home Repair	50	10
Total	**$457**	**$68**

Here's how a loan from us Solved the Problem.

Amount of Loan	$500
Amount Needed to Pay Bills	457
Extra Cash for You	43
Monthly Payment For a loan of $500	$27.81

Here's how payments of $68 were cut to $27.81

1. For how many months will the borrower have to make the monthly payments of $27.81? ________

2. If the loan were not made then:

a. In how many months will the clothing bill be paid? ________

b. In how many months will the appliance bill be paid? ________

c. In how many months will the hospital bill be paid? ________

d. In how many months will the doctor's bill be paid? ________

e. In how many months will the dentist's bill be paid? ________

f. In how many months will the car repair bill be paid? ________

g. In how many months will the home repair bill be paid? ________

Unit 1

3. In view of your answers in exercise 2, how long would it be before the last of the bills is paid? ________

4. In what way would you criticize the advertisement? ________

5. How much interest would the borrower have to pay on the loan of $500? ________

Unit 2 The Pawnshop Loan

The person who needs a relatively small amount of money but can not borrow it from either the bank or the small loan agency may turn to the pawnshop. As with other lending agencies, the interest rates on a pawnshop loan are controlled by the state. There are a few states in which the rate can run as high as 10% per month which is 120% per year! In most states, however, the rates are about the same as those charged by the small loan agency.

Although the pawnshop loan is considered a high risk loan, the great majority of them are repaid. However, whether they are repaid or not is often of little concern to the pawnbroker. All pawnshop loans are *collateral* loans. That is, the borrower must leave some article of value—jewelry, radio, TV set, watch, camera, musical instrument—until the debt is cleared. Since the value of the article is far more than the amount borrowed, in the event the debt is not repaid, the article is sold and the broker regains his money.

One major difference between the pawnshop loan and others we have examined is the fact that interest appears to be charged for *full monthly periods*. Therefore, even though the money may be returned as quickly as 5 days, the interest charge will be for 1 full month. Similarly, a person who repays a pawnshop loan in 2 months 1 day will have to pay interest on the debt for 3 full months.

Example 1 Determine the number of months on which interest will have to be paid on a pawnshop loan that was made on April 7 and paid off on August 15.

Solution

This period is considered 1 month

1	1	1	1	1
April 7 – May 7	May 7 – June 7	June 7 – July 7	July 7 – August 7	August 7 – August 15

Number of months = 5

The design of rate tables on pawnshop loans is much the same as the one below.

Monthly Interest Charge

Amount Borrowed	Monthly Rate
On first $50 or less	$2\frac{1}{2}\%$
Excess over $50	2%

It is best to determine the interest on loans over $50 as the sum of the two charges:

1. The charge on the first $50
2. The charge on the excess over $50.

Example 2 Use the table above to determine the monthly interest on a loan of $85.

Solution

$$\begin{aligned}\text{Interest on first } \$50 &= 2\tfrac{1}{2}\% \text{ of } \$50\\ &= 2.5\% \times \$50\\ &= .025 \times \$50\\ &= \$1.25\end{aligned}$$

$$\begin{aligned}\text{Excess over } \$50 &= \$85 - \$50\\ &= \$35\end{aligned}$$

$$\begin{aligned}\text{Interest on excess} &= 2\% \text{ of excess}\\ &= .02 \times \$35\\ &= \$.70\end{aligned}$$

$$\begin{aligned}\text{Total monthly interest} &= \$1.25 + \$.70\\ &= \$1.95\end{aligned}$$

Example 3 Use the table above to determine the amount repaid on a loan of \$116. The money was borrowed for 5 months 9 days.

Solution

Computation for monthly interest →

$$
\begin{aligned}
\text{Interest on first } \$50 &= \$1.25 \quad \longleftarrow \textbf{See Example 2}\\
\text{Excess above } \$50 &= \$116 - \$50\\
&= \$66\\
\text{Interest on excess} &= 2\% \text{ of excess}\\
&= .02 \times \$66\\
&= \$1.32\\
\text{Monthly interest} &= \$1.25 + \$1.32\\
&= \$2.57\\
\text{Period of loan} &= 6 \text{ months}
\end{aligned}
$$

$$
\begin{aligned}
\text{Total interest} &= \text{Number of months} \times \text{Monthly interest}\\
&= 6 \times \$2.57\\
&= \$15.42
\end{aligned}
$$

Exercises

A

1. Determine the interest period for each pawnshop loan.

	Date Borrowed	Date Returned	Number Of Months
a.	March 5	March 17	______
b.	May 18	June 2	______
c.	February 6	April 12	______
d.	September 9	December 20	______
e.	June 16	October 25	______
f.	January 9	May 5	______
g.	December 14	February 25	______
h.	October 23	January 29	______

2. Determine the monthly interest for each pawnshop loan.

	Loan	Monthly Rate	Monthly Interest
a.	\$50	2%	\$______
b.	40	3%	______
c.	15	4%	______
d.	25	$2\frac{1}{2}\%$	______
e.	65	$3\frac{1}{2}\%$	______
f.	96	$4\frac{1}{2}\%$	______

3. Determine the total interest for each pawnshop loan.

	Loan	Monthly Rate	Monthly Interest	Number Of Months	Total Interest
a.	$ 80	3%	$______	2	$______
b.	45	4%	______	5	______
c.	75	2%	______	4	______
d.	92	$3\frac{1}{2}$%	______	3	______
e.	104	$2\frac{1}{2}$%	______	6	______

4. Determine the total interest for each pawnshop loan.

	Loan	Monthly Rate	Monthly Interest	Date Borrowed	Date Returned	Number Of Months	Interest
a.	$ 20	4%	$______	May 3	June 20	______	$______
b.	68	3%	______	Feb. 21	July 10	______	______
c.	110	2%	______	June 10	Sept. 2	______	______
d.	24	$3\frac{1}{2}$%	______	Jan. 16	June 4	______	______
e.	56	$2\frac{1}{2}$%	______	Sept. 21	Jan. 28	______	______

5. Determine the amount of money that will be repaid for each pawnshop loan.

	Loan	Monthly Rate	Monthly Interest	Number Of Months	Total Interest	Amount Repaid
a.	$ 50	2%	$______	3	$______	$______
b.	28	4%	______	2	______	______
c.	41	3%	______	5	______	______
d.	58	$2\frac{1}{2}$%	______	6	______	______
e.	114	$3\frac{1}{2}$%	______	3	______	______

Unit 2

B

Use the pawnshop Monthly Interest Charge table on page 289 when doing each of the following exercises.

1. Determine the monthly interest for each loan.

	Loan	Interest on $50 or Less	Interest on Excess	Monthly Interest
a.	$35	$______	$______	$______
b.	50	______	______	______
c.	65	______	______	______
d.	78	______	______	______
e.	92	______	______	______

2. Determine the total interest for each loan.

	Loan	Interest on $50 or Less	Interest on Excess	Monthly Interest	Number Of Months	Total Interest
a.	$25	$________	$________	$________	3	$________
b.	60	________	________	________	4	________
c.	74	________	________	________	2	________
d.	91	________	________	________	3	________

C

The maximum permissible pawnshop-loan rates in one of the states are shown in the following table. Use this table when doing the exercises below.

Interest Charges
4 1/2% on the amount up to and including $25
3% on the amount in excess of $25

1. Determine the total interest for each loan.

	Loan	Interest on $25 or Less	Interest on Excess	Monthly Interest	Number Of Months	Total Interest
a.	$ 20	$________	$________	$________	2	$________
b.	30	________	________	________	5	________
c.	65	________	________	________	4	________
d.	138	________	________	________	3	________

2. Determine the amount that will be repaid for each loan.

	Loan	Monthly Interest	Date Borrowed	Date Returned	Number Of Months	Total Int.	Total Returned
a.	$ 40	$________	5/10	6/1	________	$________	$________
b.	84	________	7/6	9/4	________	________	________
c.	130	________	3/20	8/25	________	________	________
d.	185	________	11/25	2/6	________	________	________
e.	225	________	10/12	3/21	________	________	________

Unit 3 The Credit Union Loan

In general, the loan that will cost us the least is the one that is made at a credit union. Unfortunately, all of us can not borrow money from a credit union for only members are permitted to do this.

Credit unions are formed by groups of people having some common bond. They may be postal employees of the same county, teachers in the same state, employees of the same company, and so on. In many respects they operate in much the same manner as a bank. However, their only depositors are the members of their group.

Credit unions were first organized for the purpose of enabling their members to obtain relatively small loans at low interest rates. Although some credit unions are still quite small, others provide the same services as those of a fair sized bank. Their savings accounts are insured up to $40,000 by the federal government and there are even a few that grant loans as high as $100,000.

A large number of the loans made at credit unions are for the purchase of automobiles, home appliances, vacations, swimming pools, and the like. The rates vary from credit union to credit union, however, for most loans these rates appear to be between 8% and 12% per year.

Almost all their loans are installment loans. The monthly payment is determined by reference to a Table of Factors. The computation is basically the same as we used when examining bank installment loans. In view of this, some of the factors in the table below are the same as those in the table on page 267.

Monthly Installment Factors

Interest Rate	Number of Months 12	18	24	30
8%	.0869884	.0591403	.0452273	.0368833
$8\frac{1}{2}$%	.0872198	.0593687	.0454557	.0371178
9%	.0874515	.0595977	.0456847	.0373432
$9\frac{1}{2}$%	.0876835	.0598271	.0459145	.0375794
10%	.0879159	.0600571	.0461449	.0378114
11%	.0883817	.0605185	.0466078	.0382781
12%	.0888488	.0609821	.0470735	.0387481

Unit 3

Example 1 Determine the interest charge on a 9 1/2% credit union installment loan for $3,500. The debt was paid back over a 24-month period.

Solution

$$
\begin{aligned}
\text{Monthly Payment} &= \text{Loan} \times \text{Monthly factor} \\
&= \$3{,}500 \times .0459145 \\
&= \$160.70
\end{aligned}
$$

$$
\begin{aligned}
\text{Total repaid} &= \text{Number of months} \times \text{Monthly payment} \\
&= 24 \times \$160.70 \\
&= \$3{,}856.80
\end{aligned}
$$

$$
\begin{aligned}
\text{Interest} &= \text{Total repaid} - \text{Loan} \\
&= \$3{,}856.80 - \$3{,}500 \\
&= \$356.80
\end{aligned}
$$

For the people that request this, credit unions will permit them to repay the debt in a lump sum rather than in monthly installments. Under this arrangement – called a time loan – the total interest is paid at the time the money is returned.

The interest is usually computed in terms of months rather than days. Therefore, when using the interest formula, we must remember to divide the number of months by 12 to determine the fraction of a year for which the money was borrowed.

Example 2 Joan Baker borrowed $600 from her credit union to purchase a color TV set. At the end of 8 months she repaid the entire debt plus interest. If the annual interest rate was 10%, what was the total she returned?

Solution

$$\text{Interest} = P \times R \times T$$
$$= \$600 \times 10\% \times 8 \text{ months}$$

Fraction of year

$$= \$600 \times .10 \times \tfrac{8}{12}$$
$$= \$60 \times \tfrac{2}{3}$$
$$= \$40$$

$$\text{Total repaid} = \text{Loan} + \text{Interest}$$
$$= \$600 + \$40$$
$$= \$640$$

Exercises — A

1. Determine the monthly installment factor for each credit union installment loan.

	Interest Rate	Number of Months	Factor
a.	8%	18	______
b.	9%	12	______
c.	11%	30	______
d.	$8\frac{1}{2}$%	24	______
e.	$9\frac{1}{2}$%	18	______

2. Determine the monthly installment for each credit union installment loan.

	Loan	Interest Rate	Number of Months	Factor	Monthly Installment
a.	$ 500	10%	12	______	$______
b.	800	9%	18	______	______
c.	2,000	$8\frac{1}{2}$%	24	______	______
d.	1,500	$9\frac{1}{2}$%	18	______	______
e.	5,200	11%	30	______	______

3. Determine the total repaid for each credit union installment loan.

	Loan	Interest Rate	Number of Months	Factor	Monthly Installment	Total Repaid
a.	$ 600	12%	12	______	$______	$______
b.	1,000	11%	18	______	______	______
c.	1,200	9½%	24	______	______	______

4. Determine the interest charge for each credit union installment loan.

	Loan	Interest Rate	Number of Months	Factor	Monthly Installment	Total Repaid	Interest
a.	$ 700	10%	12	______	$______	$______	$______
b.	3,000	8½%	30	______	______	______	______
c.	6,400	9½%	24	______	______	______	______

B

1. Determine the interest charge for each credit union time loan.

	Loan	Annual Interest Rate	Number of Months	Interest
a.	$ 600	9%	8	$______
b.	900	10%	6	______
c.	1,000	12%	7	______
d.	1,600	11%	9	______
e.	2,500	8½%	6	______
f.	4,200	9½%	4	______

2. Determine the total repaid for each credit union time loan.

	Loan	Annual Interest Rate	Number Of Months	Interest	Total Repaid
a.	$ 800	10%	3	$______	$______
b.	750	12%	5	______	______
c.	1,800	11%	8	______	______
d.	2,600	9%	7	______	______

C

1. Steve McNair borrowed $1,600 from his credit union on a time loan at an annual interest rate of 12%. At the end of 3 months he returned $600 plus the interest that had accumulated on the

$1,600. At the end of an additional 6 months he paid off the remaining $1,000 plus interest.

a. How much did he owe the credit union before he made the $600 payment? ________

b. How much interest did he give the credit union when he made the $600 payment? ________

c. How much money did he owe the credit union after he made the $600 payment? ________

d. How much interest did he give the credit union at the time he made the last payment? ________

e. What was the total interest he paid on the debt? ________

2. $1,200 is borrowed from a credit union on a time loan at an annual interest rate of 10%. At the end of 4 months, $300 is repaid with interest that had accumulated to that point. After an additional 9 months the remainder of the debt plus the interest is repaid.

a. How much interest was paid when the first payment was made? ________

b. How much interest was paid when the final payment was made? ________

c. What was the total interest paid on this debt? ________

The Credit Card Loan Unit 4

Unit 4

Earlier in our work we examined how the cost of purchases can be charged against our account with credit companies such as VISA or Master Charge. It is also possible to borrow money from these and other credit companies. However, the total of both the purchases we have made and the amount we have borrowed can not be greater than our *credit line*. This term simply refers to the *maximum* amount of money against which we can either charge purchases or borrow money.

As an example, consider that our credit line is $1,500. If we have already charged purchases amounting to $1,000, then we can borrow no more than $500 for the total of the two is $1,500.

$1,000 + $500 = $1,500

Purchases (↑ $1,000) **Loan** (↑ $500) **Credit line** (↑ $1,500)

The annual rate of interest on a credit card loan—called a cash *advance*—is 12%. However, when computing the interest, this rate is renamed as a daily rate of .03287%.

$$12\% \div 365 = .03287\%$$

Annual rate (12%) — **Days in 1 year** (365) — **Daily rate** (.03287%)

Basically, the interest on the loan is computed on the number of days the money is borrowed.

The loan agreement with the credit company also calls for a payment each month of at least 1/24 of the cash advance balance.

Example 1 What is the smallest payment that can be made on an $800 cash advance balance?

Solution

$$\begin{aligned} \text{Smallest payment} &= \tfrac{1}{24} \text{ of balance} \\ &= \tfrac{1}{24} \times \$800 \\ &= \$33.33 \end{aligned}$$

Example 2 During a billing period of 28 days the balance on a cash advance was $600. Using the daily interest rate of .03287%, determine the interest for this period.

Solution

$$\begin{aligned} \text{Interest for 1 day} &= \text{Amount owed} \times \text{Daily rate} \\ &= \$600 \times .03287\% \\ &= \$600 \times .0003287 \\ &= \$.19722 \end{aligned}$$

Number of days (28) — **Interest per day** ($.19722)

$$\begin{aligned} \text{Total interest} &= 28 \times \$.19722 \\ &= \$5.52 \end{aligned}$$

By borrowing on a cash advance, we face some unfortunate charges because of delays in the mail. As an example, consider the diagram below.

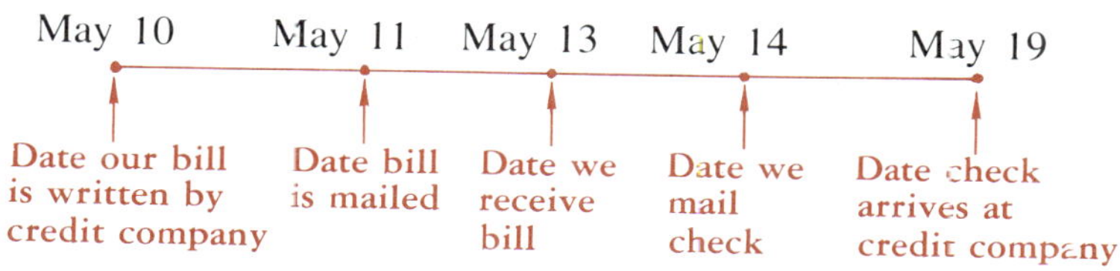

Although we paid the debt within 4 days of the date it was mailed to us, there will be an additional charge of 9 days interest on our next bill! This is the charge for the 9 days between the time the bill was written and our check was received.

Example 3 The monthly statement from a credit agency showed the balance of a cash advance to be \$450 while the daily interest rate was .04109%. The billing date of the statement was July 8, however the credit agency did not receive the payment until July 19. For what additional charge will the borrower be billed as a result of the delay in the payment?

Solution

Number of days from July 8 to July 19 = 19 − 8
= 11

Interest for 1 day = \$450 × .04109%
= \$450 × .0004109
= \$.184905

Number of days → 11; **Interest per day** → \$.184905

Total interest = 11 × \$.184905
= \$2.03

Exercises

A

Do you recall how to do the computation in the following exercises? This computation is the same as you will be doing when finding the cost of a cash advance.

Unit 3

1. Rename each percent numeral as a decimal.

a.	.04527%	________	**d.**	.03067%	________
b.	.03621%	________	**e.**	.04127%	________
c.	.02984%	________	**f.**	.04062%	________

2. Find the products.

a.	$400 × .03823%	$________	c.	$2,000 × .02506%	$________
b.	$600 × .05217%	________	d.	$1,300 × .03162%	________

B

1. The minimum monthly payment on a cash advance is 1/24 of the amount still owed. Determine the minimum payment for each balance.

	Balance	Minimum Payment		Balance	Minimum Payment
a.	$288	$________	d.	$297.60	$________
b.	600	________	e.	568.80	________
c.	888	________	f.	786.00	________

2. Determine the number of days for which the borrower will have to pay an additional interest charge because of the delay in the payment.

	Billing Date	Date of Check's Arrival	Number of Days
a.	March 12	March 18	________
b.	June 5	June 16	________
c.	April 20	April 27	________
d.	September 28	October 8	________
e.	January 27	February 5	________

3. Determine the daily interest for each cash advance. Do not round your answer.

	Cash Advance	Daily Rate	Daily Interest
a.	$500	.03287%	$________
b.	400	.03287%	________
c.	600	.04109%	________
d.	800	.04109%	________
e.	700	.04931%	________

4. Determine the interest for each cash advance for the billing period shown.

	Cash Advance	Daily Rate	Daily Interest	Billing Period (Days)	Total Interest
a.	$ 300	.03287%	$________	30	$________
b.	600	.03835%	________	30	________
c.	1,000	.04383%	________	31	________
d.	900	.04109%	________	31	________

5. Determine the additional interest that will be charged because of the delay in each of the following payments.

	Balance	Daily Rate	Daily Interest	Billing Date	Date of Check's Arrival	Number Of Days	Total Interest
a.	$600	.03287%	$________	6/5	6/10	________	$________
b.	700	.03287%	________	7/16	7/26	________	________
c.	500	.04109%	________	4/21	4/29	________	________
d.	800	.04383%	________	5/28	6/4	________	________

6. The minimum monthly payment on a cash advance is 1/24 of the new balance of the advance. Determine the minimum payment for each situation.

	Old Balance	Daily Rate	Daily Interest	Number Of Days	Interest for Billing Period	New Balance	Minimum Payment
a.	$200	.03287%	$________	30	$________	$________	$________
b.	400	.04109%	________	30	________	________	________
c.	500	.04383%	________	31	________	________	________
d.	800	.03835%	________	31	________	________	________

C

1. The daily interest rate on a credit card loan is .05479%. What is the annual rate on this loan? ________

2. Bruce Dodge repaid a debt of $800 on a credit card loan. His check was in the mail for a period of five days. How much interest did Bruce have to pay as a result of just the mailing time? The daily interest rate was .03506%. ________

3. On March 17, Martha Fuller received a credit card statement dated March 5. The balance of her loan as shown on this statement was $650. How much interest accumulated on this loan during the period of mailing if the daily rate was .04794%? ________

Review of Chapter 8

1. The Alliance Finance Corporation placed the following advertisement in a local newspaper.

LOANS TO FIT YOUR NEEDS

Cash Received By You	Amount of Monthly Payment	Number of Months
$106	$19.47	6
147	14.63	12
387	27.78	18

(annual rate of interest: 34.16%)

a. How much money will the borrower repay the loan company on the $106 loan? ________

b. How much money will the borrower repay the loan company on the $387 loan? ________

2. If Jack Sands borrowed money from a pawnshop on August 17 and paid off the debt on January 5, for how many months would he be charged interest? ________

3. Steve Billings borrowed $75 from a pawnshop that charged an interest rate of 4% per month on the unpaid balance of the debt. He returned the money with interest at the end of four months.

a. How much interest did Steve have to pay? ________

b. What amount did he return to the pawnbroker? ________

4. Julia Browne, a pawnbroker, has her shop in a state where she is permitted to charge the following monthly interest rates:

On the first $50 of the debt	$3\frac{1}{2}$%
On the amount over $50	2%

a. If a person borrows $40 for one month, how much interest will have to be paid? ________

b. If a person borrows $85 for one month, how much interest will have to be paid? ________

c. If a person borrows $125 for five months, how much interest will have to be paid? ________

5. Nancy Brookfield belongs to a credit union. In order to buy a used car she borrowed $1,500 on a 9 1/2% installment loan to be paid off monthly over a period of one year. Use the table of Monthly Installment Factors on page 293 to answer the questions below.

a. How large was each of her monthly installments? ________

b. What was the total amount of money she repaid the credit union? ________

c. How much interest did she pay on this loan? ________

6. When Phyllis Riley bought her sewing machine she borrowed $300 from her credit union at an annual rate of 9%. At the end of 8 months she repaid the entire debt plus interest in one lump sum.

a. How much interest did she pay on the loan? ________

b. Including both principal and interest, how much money did she give the credit union at the time she repaid the debt? ________

7. Harold Bentley borrowed $4,000 from his credit union at an interest rate of 10%. At the end of 9 months he repaid $1,000 plus the total interest that had accumulated during that time.

a. How much interest did he owe for that period? ________

b. How much money did he give the credit union at the time he made the payment? ________

c. How much money did he owe the credit union after making this payment? ________

d. He repaid the remainder of the debt with interest 6 months later. How large was that payment? ________

Review

8. The billing date on a credit agency statement is May 15. The borrower receives the bill on May 19 and mails a check for the entire balance. The credit agency receives the check on May 26. For how many additional days will the borrower have to pay interest because of the delay in payment? ________

9. The daily interest rate on a cash advance is .04794%. In addition, the minimum monthly payment is 1/24 of the new balance of the loan.

a. The old balance on an advance is $500. How much interest accumulated over a 30-day billing period? ________

b. What is the new balance on this advance? ________

c. What is the minimum payment that can be made on the loan for this month? ________

AMERICAN STOCK EXCHANGE

Chapter 9

Investments in Stocks and Bonds

We have learned that our money can be put to work earning more money for us if we keep it in a savings account. There are, though, many people who seek other ways to invest their money that may pay higher interest rates than those offered by banks.

In this chapter we plan to examine two such investments. It is important to realize that with these investments as with any investment, the greater the opportunity for rapid gain, the greater is the risk that we may lose either part or all of our money. On the other hand, money deposited at a bank can never be lost because it is insured – up to $40,000 – by the federal government.

The Stock Market Unit 1

Section 1 Stock Market Quotations

Let us consider for the moment a situation in which five of us decide to form a company. Our immediate needs amount to $600. To cover this each of us contributes respectively:

$200, $50, $150, $100, and $100.

So that there will be no disagreement at some later date, sheets of paper are distributed among us. Printed on each sheet is the statement that whoever holds that paper owns

$10 in the original value of the company

Each of these pieces of paper is called a *stock certificate* or a *share of stock*. Therefore, for the company we just formed, the person who contributed \$200 will receive

20 shares of stock (\$200 ÷ \$10 = 20).

Similarly, the person who contributed \$50 will receive

5 shares of stock (\$50 ÷ \$10 = 5).

And, of course, each of the others will receive comparable shares of stock making a total of

60 shares for all (\$600 ÷ \$10 = 60).

Let us now imagine that the company prospers and that its value increases from the original \$600 to \$1,800. In view of this, where originally each share was worth \$10, now each share will be worth \$30:

\$1,800 ÷ 60 = \$30.

Thus, the person owning 20 shares will have increased his investment from

\$200 to \$600 (20 × \$30).

On the other hand, if the company was to lose money, then each share of stock would decrease in value and the person owning the greatest number of shares would lose the greatest amount of money.

The situation just described is the manner in which most companies are originally organized. The shares of stock, though, are purchased through stockbrokers at markets called stock exchanges. These exchanges are located throughout the world. The largest two – New York Stock Exchange and American Stock Exchange – are in New York City.

The price of each share of stock is always stated as the sum of:

1. A whole number of dollars

and 2. One of the following fractional parts of a dollar:

$$\$\tfrac{1}{8},\ \$\tfrac{1}{4},\ \$\tfrac{3}{8},\ \$\tfrac{1}{2},\ \$\tfrac{5}{8},\ \$\tfrac{3}{4},\ \$\tfrac{7}{8}.$$

Before doing any computation with these fractions, they are renamed as decimals.

$\$\tfrac{1}{8} = \$.125$	$\$\tfrac{5}{8} = \$.675$
$\$\tfrac{1}{4} = \$.25$	$\$\tfrac{3}{4} = \$.75$
$\$\tfrac{3}{8} = \$.375$	$\$\tfrac{7}{8} = \$.875$
$\$\tfrac{1}{2} = \$.50$	

Example 1 A share of stock is quoted at 59 1/4. Rename this price as a decimal.

Explanation A quotation is understood to represent the value of *one* share. In this situation 59 1/4 means $59 1/4.

Solution

$$\begin{aligned} \$\tfrac{1}{4} &= \$.25 \\ \text{Price} &= \$59\tfrac{1}{4} \\ &= \$59.25 \end{aligned}$$

Unit 1

Example 2 Determine the cost of 75 shares of stock that sold at 46 3/8 per share.

Solution

$$\begin{aligned} \text{Cost per share} &= \$46\tfrac{3}{8} \\ &= \$46.375 \end{aligned}$$

Number of shares → 75; **Cost per share** → $46.375

$$\begin{aligned} \text{Total cost} &= 75 \times \$46.375 \\ &= \$3{,}478.125 \text{ or } \$3{,}478.13 \end{aligned}$$

Exercises

1. Rename each mixed number as a decimal.

a.	$26\frac{1}{2}$	________	**e.**	$65\frac{3}{8}$	________
b.	$15\frac{1}{4}$	________	**f.**	$127\frac{5}{8}$	________
c.	$86\frac{3}{4}$	________	**g.**	$104\frac{7}{8}$	________
d.	$34\frac{1}{8}$	________	**h.**	$10\frac{3}{8}$	________

2. Rename each stock quotation in terms of its dollar value.

	Quotation	Dollar Value		Quotation	Dollar Value
a.	$15\frac{1}{4}$	$________	e.	$84\frac{5}{8}$	$________
b.	$43\frac{1}{2}$	________	f.	$29\frac{3}{4}$	________
c.	$26\frac{1}{8}$	________	g.	$117\frac{1}{8}$	________
d.	$57\frac{3}{8}$	________	h.	$226\frac{7}{8}$	________

3. Determine the total cost for each stock purchase.

	Quotation	Dollar Value	Number of Shares	Total Cost
a.	27	$________	40	$________
b.	$14\frac{1}{2}$	________	60	________
c.	$31\frac{1}{2}$	________	50	________
d.	$20\frac{1}{4}$	________	100	________
e.	$56\frac{1}{4}$	________	80	________
f.	$31\frac{3}{4}$	________	70	________
g.	$67\frac{3}{4}$	________	200	________
h.	$11\frac{1}{8}$	________	30	________
i.	$23\frac{1}{8}$	________	100	________
j.	$42\frac{3}{8}$	________	200	________
k.	$29\frac{3}{8}$	________	300	________
l.	$12\frac{5}{8}$	________	25	________
m.	$53\frac{5}{8}$	________	45	________
n.	$36\frac{7}{8}$	________	150	________
o.	$74\frac{7}{8}$	________	250	________

Section 2 The Stock Market Report

To keep people informed of the value of their investments, many newspapers publish the daily sales transactions of at least one stock exchange. These reports are usually in the form of the one shown on page 309.

STOCK MARKET QUOTATIONS

Year To Date High	Year To Date Low	Stocks & Dividend	Sales In 100's	Open	High	Low	Close	Net Change
19½	12	Abacus Corp. 2	16	15½	16	15	15¼	+½
85	72	Abbott Markets 5	4	75	75½	75	75	−1
16¾	12½	Adams Department Store 1	29	15¼	15¾	14⅞	15	+¼
39	25¼	Alabama Electric Co. 2	1	36⅞	36⅞	36⅞	36⅞	+½
16⅝	12⅞	Allen, Inc. 1.20	32	13¼	13⅜	13⅛	13¼	+⅛
78	64	American Bakers pf. 2½	6	64	64¾	64	64¼	−½
96⅞	71½	American Corporation 2	58	84½	85¼	83¼	84½	+1½
12¼	5¾	Atlas Industries	24	8	8¾	8	8½	
48⅝	41⅛	Baldwin Tools 2.50	26	44⅞	45¼	44¾	45	+⅜
58¼	44⅞	Baltimore Credit 1	31	47½	48¼	47⅛	47⅝	−½
65⅜	50¾	Baxter, Inc., pf. 3	7	62⅝	63⅞	62⅜	63¾	+1¼
5⅞	3⅛	Beech Instruments .15	81	4¼	4⅜	4⅛	4⅜	+⅛
17¾	13	Bliss Petroleum 1	9	15¼	15⅝	15¼	15½	+⅞
55½	41½	Bond Foods pf. 2½	14	48½	49	48¼	48¾	−¼
40⅜	23¼	Bullard and Sons 1.70	38	27¾	28	27½	28	+¼
36⅞	29⅝	California Breweries 2	6	30⅝	30⅝	30⅝	30⅝	
46½	37½	Canadian Gas Co. 2	14	38¼	39½	38	38	−1
81⅞	62¼	Carpenter Corp. 1.80	55	78¾	79½	78¼	79	+⅞
43½	35¼	Cenco Chemicals 1.50	26	42	42½	41	42¼	−½
31½	27	Champion Soup Co. 1.06	18	28	28	27	28	
46¾	37⅞	City R. R. Services 1½	16	39⅛	39⅞	38¾	39½	−⅛
82¼	75	Colonial, Inc. 3.25	88	78½	78¾	78	78½	+2
20⅝	14¾	Crane Industries pf. 1½	75	19⅛	19¼	17⅝	18⅜	+¾
33	29⅝	Dana Power and Light Co. 1.60	9	30½	30½	30¼	30½	+⅛
34½	19⅞	Diamond Airlines	68	28⅜	28½	26	28	+1⅛
20¼	13½	Dobbs Productions .37	14	14	14¼	13⅞	13⅞	−⅞
36¾	30¾	Dow Aircraft .84	92	34	35	33	34½	+1
36	25¼	Drexel Mines .70	44	28	28½	28	28	−⅜
23⅜	18¼	Dunhill Limited pf. 1.05	27	23	23⅛	22½	22⅞	+¼
24¼	16½	Dynamics, Inc. 1.20	16	18¼	18½	17½	18	+⅝

Unit 1

The diagram on page 310 gives an explanation of the quotations for the Abacus Corporation. This is the first company listed in the stock market report. For reasons of simplicity, the information in the diagram has been rearranged on two lines.

Example 1 Determine the highest price paid for a share of Atlas Industries during the day shown in the report on this page.

Year To Date High	Year To Date Low	Stocks & Dividend	Sales In 100's
$19\frac{1}{2}$	12	Abacus Corp. 2	16
This is the highest price paid for a share of stock thus far *that year*.	This is the lowest price paid for a share of stock thus far *that year*.		This is the number of hundreds of shares sold *that day*.

Open	High	Low	Close	Net Change
$15\frac{1}{2}$	16	15	$15\frac{1}{4}$	$+\frac{1}{2}$
This is the first price paid for a share *that day*	This is the highest price paid for a share *that day*.	This is the lowest price paid for a share *that day*.	This is the last price paid for a share *that day*.	This is the amount the last price gained (+) or lost (−) over the last price of the previous day.

Explanation

Stocks & Dividend | **Open** | **High**

Read down → Atlas Industries — — — — — — → $8\frac{3}{4}$

Read across

Solution

$$\text{Highest price paid that day} = 8\tfrac{3}{4}$$
$$= \$8.75$$

Example 2 Determine the lowest price paid thus far that year for a share of Crane Industries.

Explanation

Year to Date | | **Stocks & Dividend**
High **Low**

Read down

$14\frac{3}{4}$ ← — — — — — — — — — Crane Industries

Read across

Solution

$$\text{Lowest price paid that year} = 14\tfrac{3}{4}$$
$$= \$14.75$$

Example 3 Joe Manning purchased 400 shares of Baxter, Inc. stock at the Low of the day shown in the stockmarket quotations. Determine the amount paid for these shares.

Solution

$$\text{Low of day} = 62\tfrac{3}{8}$$
$$= \$62.375$$

Number of shares ↓ ↓ **Cost per share**

$$\text{Total cost} = 400 \times \$62.375$$
$$= \$24{,}950$$

Example 4 Determine the last price paid for a share of Allen, Inc. on the day prior to that of the quotations on page 309.

Explanation

Stocks & Dividend **Net Change**

Allen, Inc. 1.20 ——————————→ $+\frac{1}{8}$

This sign means a gain over the prior day.

Since the value increased \$1/8 over its price of the previous day, then to determine the last price of the previous day we will have a subtract \$1/8 from the last price of the day of the quotations.

Solution

$$\text{Last price that day} = 13\tfrac{1}{4}$$
$$\text{Net change} = +\tfrac{1}{8}$$
$$\begin{aligned}\text{Last price of previous day} &= 13\tfrac{1}{4} - \tfrac{1}{8}\\ &= \$13.25 - \$.125\\ &= \$13.125\end{aligned}$$

Exercises A

Use the stockmarket quotations on page 309 when doing the exercises of B, C, and D.

Do you recall how to do the computation in the following exercises? This computation is the same as you will be doing when finding the price of a share of stock.

1. Find the sums. Rename each numeral as a decimal before doing the computation.

a. $\$16 + \$\frac{1}{2} =$ \$________

b. $\$27 + \$\frac{3}{4} =$ ________

c. $\$41 + \$\frac{5}{8} =$ ________

d. $\$8\frac{1}{2} + \$\frac{1}{4} =$ ________

e. $\$5\frac{3}{4} + \$\frac{1}{2} =$ ________

j. $\$16\frac{1}{8} + \$\frac{1}{4} =$ \$________

k. $\$23\frac{3}{8} + \$\frac{1}{2} =$ ________

l. $\$15\frac{3}{4} + \$\frac{5}{8} =$ ________

m. $\$29\frac{7}{8} + \$\frac{1}{4} =$ ________

n. $\$76\frac{5}{8} + \$\frac{3}{8} =$ ________

f. $\$14\frac{1}{4} + \$\frac{3}{4} =$ ________ o. $\$41\frac{1}{2} + \$\frac{7}{8} =$ ________

g. $\$27\frac{1}{2} + \$\frac{3}{4} =$ ________ p. $\$59\frac{1}{8} + \$1\frac{1}{4} =$ ________

h. $\$85\frac{1}{4} + \$1\frac{1}{2} =$ ________ q. $\$62\frac{3}{8} + \$1\frac{5}{8} =$ ________

i. $\$31\frac{3}{4} + \$1\frac{3}{4} =$ ________ r. $\$80\frac{1}{8} + \$2\frac{5}{8} =$ ________

2. Find the differences. Rename each numeral as a decimal before doing the computation.

a. $\$14 - \$\frac{1}{2} =$ $________ f. $\$45\frac{1}{8} - \$\frac{3}{4} =$ $________

b. $\$16 - \$\frac{3}{8} =$ ________ g. $\$71 - \$1\frac{1}{4} =$ ________

c. $\$27\frac{3}{4} - \$\frac{1}{2} =$ ________ h. $\$83 - \$2\frac{7}{8} =$ ________

d. $\$39\frac{5}{8} - \$\frac{3}{8} =$ ________ i. $\$94\frac{5}{8} - \$1\frac{3}{4} =$ ________

e. $\$62\frac{1}{4} - \$\frac{5}{8} =$ ________ j. $\$50\frac{1}{4} - \$3\frac{5}{8} =$ ________

B

1. Find the highest price paid so far this year for a share of stock in each company.

	Company	Price per Share
a.	Alabama Electric Co.	$________
b.	Bond Foods	________
c.	Baltimore Credit	________
d.	Bliss Petroleum	________
e.	Baxter, Inc.	________
f.	California Breweries	________

2. Find the first price paid that day for a share of stock in each company.

	Company	Price per Share
a.	Abbott Markets	$________
b.	American Corporation	________
c.	Bullard and Sons	________
d.	City R. R. Services	________
e.	Diamond Airlines	________
f.	Baldwin Tools	________

Unit 1

3. Find the lowest price paid that day for a share of stock in each company.

	Company	Price per Share
a.	Atlas Industries	$________
b.	Drexel Mines	________
c.	Dynamics, Inc.	________
d.	Dana Power and Light Co.	________
e.	City R. R. Services	________
f.	Adams Department Store	________

4. Determine the net change of a share of stock in each company. Indicate whether the net change is an increase or a decrease.

	Company	Net Change	Increase or Decrease
a.	Colonial, Inc.	$________	________
b.	Abbott Markets	________	________
c.	Alabama Electric Co.	________	________
d.	Beech Instruments	________	________
e.	Drexel Mines	________	________
f.	Diamond Airlines	________	________

5. Determine the last price paid for a share of stock in each company on the day prior to that of the stockmarket quotations.

	Company	Last Price For Day Shown	Net Change	Last Price of Previous Day
a.	Dow Aircraft	$________	$________	$________
b.	Canadian Gas Co.	________	________	________
c.	Champion Soup Co.	________	________	________
d.	Adams Department Store	________	________	________
e.	Bond Foods	________	________	________
f.	Cenco Chemicals	________	________	________
g.	Baldwin Tools	________	________	________
h.	Bliss Petroleum	________	________	________
i.	Baltimore Credit	________	________	________
j.	Crane Industries	________	________	________

C

1. Determine the cost of each stock purchase.

	Company	Number of Shares	Price per Share	Price per Share in Dollars	Total Price
a.	Atlas Industries	100	Open	$________	$________
b.	Abbott Markets	100	Low	________	________
c.	Bullard and Sons	100	Low	________	________
d.	Baxter, Inc.	200	Close	________	________
e.	Allen, Inc.	200	High	________	________
g.	Crane Industries	200	Open	________	________
h.	Dobbs Productions	300	Low	________	________
i.	Baltimore Credit	400	Low	________	________

D

1. Timothy Crane purchased 200 shares of Adams Department Store stock at the closing price.

a. How much did he pay for the stock? ________

b. How much would the stock have cost him had he made the purchase at the closing price of the previous day? ________

c. How much could he have saved had he been able to make the purchase at the closing price of the previous day? ________

2. Frank Delman purchased 300 shares of Baxter, Inc., preferred stock at the "Low" of the day.

a. How much did he save on each share of stock by not having to buy it at the "High" of the day? ________

b. How much did he save on the 300 shares by buying them at the "Low" of the day rather than the "High"? ________

3. Patricia Trent was lucky enough to have been able to purchase 100 shares of Bullard and Sons at the "Low" of the year.

a. How much did these stocks cost her? ________

b. Had she been equally lucky and sold them at the "High" of the year, how much would she have received? ________

c. How much profit would she have made on the purchase and sale of this stock? ________

Section 3 Stock Dividends

Unit 1

In the event a company earns a profit during the year, that profit is divided among the owners—these are the stockholders. To illustrate the amount each stockholder receives let us return to the example described at the beginning of the chapter where five of us had organized a company. As you recall, there were 60 shares in all. Therefore, if the company had earned a profit of $120, then $2 of this profit would be designated for each share.

$$\$120 \div 60 = \$2$$

The amount of profit assigned to each share is called a *dividend*. In this illustration the person who owns 5 shares will receive a total dividend payment of $10.

Number of shares → 5; **Dividend per share** → $2

Total dividends = 5 × $2

Example 1 A company declares a dividend of 27 1/2¢. Determine the total dividends a stockholder will receive for 350 shares of stock.

Solution

Number of shares → 350; **Dividend per share** → $27\frac{1}{2}$¢

Total dividends = 350 × $27\frac{1}{2}$¢
= 350 × 27.5¢
= 350 × $.275
= $96.25

There are many varieties of stock that a company can issue. Basically, though, they fall into two groups:

1. Common Stock
2. Preferred Stock

The manner in which the company profit is divided as described above applies only to common stock. The owners of preferred stock are guaranteed a fixed dividend each year. That dividend is based on a specified percent of the *par value* of the stock. The par value of a share can actually be any value a company wants it to be. Companies will frequently assign the original value of the stock as the par value. In the event the entire profit is needed to pay the preferred stockholders their guaranteed dividend, then the common stockholders will receive no dividend.

Example 2 Peter Schilling owns 300 shares of 5% preferred stock of Graphic Central Corporation. The par value of the stock is $40; however, Mr. Schilling paid $62.50 per share at the time he purchased the stock. What were the dividends he received each year for these shares?

Explanation Since the dividend is based on the par value, we ignore the \$62.50. The price Mr. Schilling paid for the stock has no bearing on the dividend.

Solution

Dividend on each share = 5% of par value
= .05 × \$40
= \$2.00

Number of shares → 300; Dividend per share → \$2

Total dividends = 300 × \$2
= \$600

In the example above the \$600 represents the total dividends that were paid to Mr. Schilling for the entire year. There are many companies, however, that pay dividends either:

1. *Semiannually*—Twice each year

or 2. *Quarterly*—Four times each year.

Quite apparently:

semiannual dividend = 1/2 of annual dividend
quarterly dividend = 1/4 of annual dividend.

Example 3 Mary Ryan owns 80 shares of Texas Oil Company's 6 1/2% preferred stock. If the par value of the stock is \$100, how large is Mary's quarterly dividend?

Solution

Dividend on each share = $6\frac{1}{2}\%$ of par value
= 6.5% × \$100
= .065 × \$100
= \$6.50

Number of shares → 80; Dividend per share → \$6.50

Total annual dividends = 80 × \$6.50
= \$520

Quarterly dividends = $\frac{1}{4}$ of annual dividend
= $\frac{1}{4}$ × \$520
= \$130

Unit 1

One of the pieces of information we ignored at the time we examined the stock market quotations was the numeral immediately to the right of the name of the company.

Year To Date		Stocks & Dividend		Sales in 100's
High	Low			
$19\frac{1}{2}$	12	Abacus Corp.	2	16

This numeral shows the dividend paid by this company during the previous year.

The purpose of including the dividend with the stockmarket quotations is to give us some indication of how much our money will earn if we purchased this stock.

Example 4 Raymond Da Silva owns 200 shares of Bullard and Sons stock. How much did Raymond receive in dividends last year for these shares? (See quotations on page 309.)

Explanation

Stocks & Dividend

Bullard and Sons 1.70 — **Last year's dividend**

Solution

Dividend per share = \$1.70

Number of shares (200), **Dividend per share** (\$1.70)

$$\text{Total dividends} = 200 \times \$1.70$$
$$= \$340$$

Exercises A

Whenever necessary refer to the quotations on page 309 in order to do the exercises below.

1. Determine the dividend paid by each company for the year prior to the quotations.

	Company	Dividend		Company	Dividend
a.	Baltimore Credit	$______	e.	Diamond Airlines	$______
b.	Colonial, Inc.	______	f.	California Breweries	______
c.	Dunhill Limited pf.	______	g.	Dobbs Productions	______
d.	Crane Industries pf.	______	h.	Beech Instruments	______

2. Determine the total annual dividends paid to each stockholder.

	Number of Shares	Annual Dividend Per Share	Total Dividends
a.	100	50¢	$______
b.	100	75¢	______
c.	80	65¢	______
d.	40	$1.20	______
e.	75	$2.25	______
f.	200	$62\frac{1}{2}$¢	______
g.	250	$18\frac{1}{2}$¢	______
h.	300	$67\frac{1}{2}$¢	______

3. Determine the dividends paid to each stockholder during the year prior to the quotations.

	Number Of Shares	Company	Annual Dividend Per Share	Total Dividends
a.	100	Albany Electric Co.	$______	$______
b.	100	Baldwin Tools	______	______
c.	200	Bullard and Sons	______	______
d.	800	Dow Aircraft	______	______
e.	75	Bond Foods	______	______
f.	35	City R. R. Service	______	______

4. Determine the annual dividend on each preferred stock.

	Par Value	Dividend Rate	Annual Dividend Per Share
a.	$ 50	5%	$______
b.	100	5%	______
c.	100	$5\frac{1}{2}$%	______
d.	50	$5\frac{1}{2}$%	______
e.	50	$6\frac{1}{4}$%	______
f.	100	$4\frac{3}{4}$%	______
g.	50	$4\frac{3}{4}$%	______
h.	25	5%	______
i.	25	$4\frac{1}{2}$%	______
j.	10	8%	______
k.	10	$5\frac{1}{2}$%	______

5. Determine the total dividends that will be paid annually to each preferred stockholder.

	Number of Shares	Par Value	Dividend Rate	Annual Dividend Per Share	Total Dividends
a.	100	$ 50	6%	$______	$______
b.	200	50	7%	______	______
c.	100	100	8%	______	______
d.	50	50	$6\frac{1}{2}$%	______	______
e.	200	100	$5\frac{1}{2}$%	______	______
f.	200	40	4.75%	______	______
g.	500	10	8.3%	______	______

6. Determine the total dividends that will be paid semiannually to each preferred stockholder.

	Number Of Shares	Par Value	Dividend Rate	Annual Dividend Per Share	Total Annual Dividends	Total Semiannual Dividends
a.	100	$100	7%	$______	$______	$______
b.	50	100	3%	______	______	______
c.	40	50	$4\frac{1}{2}$%	______	______	______
d.	70	50	5.8%	______	______	______
e.	150	50	6.45%	______	______	______

7. Determine the total dividends that will be paid quarterly to each preferred stockholder.

	Number Of Shares	Par Value	Dividend Rate	Annual Dividend Per Share	Total Annual Dividends	Total Quarterly Dividends
a.	200	$100	7%	$______	$______	$______
b.	100	50	4.2%	______	______	______
c.	30	80	5.3%	______	______	______
d.	150	60	7.4%	______	______	______

B

1. Bill Stanton purchased 65 shares of common stock and 200 shares of 5% preferred stock of the Manhattan Business Corporation.

a. If the company has been paying an annual dividend of $1.07 per share on its common stock, how much in dividends should Bill expect to receive over the period of one year on his common stock? ______

b. If the par value of the preferred stock is $50, how much in dividends should he expect to receive over the period of one year on his preferred stock? ________

c. What total dividends should he expect to receive from the Manhattan Business Corporation over the period of one year? ________

2. For some years now, Mona Feret has owned 200 shares of Dynamics, Inc., stock and 350 shares of Cenco Chemicals stock. The stock of both of these companies is sold at the exchange whose quotations appear on page 309.

a. How much did she receive in dividends from the Dynamics company last year? ________

b. How much did she receive in dividends from Cenco Chemicals last year? ________

c. What total dividends did she receive from both of these companies last year? ________

3. The 5% preferred stock of the Trans-Atlantic Airlines Company is selling for $75 per share, but it has a par value of $100. The 6% preferred stock of the Southern Airlines Company is also selling for $75 per share, but it has a par value of $50.

a. What is the annual dividend paid per share by the Trans-Atlantic Company? ________

b. What is the annual dividend paid per share by the Southern Airlines Company? ________

c. When you consider only the dividend and cost of each of these stocks, which is the better buy? ________

Section 4 Profit or Loss on Stock Investments

Many people are under the impression that a great deal of money can be made by continually buying and selling stock after a slight rise. Doing this may make the broker wealthy, but not the investor. On small investments, the broker's fee alone can run as high as 10%!

The person investing only $300 to $400 will find that his broker's fee is approximately 8% of the value of the stock. On the other-hand, the person investing $5,000 will probably pay his broker somewhere between 1 1/2% to 2% of the investment. This charge does not include a city tax, a state tax, and a federal income tax that will also have to be paid!

In general, the average person buying or selling stock can expect to pay at least

2% of the stock value

as a broker's fee. This is the rate we will use for the computation in this section.

Example 1 Determine the approximate broker's fee on the purchase of 100 shares of Hayes Products, Inc., at 23 1/2.

Solution

$$\begin{aligned}
\text{Cost per share} &= 23\tfrac{1}{2} \\
&= \$23.50 \\
\text{Total stock cost} &= 100 \times \$23.50 \\
&= \$2{,}350 \\
\text{Broker's fee} &= 2\% \text{ of stock value} \\
&= .02 \times \$2{,}350 \\
&= \$47
\end{aligned}$$

When buying stock through a broker, the bill for the purchase will include both the total stock cost and the broker's fee. Therefore, the total cost of the purchase is the sum of these quantities.

Add

Total purchase price = Total stock cost + Broker's fee

However, at the time we sell stock, the broker will subtract his fee from the selling price. The difference is turned over to us.

Subtract

Net selling price = Total selling price − Broker's fee

Amount remaining

Example 2 Approximately how much money did Marie Larson receive from her broker at the time she sold 50 shares of Howard Nickle Company stock at 42 3/8?

Solution

$$\begin{aligned}\text{Selling price per share} &= 42\tfrac{3}{8}\\ &= \$42.375\\ \text{Total selling price} &= 50 \times \$42.375\\ &= \$2{,}118.75\\ \text{Broker's fee} &= 2\% \text{ of stock value}\\ &= .02 \times \$2{,}118.75\\ &= \$42.38\\ \text{Net selling price} &= \text{Total selling price} - \text{Broker's fee}\\ &= \$2{,}118.75 - \$42.38\\ &= \$2{,}076.37\end{aligned}$$

Example 3 Determine Michael O'Hara's profit or loss on an investment in 100 shares of Reliable Products, Inc. which he purchased at 37 and sold at 38 1/8.

Solution

Computation on Purchase	Computation on Sale
Cost = 100 × \$37 = \$3,700	Selling price = 100 × \$38.125 = \$3,812.50
Broker's fee = 2% of \$3,700 = .02 × \$3,700 = \$74	Broker's fee = 2% of \$3,812.50 = .02 × \$3,812.50 = \$76.25
Total purchase price = \$3,700 + \$74 = \$3,774	Net selling price = \$3,812.50 − \$76.25 = \$3,736.25

$$\begin{aligned}\text{Loss} &= \text{Total purchase price} - \text{Net selling price}\\ &= \$3{,}774 - \$3{,}736.25\\ &= \$37.75\end{aligned}$$

Unit 1

Explanation Although each share was sold for more money than it was bought, Michael lost \$37.75 on this transaction because of the broker's fees that had to be paid!

Exercises A

1. Determine the broker's fee for each stock purchase.

	Total Purchase Price	Broker's Fee		Total Purchase Price	Broker's Fee
a.	$2,000	$________	d.	$875	$________
b.	3,000	________	e.	618	________
c.	1,500	________	f.	746	________

2. Determine the broker's fee for each stock sale.

	Number Of Shares	Quotation	Total Selling Price	Broker's Fee
a.	100	25	$________	$________
b.	100	$27\frac{1}{2}$	________	________
c.	100	$16\frac{1}{8}$	________	________
d.	50	$22\frac{5}{8}$	________	________
e.	40	$62\frac{3}{8}$	________	________
f.	60	$71\frac{7}{8}$	________	________

3. Use the stock quotations on page 309 to determine the broker's fee for each stock sale.

	Number Of Shares	Company	Selling At:	Total Selling Price	Broker's Fee
a.	100	Atlas Industries	Open	$________	$________
b.	100	Bond Foods	High	________	________
c.	200	Abacus Corp.	Low	________	________
d.	200	Cenco Chemicals	Low	________	________
e.	300	American Bakers	Close	________	________

4. Determine the total purchase price including the broker's fee paid by the buyer for each stock purchase.

	Cost	Broker's Fee	Total Purchase Price
a.	$1,000	$________	$________
b.	3,600	________	________
c.	1,250	________	________
d.	2,375	________	________

5. Determine the net selling price received by the seller for each stock sale.

	Selling Price	Broker's Fee	Net Selling Price
a.	$5,000	$______	$______
b.	4,300	______	______
c.	2,450	______	______
d.	1,325	______	______

6. Determine the total purchase price including broker's fee paid by the buyer for each stock purchase.

	Number Of Shares	Quotation	Cost	Broker's Fee	Total Purchase Price
a.	100	15	$______	$______	$______
b.	100	$21\frac{1}{2}$	______	______	______
c.	200	$34\frac{1}{4}$	______	______	______
d.	200	$26\frac{3}{4}$	______	______	______
e.	400	$12\frac{5}{8}$	______	______	______

7. Determine the net selling price received by the seller for each stock sale. Use the quotations on page 309.

	Number Of Shares	Company	Selling at:	Selling Price	Broker's Fee	Net Selling Price
a.	100	Bullard and Sons	Close	$______	$______	$______
b.	100	Abbott Markets	Low	______	______	______
c.	200	Dobbs Productions	High	______	______	______
d.	200	Diamond Airlines	Open	______	______	______
e.	300	City R. R. Services	High	______	______	______

Unit 1

B

1. Determine the profit or loss for each stock transaction. Do not include broker's fees in your computation.

	Number Of Shares	Purchase Quotation	Cost	Selling Quotation	Selling Price	Profit Or Loss
a.	100	50	$______	56	$______	$______
b.	40	28	______	37	______	______
c.	100	$31\frac{3}{4}$	______	$47\frac{5}{8}$	______	______
d.	65	$12\frac{1}{2}$	______	$9\frac{3}{4}$	______	______
e.	30	$54\frac{1}{8}$	______	$42\frac{1}{4}$	______	______

2. Determine the profit or loss for each stock transaction. Include broker's fees in your computation.

	Number Of Shares	Purchase Quotation	Total Purchase Price	Selling Quotation	Net Selling Price	Profit Or Loss
a.	100	24	$________	30	$________	$________
b.	100	32	________	46	________	________
c.	100	16	________	17	________	________
d.	300	50	________	51	________	________
e.	200	20	________	16	________	________

Section 5 Rate of Return on Stock Investments

People invest their money in stocks for either one or both of the following reasons:

1. They feel that the rate of interest their money will earn in a stock investment will be greater than the rate of interest paid by the banks.
2. They hope that during the time they own the stock its value will increase. This will enable them to sell their shares at a profit.

In this section we are going to examine how the rate of interest (actually called the rate of return) is determined on a stock investment. To show how this is done, we must recall a point developed earlier. Thus, if

First Number **Second Number** **Product**

7 × 8 = 56

then,

Second Number **Product** **First Number**

8 = 56 ÷ 7

A similar situation exists when finding the yearly interest on $500 at an annual interest rate of 6%.

Principal **Annual Rate** **Interest**

$500 × 6% = $30

First Number **Second Number** **Product**

Therefore, we can say that

Second Number		Product		First Number
6%	=	$30	÷	$500

However, since

6% is the annual interest rate;
$30 is the interest;
and $500 is the principal;

we have just shown that

Annual interest rate = Interest ÷ Principal.

When the formula is translated into the language of stock investments it becomes

Annual rate of return = Annual dividend ÷ Cost of one share

Notice that the rate of return is computed on just one share of stock. Whatever the rate is on one share, it will be exactly the same on all shares.

Example 1 What rate of return can a person expect on the purchase of 100 shares of Bullard and Sons at the High of the day shown in the quotations on page 309?

Solution

Dividend = $1.70
Cost of one share = $28
Annual rate of return = Annual dividend ÷ Cost of one share
= $1.70 ÷ $28
= .061 or 6.1%

Unit 1

Example 2 The 4% preferred stock of the Pontiac Glass Company is selling at 37 1/2. What rate of return will an investor receive if the par value of the stock is $50?

Solution

$$\text{Dividend} = 4\% \text{ of par value} = .04 \times \$50 = \$2$$

$$\text{Annual rate of return} = \text{Annual dividend} \div \text{Cost of one share} = \$2 \div \$37.50 = .053 \text{ or } 5.3\%$$

Example 3 Stock selling at 52 3/4 paid a quarterly dividend of \$1.75. What rate of return is this stock paying?

Solution

$$\text{Quarterly dividend} = \$1.75$$

Number of quarters in a year (4)

$$\text{Annual Dividend} = 4 \times \$1.75 = \$7$$

$$\text{Annual rate of return} = \text{Annual dividend} \div \text{Cost of one share} = \$7 \div \$52.75 = .133 \text{ or } 13.3\%$$

Exercises A

In each of the exercises below the rate of return should be computed to the nearest tenth of a percent.

1. Determine the rate of return for each stock purchase.

	Annual Dividend	Cost Per Share	Rate Of Return
a.	\$3.00	\$52	______%
b.	4.00	65	______%
c.	2.50	46	______%
d.	1.80	30.50	______%
e.	2.30	28.75	______%
f.	4.75	83.75	______%

2. Determine the rate of return that can be expected on each stock purchase. Use the quotations on page 309.

Company	Purchased At:	Cost Per Share	Annual Dividend	Rate Of Return
a. Canadian Gas Co.	Close	$________	$________	________%
b. Colonial, Inc.	Low	________	________	________%
c. American Bakers	Low	________	________	________%
d. Allen, Inc.	Open	________	________	________%
e. Abacus Corp.	Close	________	________	________%
f. Dobbs Products	Open	________	________	________%

3. Determine the rate of return that can be expected on the purchase of these preferred stocks.

	Interest Rate	Par Value	Dividend	Purchase Price	Rate Of Return
a.	5%	100	$________	$110	________%
b.	6%	100	________	112	________%
c.	$4\frac{1}{2}$%	50	________	38	________%
d.	$5\frac{1}{2}$%	50	________	55	________%
e.	4.8%	100	________	82	________%
f.	6.1%	100	________	95	________%

4. Determine the rate of return that can be expected on each stock purchase. Dividends are paid semiannually.

	Semiannual Dividend	Annual Dividend	Purchase Price	Rate Of Return
a.	$1.50	$________	$ 40	________%
b.	.40	________	14	________%
c.	1.68	________	52	________%
d.	2.37	________	63	________%
e.	3.69	________	107	________%

Unit 1

5. Determine the rate of return that can be expected on each stock purchase. Dividends are paid quarterly.

	Quarterly Dividend	Annual Dividend	Purchase Price	Rate Of Return
a.	$ .28	$________	$ 20	________%
b.	.73	________	42	________%
c.	1.30	________	75	________%
d.	2.50	________	180	________%
e.	1.86	________	135	________%

B

1. Bernard Coswell purchased 100 shares of Drexel Mines at the High for the year shown in the quotations on page 309.

a. What rate of return can he expect to receive on his investment? ________

b. If he purchased the stock at the Low of the year, what rate of return can he expect to receive? ________

c. Considering only the rate of return, should Mr. Coswell have invested his money in stock or should he have deposited it in a bank? ________

2. June Farley purchased 100 shares of Atlas Industries at the opening price shown in the quotations on page 309.

a. What rate of return can she expect to receive on her investment? ________

b. Why would she want to invest her money in Atlas Industries? ____________________

Unit 2 The Bond Market

Section 1 The Bond Market Report

Large companies are often in need of money in order to expand their business. On these occasions they borrow this money either from a bank or from a private individual such as ourself. As evidence of the fact that the company owes us money, we will receive a piece of paper called a *bond.* Printed on the bond is the date on which the debt will be paid. Usually, the period of the loan is for many years.

Appearing also on the bond is a statement concerning the amount borrowed (*the denomination of the bond*) and the annual interest rate that will be paid on the loan. The denomination of most bonds is $1,000.

Before the company profits can be shared by the stockholders, the bondholders must be paid the interest on the money borrowed from them by the company.

A stockholder is one of the owners of the company.

A bondholder is someone from whom the owners have borrowed money.

BOND MARKET QUOTATIONS

Year to Date High	Year to Date Low	Bonds & Rate	Sales In $1,000	High	Low	Last	Net Change
106½	85	Eastern Manufacturing Co. 5s 92	85	89	88¼	88⅜	+⅛
155	99¾	Electronic Associates 3½s 87	22	116	115½	116	+1
135	98⅛	Emerson-Brody Corp. 5s 89	17	108	106	107	−½
104½	97	Equitable Light Co. 6s 93	33	103	102½	102½	
96	92	Evans Storage Battery Co. 6s 94	5	94¾	94¾	94½	−1⅛
27	18	Falstaff Finance Corp. 5s 2020	1	20¼	20¼	20¼	+¼
29	20⅛	Family Food Stores 4½s 2015	47	24½	24	24	−1
40	36	Federal Mineral Co. 3⅛s 2000	2	38	38	38	−½
102	71⅜	Florida Products, Inc. 4⅜s 92	17	77½	76⅞	77¼	+1
124	104	Foremost Tire Co. 5½s 87	15	109	108	108½	−1
80¾	74	Franklin Sulphur Co. 3¼s 88	12	80¾	79½	80¾	+1½
97⅞	59⅝	General Department Stores 5½s 2011	58	90	84¼	89⅞	+8⅛
73	68½	Gibraltar Power Co. 5s 2001	2	72¾	72⅜	72⅜	
84½	70	Glenn-Holmes Corp. 4.80s 89	31	78	77	77	+1
163¼	124	Gulf National, Inc. 5½s 90	25	161⅛	159	160⅝	−4⅛
90	82½	Harvey Grace Motors, Inc. 7s 88	137	84¾	83⅜	84	+⅛
129½	109⅜	Hayes Products 5.30s 92	15	120⅞	120⅛	120½	+2⅛
145	112	Hotels International 4¼s 90	13	125	124¼	124⅝	+6
120	98¾	Household Services, Ltd. 6¼s 87	3	105	104½	104½	−½
124	104	Hunt-Portland Engineering Company 4¼s 92	8	116	114	115	−⅛
74	70¼	Idaho Paper Co. 3¼s 86	5	74	72½	73	−1
85½	80	Ideal Powder Co. 5½s 92	16	85¼	84⅜	85	
105	101⅛	Illinois Telephone Co. 4¼s 93	62	103½	102½	102⅝	−⅜
144¾	112	International Sugar Co. 4¾s 91	37	117½	116½	117	−½
103	94⅝	Interstate Transportations 4s 92	36	95½	94¾	95⅜	−2¼
131	109¾	Iowa Refrigerator Co. 5¾s 88	39	125⅞	123¼	124	−⅞
58½	49	Johnson Manufacturing Co. 4s 91	25	54¾	54	54¼	+1¼
106⅝	103⅝	Jonathon and Sons, Inc. 7⅜s 88	54	104½	104	104⅛	−¼
84	74⅞	Jones and Kendall, Inc. 4⅞s 90	10	81⅞	80⅝	81¾	+¼
100¾	93⅞	Joy Steel Corp. 6½s 87	16	98	97¼	97⅝	−1

Since the period of the loan covers so many years, bondholders often sell their bonds prior to the *redemption date* (this is the date on which the company has promised to repay the loan.) As in the case of stock, bonds are sold at a bondmarket through bond brokers. Reports of these transactions appear in newspapers in much the same way as those of the stockmarket. However, there is one major difference - the decimal point in a quotation is *one place further to the left* than it should be. As an example, the quotation 95 implies that the price of the bond is actually

$95.0. or $950.

Move decimal point one place to the right.

Unit 2

Example 1 Determine the value of a bond that is quoted at 107 1/4.

Solution

$$\text{Quotation} = 107\tfrac{1}{4}$$
$$= 107.250$$

Attach extra zero

$$\text{Value of bond} = 107.2.50$$

Move decimal point one place to right

$$= \$1{,}072.50$$

A small section of a bond market report appears on page 331. The heading of each column has exactly the same interpretation as the comparable heading of a stockmarket report - except for the few differences shown below.

Bonds & Rate	**Sales in $1,000**
Eastern Manufacturing Co. 5s 92	85

This is the annual rate of interest on the bond - 5%. **This is the year in which the loan will be repaid - 1992.** **This is the number of bonds sold that day.**

Example 2 Determine the cost of 5 bonds of Florida Products, Inc. that were purchased at the Low of the day.

Solution

$$\text{Quotation} = 76\tfrac{7}{8}$$
$$= 76.875$$
$$\text{Cost per bond} = \$76.8.75$$

Decimal point moved one place to right

$$= \$768.75$$

Number of bonds **Cost per bond**

$$\text{Total cost} = 5 \times \$768.75$$
$$= \$3{,}843.75$$

Example 3 Determine the last price paid for a bond of Evans Storage Battery Co. on the day prior to that of the quotations on page 331.

Explanation

Loss

$$\text{Net change} = -\ 1\tfrac{1}{8}$$
$$= -\ 1.125$$
$$= -\ \$1\odot1.25$$

Move decimal point

$$= -\ \$11.25$$

Since the value decreased $11.25 from its price of the previous day, then to determine the last price of the previous day we will have to add $11.25 to the last price of that day.

Solution

$$\text{Last price that day} = 94\tfrac{1}{2}$$
$$= 94.5$$
$$= \$94\odot5. \text{ or } \$945$$

$$\text{Net change} = 1\tfrac{1}{8}$$
$$= -\ \$11.25$$

$$\text{Last price of previous day} = \$945 + \$11.25$$
$$= \$956.25$$

Exercises A

Use the bond market quotations on page 331 when doing the exercises in B, C, and D.

Unit 2

1. Determine the actual price of the bond from each quotation.

	Quotation	Price		Quotation	Price
a.	$96\frac{1}{2}$	$______	h.	$114\frac{1}{8}$	$______
b.	$117\frac{1}{2}$	______	i.	$76\frac{3}{8}$	______
c.	$52\frac{1}{4}$	______	j.	$127\frac{3}{8}$	______
d.	$123\frac{1}{4}$	______	k.	$69\frac{5}{8}$	______
e.	$85\frac{3}{4}$	______	l.	$103\frac{5}{8}$	______
f.	$104\frac{3}{4}$	______	m.	$89\frac{7}{8}$	______
g.	$80\frac{1}{8}$	______	n.	$116\frac{7}{8}$	______

2. Determine the actual price change for each net change.

	Net Change	Price Change		Net Change	Price Change
a.	$+\frac{1}{4}$	$____	f.	$+ 1\frac{1}{2}$	$____
b.	$+\frac{3}{8}$	____	g.	$- 7$	____
c.	$-\frac{5}{8}$	____	h.	$- 2\frac{3}{4}$	____
d.	$+\frac{3}{4}$	____	i.	$+ 4\frac{1}{8}$	____
e.	$-\frac{1}{2}$	____	j.	$- 5\frac{3}{8}$	____

B

1. Determine the year in which the bonds for each company will be redeemed.

	Company	Year
a.	Equitable Light Co.	____
b.	Hotels International	____
c.	Iowa Refrigerator Co.	____
d.	General Department Stores	____
e.	Federal Mineral Co.	____

2. Determine the lowest price paid that year for these bonds.

	Company	Quotation	Price
a.	Gibraltor Power Co.	____	$____
b.	Iowa Refrigerator Co.	____	____
c.	Idaho Paper Co.	____	____
d.	Florida Products, Inc.	____	____
e.	General Department Stores	____	____

3. Determine the highest price paid that day for these bonds.

	Company	Quotation	Price
a.	Franklin Sulphur Co.	______	$______
b.	Ideal Powder Co.	______	______
c.	International Sugar Co.	______	______
d.	Household Services, Ltd.	______	______

4. Determine the last price paid that day for these bonds.

	Company	Quotation	Price
a.	Evans Storage Battery Co.	______	$______
b.	General Department Stores	______	______
c.	Illinois Telephone Co.	______	______
d.	Joy Steel Corp.	______	______

5. Determine the net change in the price of these bonds. Indicate whether the price had decreased or increased over the closing price of the previous day.

	Company	Net Change	Price Change	Decrease or Increase
a.	Falstaff Finance Corp.	______	$______	______
b.	Family Food Stores	______	______	______
c.	Interstate Transportations	______	______	______
d.	Hotels International	______	______	______
e.	Ideal Powder Co.	______	______	______

6. Determine the last price paid for these bonds on the day prior to that of the bond quotations.

Unit 2

	Company	Last Price For Day Shown	Net Change	Price Change	Last Price of Previous Day
a.	Emerson-Brody Corp.	$______	______	$______	$______
b.	Idaho Paper Co.	______	______	______	______
c.	Glenn-Holmes Corp.	______	______	______	______
d.	Gibraltor Power Co.	______	______	______	______
e.	Franklin Sulphur Co.	______	______	______	______
f.	Jonathon and Sons Inc.	______	______	______	______
g.	Hayes Products	______	______	______	______

C

1. Determine the cost for each bond purchase.

	Number Of Bonds	Bond Quotation	Price Per Bond	Total Price
a.	2	87	$______	$______
b.	3	$96\frac{1}{2}$	______	______
c.	5	$75\frac{3}{4}$	______	______
d.	5	$103\frac{5}{8}$	______	______
e.	4	$107\frac{1}{8}$	______	______
f.	4	$100\frac{7}{8}$	______	______
g.	8	$95\frac{1}{4}$	______	______

2. Determine the cost for each bond purchase.

Company	Number Of Bonds	Bond Quotation	Price Per Bond	Total Price
a. Equitable Light Co.	3	High	$______	$______
b. Family Food Stores	10	High	______	______
c. Jones & Kendall, Inc.	6	Last	______	______
d. Interstate Transportations	20	Low	______	______
e. Hayes Products	8	Low	______	______
f. Florida Products, Inc.	7	Low	______	______
g. Iowa Refrigerator Co.	16	High	______	______

D

1. Robert Perle purchased 5 bonds of Hunt-Portland Engineering Company.

a. How much would he have had to pay had he purchased them at the High of the year? ______

b. How much would he have had to pay had he purchased them to the Low of the year? ______

c. How much would he have saved had he been able to purchase them at the Low of the year rather than at the High? ______

2. Linda Turner purchased 10 bonds of Hotels International at the closing price. How much more did she pay for them than she would have had they been purchased at the closing price of the previous day? ______

Section 2 Interest Payments

The interest on bond loans is computed in exactly the same manner as we would compute the interest on any time loan. These payments are made in one of the following ways:

1. Annually—once each year
2. Semiannually—twice each year
3. Quarterly—four times each year

Many companies mail the interest payments directly to the bondholder. However, there is a variety of bonds, called *coupon bonds,* in which this is not necessary. Each of these bonds have a great number of coupons attached to it, and each of these coupons has a date printed upon it. On each of these dates the bondholder takes the coupon bearing that date to his bank. The bank gives him the interest payment in return for the coupon.

It is important that we realize that the interest on a bond is computed on the original value (par value) of the bond and not on what we may have paid for it. Although the purchase price may have been only \$837, the interest computation is based on \$1,000 because that is the amount that had been borrowed from the original lender.

In the examples and exercises that follow, any reference to company names implies those companies listed in the quotations on page 331.

Example 1 Determine the annual interest payment on an Illinois Telephone Company bond.

Explanation

Bonds & Rate
Illinois Telephone Co. $4\frac{1}{4}$s 93

Annual interest rate (arrow to $4\frac{1}{4}$)

Solution

Amount borrowed (arrow to \$1,000) **Interest rate** (arrow to $4\frac{1}{4}\%$)

Annual interest $= \$1,000 \times 4\frac{1}{4}\%$
$= \$1,000 \times 4.25\%$
$= \$1,000 \times .0425$
$= \$42.50$

Unit 2

Example 2 Fred Schaeffer owns 6 bonds of Florida Products, Inc. Determine the quarterly interest payment on these bonds.

Explanation

Interest rate → $4\frac{3}{8}$

Florida Products, Inc. $4\frac{3}{8}$s 92

Solution

Amount borrowed → \$1,000; **Interest rate** → $4\frac{3}{8}\%$

Annual interest on 1 bond $= \$1,000 \times 4\frac{3}{8}\%$
$= \$1,000 \times 4.375\%$
$= \$1,000 \times .04375$
$= \$43.75$

Number of bonds → 6; **Interest per bond** → \$43.75

Annual interest on 6 bonds $= 6 \times \$43.75$
$= \$262.50$

Fraction of year → $\frac{1}{4}$; **Annual interest** → \$262.50

Quarterly interest on 6 bonds $= \frac{1}{4} \times \$262.50$
$= \$65.63$

Exercises A

1. Rename each interest rate as a decimal.

	Interest Rate	Decimal		Interest Rate	Decimal
a.	4.80%	________	f.	$6\frac{3}{4}\%$	________
b.	5.30%	________	g.	$7\frac{1}{8}\%$	________
c.	6.25%	________	h.	$5\frac{3}{8}\%$	________
d.	$4\frac{1}{2}\%$	________	i.	$4\frac{7}{8}\%$	________
e.	$5\frac{1}{4}\%$	________	j.	$6\frac{5}{8}\%$	________

2. Determine the annual interest payment for each \$1,000 bond.

	Interest Rate	Annual Interest		Interest Rate	Annual Interest
a.	4%	\$________	f.	$5\frac{1}{2}\%$	\$________
b.	5%	________	g.	$6\frac{3}{4}\%$	________
c.	7%	________	h.	$4\frac{3}{8}\%$	________
d.	5.80%	________	i.	$5\frac{7}{8}\%$	________
e.	6.30%	________	j.	$7\frac{1}{8}\%$	________

3. Determine the semiannual interest payment for each $1,000 bond.

	Interest Rate	Annual Interest	Semiannual Interest
a.	5%	$________	$________
b.	4.20%	________	________
c.	$6\frac{1}{2}$%	________	________
d.	$7\frac{1}{4}$%	________	________

4. Determine the quarterly interest payment for each $1,000 bond.

	Interest Rate	Annual Interest	Quarterly Interest
a.	8%	$________	$________
b.	7.40%	________	________
c.	$7\frac{1}{2}$%	________	________
d.	$6\frac{3}{4}$%	________	________

5. Determine the total interest payment made to each bondholder for the period of time indicated. The par value of each bond is $1,000.

	Number Of Bonds	Interest Rate	Payments Made	Annual Interest Per Bond	Total Interest
a.	3	6.7%	Annually	$________	$________
b.	5	7.4%	Annually	________	________
c.	4	$6\frac{1}{2}$%	Semiannually	________	________
d.	2	$5\frac{3}{4}$%	Semiannually	________	________
e.	8	5%	Quarterly	________	________
f.	7	8%	Quarterly	________	________

B

1. Determine the annual interest payment for each bond.

	Company	Interest Rate	Interest
a.	Emerson-Brody Corp.	________%	$________
b.	Family Food Stores	________%	________
c.	International Sugar Co.	________%	________
d.	Jonathon and Sons, Inc.	________%	________

2. Determine the semiannual interest payment for each bond.

Company	Interest Rate	Annual Interest	Semiannual Interest
a. Eastern Manufacturing Co.	______%	$______	$______
b. Hayes Products	______%	______	______
c. Foremost Tire Co.	______%	______	______
d. Franklin Sulphur Co.	______%	______	______

3. Determine the quarterly interest payment for each bond.

Company	Interest Rate	Annual Interest	Quarterly Interest
a. Gibraltor Power Co.	______%	$______	$______
b. General Depart. Stores	______%	______	______
c. Iowa Refrigerator Co.	______%	______	______
d. Jones and Kendall, Inc.	______%	______	______

4. Determine the total interest payment that will be received for each bond investment.

	Company	Number Of Bonds	Payments Made	Interest Rate	Total Interest
a.	Falstaff Finance Corp.	4	Annually	______%	$______
b.	Harvey Grace Motors, Inc.	5	Semiannually	______%	______
c.	Gulf National, Inc.	10	Semiannually	______%	______
d.	Household Services, Ltd.	3	Semiannually	______%	______
e.	Glenn-Holmes Corp.	8	Quarterly	______%	______
f.	Joy Steel Corp.	16	Quarterly	______%	______

Section 3 Rate of Return on Bond Investments

People invest in bonds hoping they will receive a greater rate of interest on their money than the rate paid by banks. In view of this, it is important that we develop some way to determine the interest rate on a bond investment. Knowing that rate, we can compare it with the bank rate and then decide whether the investment is a wise one.

Actually, the annual interest rate (rate of return) on a bond investment is determined in exactly the same manner as the rate of return on a stock investment; only the names are changed:

"annual dividend"	becomes	"annual interest"
"cost of one share"	becomes	"cost of one bond."

Stock investment

Annual rate of return = Annual dividend ÷ Cost of one share

Bond investment

Annual rate of return = Annual interest ÷ Cost of one bond

Unit 2

As emphasized earlier, the annual interest on a bond is paid on the par value of the bond. Therefore, the annual interest itself is computed on the original $1,000 that was borrowed. However, the annual rate of return is computed on the *amount we paid for the bond*.

In the examples and exercises that follow, any reference to company names implies the companies listed in the quotations on page 331. Also, the par value of every bond is $1,000 and the interest rates will be computed to the nearest tenth of a percent.

Example 1 Determine the rate of return on a General Department Stores bond that was purchased at the Low of the day of the quotations.

Explanation

Bonds & Rate	**Low**
General Department Stores $5\frac{1}{2}$s 2011	$84\frac{1}{4}$
Interest rate	**Cost of bond – $842.50**

Solution

$$\text{Annual interest} = \$1{,}000 \times 5\tfrac{1}{2}\% = \$1{,}000 \times .055 = \$55$$

$$\text{Annual rate of return} = \text{Annual interest} \div \text{Cost of one bond}$$
$$= \$55 \div \$842.50$$
$$= .065$$
$$= 6.5\%$$

Example 2 Julia Carslaw received a quarterly interest payment of $27. The quotation of the bond at the time she purchased it was 106 3/8. Determine the rate of return on this investment.

Solution

$$\text{Quarterly interest} = \$27$$

$$\text{Annual interest} = 4 \times \$27 = \$108$$

$$\text{Cost of bond} = 106\tfrac{3}{8} = 106.375 = \$1{,}063.75$$

$$\text{Rate of return} = \text{Annual interest} \div \text{Cost of one bond}$$
$$= \$108 \div \$1{,}063.75$$
$$= .102$$
$$= 10.2\%$$

Exercises

1. Determine the annual rate of return on each bond purchase.

	Annual Interest	Cost	Rate Of Return
a.	$53	$ 950	________%
b.	65	870	________%
c.	72	1,075	________%
d.	68	1,052	________%

2. Determine the annual rate of return on each $1,000 bond.

	Interest Rate	Annual Interest	Quotation	Cost	Rate Of Return
a.	5%	$________	95	$________	________%
b.	8%	________	104	________	________%
c.	6.4%	________	$89\frac{1}{2}$	________	________%
d.	$5\frac{1}{2}$%	________	$92\frac{1}{2}$	________	________%
e.	$7\frac{1}{4}$%	________	$106\frac{1}{2}$	________	________%

3. Determine the annual rate of return on each bond purchase.

	Interest	Annual Interest	Quotation	Cost	Rate Of Return
a.	$25—semiannually	$________	84	$________	________%
b.	$32—semiannually	________	96	________	________%
c.	$36—semiannually	________	$92\frac{1}{2}$	________	________%
d.	$13—quarterly	________	72	________	________%
e.	$33—quarterly	________	$107\frac{1}{2}$	________	________%

4. Determine the annual rate of return on each bond purchase.

	Company	Interest Rate	Annual Interest	Quotation	Cost	Rate Of Return
a.	Eastern Manufacturing	______%	$______	High	$______	______%
b.	Harvey Grace Motors, Inc.	______%	______	Last	______	______%
c.	Emerson-Brody Corp.	______%	______	Low	______	______%
d.	Family Food Stores	______%	______	High	______	______%
e.	Foremost Tire Co.	______%	______	Last	______	______%
f.	Franklin Sulphur Co.	______%	______	High	______	______%

Section 4 Government Bonds

Federal, state, and local governments follow the pattern of large companies; they, too, raise money by selling bonds. City governments often need a great deal of money to build schools, or roads, or municipal buildings, and the like. If they obtain the money through taxes during a period of one or two years, the taxpayers would be unable to pay that large a sum. Therefore, they sell bonds which they can repay after a long period of time.

An important advantage in purchasing municipal bonds is the fact that no federal income tax has to be paid on the interest earned by these bonds. For many people this proves to be quite a savings. Investors whose income is somewhat above that of the average person often find that at least half the interest their money earns in normal investments must be passed on to the government in payment of taxes.

Federal bonds, too, have an advantage over corporation bonds. The owner of a federal bond pays no tax on the bond until the bond is redeemed — that is until his money plus the accumulated interest is returned to him. Often the bondholder is retired at that point and may therefore have a much smaller income that is taxable.

Although the federal government issues many different bonds, the two most commonly purchased by private individuals are the Series HH and Series EE bonds. Just as interest on municipal bonds is exempt from federal tax, so too is interest on federal bonds exempt from state income tax.

Series HH Bonds

Denominations: \$500, \$1,000, \$5,000, \$10,000
Annual interest rate: 6%
Interest payments: semiannually. Mature in 10 years.

Series EE Bonds

Denominations: \$50, \$75, \$100, \$200, \$500,
\$1,000, \$5,000, \$10,000

Annual interest rate if redeemed at end of:
1 Year–4.53% 3 Years–4.95%
2 Years–4.76% 4 Years–5.14%

5 Years to Maturity–6.00%
Mature in 11 years 9 months.

The denominations shown above for the Series EE bonds are actually the values of these bonds at maturity. The amount we pay for these bonds at the time of purchase is exactly

one-half of the maturity value.

Example 1 Determine the cost of a $500 Series EE bond.

Solution

$$\begin{aligned} \text{Cost} &= \text{one-half of maturity value} \\ &= \tfrac{1}{2} \times \$500 \\ &= \$250 \end{aligned}$$

The Series EE bond can be redeemed at any time after the first six months from the date of purchase. The table below shows the redemption value of a $50 bond at the end of annual periods.

Redemption Value - $50 Series EE Bond			
If redeemed at end of:	**Redemption Value**	**If redeemed at end of:**	**Redemption Value**
1 Year	$26.15	7 Years	$37.81
2 Years	27.47	8 Years	40.12
3 Years	28.95	9 Years	42.56
4 Years	30.63	10 Years	45.15
5 Years	33.60	11 Years	47.90
6 Years	35.64	11 Years 9 Months	50.00

Example 2 Determine the amount of money a person will receive for redeeming a $200 Series EE bond at the end of 4 years.

Explanation

If redeemed at end of:	**Redemption Value**
4 Years →	30.63 ← **Value of $50 Bond**

Solution

Number of $50 amounts in $200 = $200 ÷ $50
= 4

Number of $50 amounts → 4; **Value of $50 bond** → $30.63

Redemption value of $200 bond = 4 × $30.63
= $122.52

Unit 2

Exercises

A

1. Determine the cost of these Series EE bonds

	Bond	Cost		Bond	Cost
a.	$ 200	$______	c.	$5,000	$______
b.	1,000	______	d.	75	______

2. Determine the redemption value of these $50 Series EE bonds.

	Years Held	Redemption Value		Years Held	Redemption Value
a.	3	$______	c.	10	$______
b.	7	______	d.	5	______

3. Determine the redemption value of these Series EE bonds.

	Bond	Years Held	Redemption Value For $50 Bond	Redemption Value
a.	$ 100	4	$______	$______
b.	100	7	______	______
c.	500	9	______	______
d.	200	6	______	______
e.	1,000	8	______	______
f.	5,000	10	______	______

4. Determine the interest received at the time these Series EE bonds were redeemed.

	Bond	Years Held	Redemption Value For $50 Bond	Redemption Value	Interest
a.	$ 100	3	$______	$______	$______
b.	200	5	______	______	______
c.	500	4	______	______	______
d.	1,000	7	______	______	______
e.	10,000	6	______	______	______

B

1. Determine the interest paid annually for each municipal bond.

	Par Value	Annual Interest Rate	Interest
a.	$ 500	4%	$______
b.	500	4.8%	______
c.	1,000	5.2%	______
d.	5,000	$5\frac{1}{2}\%$	______
e.	10,000	$6\frac{1}{2}\%$	______

2. Determine the semiannual interest payments that each bondholder will receive.

	Value	Of Bonds	Interest Rate	Annual Interest Per Bond	Total Annual Interest	Total Semiannual Interest
a.	$ 500	2	5%	$______	$______	$______
b.	500	4	5.3%	______	______	______
c.	1,000	5	6.2%	______	______	______
d.	5,000	7	$6\frac{1}{2}$%	______	______	______
e.	10,000	8	$7\frac{1}{4}$%	______	______	______

Review of Chapter 9

When doing the computation for the exercises that follow it will be necessary to refer to the tables in this chapter.

1. Determine the total cost for each stock purchase.

	Quotation	Dollar Value	Number Of Shares	Total Cost
a.	$16\frac{1}{2}$	$______	20	$______
b.	$23\frac{1}{4}$	______	70	______
c.	$9\frac{1}{8}$	______	50	______
d.	$34\frac{5}{8}$	______	100	______

2. Use the stock-market quotations on page 309 to answer each of the following questions.

a. What was the lowest price paid that year prior to the day shown for a share of Bullard and Sons stock? ______

b. What was the lowest price paid that day for a share of Carpenter Corp. stock? ______

c. What was the last price paid that day for a share of Dow Aircraft stock? ______

d. How much more did the last buyer of that day have to pay for each share of Colonial, Inc., stock than the last buyer of the previous day? ______

Review

3. What was the last price paid per share by the buyer of American Bakers preferred stock on the day prior to the one shown in the quotations? ______

4. If the broker's fees are not considered, how much will each of the following purchases cost?

a. The purchase of 100 shares of stock quoted at 17 3/4.

______

b. The purchase of 300 shares of Bliss Petroleum stock purchased at the High of the day shown in the stock-market quotations ________

5. If broker's fees are not considered, how much profit would Joyce Blake have made if she purchased 100 shares of Diamond Airlines at the Low of the year and sold them at the High of the year shown in the quotations. ________

6. Approximately how much would Carla Sims have to pay in fees when she purchased stock that cost her $2,350? ________

7. Thomas Paterno owns 300 shares of Allen, Inc., stock. If he owned them the year prior to the quotations shown, how much would he have received in dividends from the company? ________

8. What annual dividend payment will the holder of 200 shares of a 5 1/2% preferred stock receive if the par value of the stock is $100? ________

9. Joe Skinner owns 100 shares of 4.40% preferred stock of the Reed Chain Corporation. The par value of the stock is $50, but Joe paid only $34 3/4 for each share. How much will he receive at each quarterly dividend period? ________

10. Determine the rate of return to the nearest tenth of a percent on each of the following stock purchases:

a. A stock that is paying an annual dividend of $3.50 and was purchased for $48. ________

b. A share of stock of Cenco Chemicals that was purchased at the High of the day shown in the quotations. ________

c. A share of 4 1/2% preferred stock that was purchased at 44 3/4, but has a par value of $50. ________

11. Change each of the following bond quotations to actual prices.

	Quotation	Price
a.	$87\frac{1}{2}$	$________
b.	$124\frac{5}{8}$	________

12. Use the bond-market quotations on page 331 to answer each of the following questions.

a. What was the lowest price paid that day for a bond of the Emerson-Brody Corporation?

Quotation:________ Price: $________

b. What was the lowest price paid that year to the day shown for a bond of the Gibraltar Power Company?

Quotation:________ Price: $________

c. How much less did the last buyer of a bond of the Joy Steel Corporation have to pay for it the day shown than he would have had to pay for it had it been purchased the previous day? Amount less: $________

13. Robert Kooms purchased 5 bonds of the Franklin Sulphur Co., at the closing price shown in the quotations. How much did he have to pay for these bonds? ________

14.

a. What annual interest payment will the holder of two 5 1/4% bonds receive if the par value of the bonds is $1,000? ________

b. What semiannual interest payment will the holder of six Iowa Refrigerator Co. bonds receive? (See the bond-market quotations.) ________

15.

a. To the nearest tenth of a percent, what rate of return will an investor receive on a bond for which he paid $940. and which gives an annual interest of $62? ________

b. To the nearest tenth of a percent, what rate of return will the purchaser receive on a bond of the Equitable Light Co. that was purchased at the low price of the day shown in the quotations? ________

Review

16.

a. What is the cost of a $50 Series EE bond? ________

b. What is the cost of a $200 Series EE bond? ________

17.

a. How much will a person receive if he returns a $200 Series EE bond after 4 years? ________

b. How much will a person receive if he returns a $75 Series EE bond after 9 years? ________

Chapter 10

Insurance

In the last few chapters we have been concerned with the ways in which our earnings can earn yet more money for us. In this chapter our concern turns toward an examination of the ways in which we can protect the money we accumulate over the years – not only for ourself but also for our family in the event of our death. The topic we plan to investigate is that of insurance.

People insure either themselves or their property because they are afraid their savings will not carry them through a sudden and unexpected loss. As an example, let us consider that 100 of us own motorcycles valued at $2,000 each. From past experience we know that 1 of these motorcycles will be completely destroyed each year.

This implies that each year one of us will have to withstand a loss of $2,000. Since this is far more money than any of us can afford to lose, each of us contributes $20 to a fund to cover the yearly cost of $2,000.

$$100 \times \$20 = \$2{,}000$$

Were we certain that our motorcycle is not the one that will be destroyed, then there would be no need to pay even the small sum of $20 to protect ourself. Unfortunately though, since none of us can predict what tomorrow will bring, each of us pays the $20 for the peace of mind that it brings.

Life Insurance Unit 1

Section 1 Mortality Tables

For many years, life insurance companies have been collecting information that enables them to predict the number of people that will die each year. Also, they are able to predict what our life

MORTALITY TABLE

Commissioners 1958 Standard Ordinary (1950-1954)

Age	Deaths Per 1,000	Expection of Life (Years)	Age	Deaths Per 1,000	Expection of Life (Years)	Age	Deaths Per 1,000	Expection of Life (Years)
0	7.08	68.30	35	2.51	36.69	70	49.79	10.12
1	1.76	67.78	36	2.64	35.78	71	54.15	9.63
2	1.52	66.90	37	2.80	34.88	72	58.65	9.15
3	1.46	66.00	38	3.01	33.97	73	63.26	8.69
4	1.40	65.10	39	3.25	33.07	74	68.12	8.24
5	1.35	64.19	40	3.53	32.18	75	73.37	7.81
6	1.30	63.27	41	3.84	31.29	76	79.18	7.39
7	1.26	62.35	42	4.17	30.41	77	85.70	6.98
8	1.23	61.43	43	4.53	29.54	78	93.06	6.59
9	1.21	60.51	44	4.92	28.67	79	101.19	6.21
10	1.21	59.58	45	5.35	27.81	80	109.98	5.85
11	1.23	58.65	46	5.83	26.95	81	119.35	5.51
12	1.26	57.72	47	6.36	26.11	82	129.17	5.19
13	1.32	56.80	48	6.95	25.27	83	139.38	4.89
14	1.39	55.87	49	7.60	24.45	84	150.01	4.60
15	1.46	54.95	50	8.32	23.63	85	161.14	4.32
16	1.54	54.03	51	9.11	22.82	86	172.82	4.06
17	1.62	53.11	52	9.96	22.03	87	185.13	3.80
18	1.69	52.19	53	10.89	21.25	88	198.25	3.55
19	1.74	51.28	54	11.90	20.47	89	212.46	3.31
20	1.79	50.37	55	13.00	19.71	90	228.14	3.06
21	1.83	49.46	56	14.21	18.97	91	245.77	2.82
22	1.86	48.55	57	15.54	18.23	92	265.93	2.58
23	1.89	47.64	58	17.00	17.51	93	289.30	2.33
24	1.91	46.73	59	18.59	16.81	94	316.66	2.07
25	1.93	45.82	60	20.34	16.12	95	351.24	1.80
26	1.96	44.90	61	22.24	15.44	96	400.56	1.51
27	1.99	43.99	62	24.31	14.78	97	488.42	1.18
28	2.03	43.08	63	26.57	14.14	98	668.15	.83
29	2.08	42.16	64	29.04	13.51	99	1000.00	.50
30	2.13	41.25	65	31.75	12.90			
31	2.19	40.34	66	34.74	12.31			
32	2.25	39.43	67	38.04	11.73			
33	2.32	38.51	68	41.68	11.17			
34	2.40	37.60	69	45.61	10.64			

expectancy is at any particular age. Thus, from their records they know that the *average* 15-year old person will live another 55 years. This simply means that half of the 15-year olds will live less than 55 years while the other half will live more than 55 years.

The tables containing these records are called *mortality tables*. Although the one on page 352 is not the most current table, it is the one used by many life insurance companies. It is interesting to note that when the first mortality information was collected (1843-1858) a person had a life expectancy at birth of 41 years The table on page 352 gives the life expectancy at birth as 68 years. And it is very likely that in the years since the information for this table was gathered, the life expectancy has increased even more.

Example 1 Determine the age to which the average 18-year old person can expect to live.

Explanation

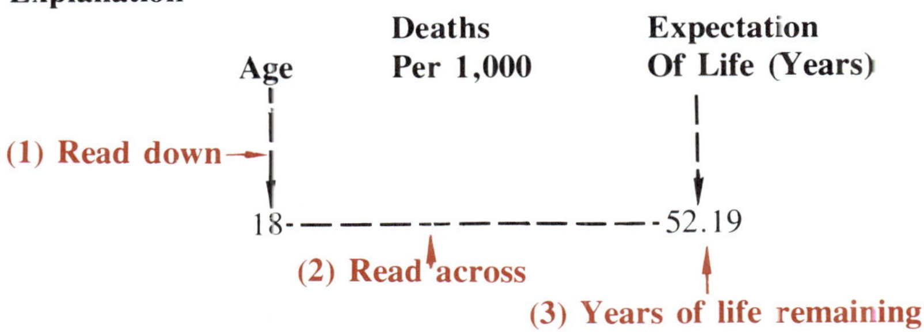

Solution

Expectation of life at 18 = 52.19 years

Present age → 18, **Future years** → 52.19

Age a person of 18 will reach = 18 + 52.19
= 70.19

Unit 1

Example 2 Determine the number of people per 1,000 at age 30 who will die before reaching the age of 31.

Explanation

Age	Deaths Per 1,000	Expectation Of Life (Years)
30 ⟶	2.13	

2.13 ← **Number per 1,000 who will die**

Solution

Number per 1,000 who will die = 2.13

Based on Example 2 above we know that 2.13 persons per 1,000 will die during the year after their 30th birthday. An insurance company, however, may have as many as 80,000 people at age 30 who are insured. Their interest is in knowing how many of those 80,000 people will die that year. To determine this they will have to find the number of 1,000's in 80,000.

To show how this is done, we will establish a pattern as we have in the past.

Number of 1,000's in 2,000 is 2:	$2{,}000 \div 1{,}000 = 2$
Number of 1,000's in 3,000 is 3:	$3{,}000 \div 1{,}000 = 3$
Number of 1,000's in 4,000 is 4:	$4{,}000 \div 1{,}000 = 4$

Therefore, to determine the number of 1,000's in a number, divide the number by 1,000.

Example 3 Determine the number of people in a group of 70,000 at age 20 who will die before reaching the age of 21.

Solution

Number per 1,000 who will die = 1.79

Number of 1,000's in 70,000 = 70,000 ÷ 1,000
= 70

Number of 1,000's (↓ 70) **Deaths per 1,000** (↓ 1.79)

Total number who will die = 70 × 1.79
= 125.3 or 125

Exercises

A

1. Determine the life expectancy of a person for each age.

	Age	Life Expectancy (Years)
a.	10	________
b.	14	________
c.	17	________
d.	28	________

	Age	Life Expectancy (Years)
e.	34	________
f.	56	________
g.	65	________
h.	93	________

2. Determine the age to which each person can expect to live. Their present age is shown below.

	Present Age	Life Expectancy (Years)	Probable Age At Death
a.	1	________	________
b.	5	________	________
c.	14	________	________
d.	21	________	________
e.	43	________	________
f.	57	________	________
g.	74	________	________
h.	86	________	________

3. Determine the number of people per 1,000 at each age who will die before reaching their next birthday.

	Age	Deaths Per 1,000
a.	2	________
b.	7	________
c.	16	________
d.	21	________

	Age	Deaths Per 1,000
e.	46	________
f.	59	________
g.	68	________
h.	75	________

4. Determine the number of people in each group below who will die before reaching their next birthday.

	Age	Deaths Per 1,000	Number Of People	Total Number Who Die
a.	3	________	100,000	________
b.	6	________	400,000	________
c.	17	________	500,000	________
d.	45	________	70,000	________
e.	70	________	60,000	________

Unit 1

5. Determine the number of people in each group below who will live to reach their next birthday.

	Age	Deaths Per 1,000	Number Of People	Total Number Who Die	Total Number Who Live
a.	4	________	300,000	________	________
b.	12	________	600,000	________	________
c.	20	________	500,000	________	________
d.	80	________	20,000	________	________

B

1. In 1850 the number of deaths per 1,000 during the first year of life was 154.70.

a. Of 200,000 babies born in 1850, how many of them probably died before reaching the age of 1? ________

b. According to the life expectancy table on page 352, of 200,000 babies born this year, how many of them will probably die before reaching the age of 1? ________

2. The life expectancy of a 15-year old in 1850 was 45.50 years. How many more years can a 15-year old be expected to live at the present time? ________

Section 2 Term Insurance

Knowing that death will eventually come to each of us, we try to arrange for the financial support of our dependents when we can no longer provide it. Most of us will turn to some form of life insurance for this protection.

Basically, life insurance coverage falls into four major groups. Of these, the least expensive is *term insurance*. This type of insurance is usually purchased for either a 5-year period, 10-year period, or 20-year period.

As an illustration, let us consider a 5-year term policy whose *face value* is $10,000. The face value or *face amount* of a policy is the amount for which we are insured. The distinguishing features of this policy are as follows:

1. In the event we live for the entire 5-year period:
   - **a.** We make payments – pay *premiums* – to the insurance company for the entire period.
   - **b.** We are no longer insured at the end of the period.
   - **c.** We receive no money from the insurance company.

2. In the event we die during the 5-year period:
   - **a.** The person whom we have named in the policy – the *beneficiary* – will receive the face value of the policy, $10,000.
   - **b.** No further premiums are paid to the insurance company after our death.

One of the selling points of most term insurance contracts is the fact that no physical examination is required should we want to renew the insurance. This is particularly important to people who are seriously ill at the time their insurance expires.

The rate table on page 358 shows the typical cost of term insurance for people at various ages. Notice that as the face value of the policy increases, the cost per $1,000 decreases.

These amounts increase

Male	Female	$10,000 - $19,999	$20,000 - $39,999	$40,000 and over
20	23	$3.86	$3.36	$3.11

These amounts decrease

Of interest also is the fact that the cost of insurance for a man is more than that for a woman at the same age. This is because the average woman has a life expectancy that is three years more than the average man. In view of this, the average woman is more likely to survive the five-year period.

ANNUAL PREMIUM PER $1,000 OF FIVE-YEAR TERM INSURANCE				
Age		Face Value		
Male	Female	$10,000 - $19,999	$20,000 - $39,999	$40,000 and over
20	23	$ 3.86	$ 3.36	$ 3.11
25	28	4.13	3.65	3.38
30	33	4.65	4.15	3.90
35	38	5.55	5.05	4.80
40	43	7.29	6.79	6.54
45	48	9.99	9.49	9.24
50	53	14.22	13.72	13.27
55	58	20.71	20.21	19.96
60	63	32.86	32.36	32.11

Example 1 Determine the annual premium on a $25,000 5-year term policy taken out at age 33 by a woman.

Explanation

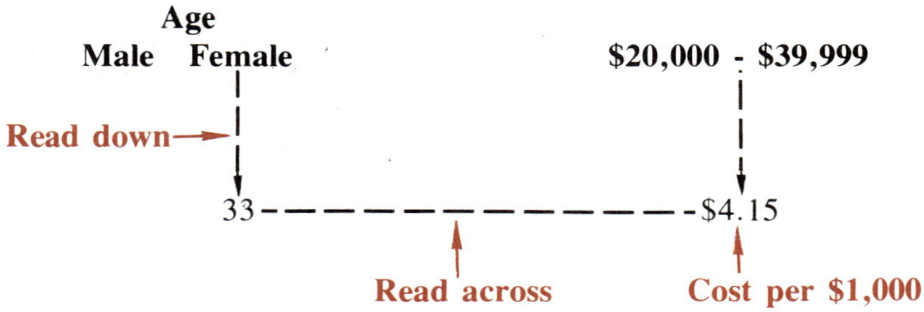

Solution

Annual cost per $1,000 = $4.15

Number of $1,000's in $25,000 = $25,000 ÷ $1,000
= 25

Number of $1,000's — Cost per $1,000

Annual premium = 25 × $4.15
= $103.75

During the early years of married life people have the need for a great deal of financial protection. However, as we become older, the need for this protection is reduced with each succeeding year. Unfortunately, although the need for protection is greatest during the early years, the earnings are the least. In view of this, a form of term insurance called *decreasing-term insurance* was designed to cover this situation. As each year passes, the amount for which the person is insured grows less. As an example, let us say that

	the coverage the first year is	$30,000;
then	the coverage the second year will probably be	$27,750;
and	the coverage the third year will probably be	$25,350;

and so on. Since the amount for which we are insured decreases each year, the cost of the insurance is less than that of a normal $30,000 term policy.

ANNUAL PREMIUM FOR $1,000 INITIAL COVERAGE OF DECREASING TERM INSURANCE—MINIMUM COVERAGE: $20,000

Age		Term	
Male	Female	10-Year	20-Year
20	23	$ 2.72	$ 2.62
25	28	2.85	2.78
30	33	3.14	3.03
35	38	3.77	3.89
40	43	4.96	5.38
45	48	6.94	7.92
50	53	10.20	11.90
55	58	15.36	18.06
60	63	23.45	

DECREASING TERM INSURANCE
COVERAGE PER $1,000 OF INITIAL INSURANCE

10-Year Term

Year	Coverage
1	$1,000
2	925
3	845
4	760
5	670
6	575
7	473
8	365
9	251
10	129

20-Year Term

Year	Coverage	Year	Coverage
1	$1,000	11	$646
2	974	12	597
3	945	13	546
4	915	14	491
5	883	15	433
6	849	16	371
7	813	17	306
8	775	18	236
9	735	19	162
10	672	20	84

Example 2 Tom Peters bought a $24,000 20-year decreasing term insurance policy. What is his coverage during the 6th year after the date of the policy?

Explanation

Decreasing Term Insurance
20-Year Term

Year → 6 → 849 ← Coverage

Coverage per $1,000 of initial insurance (849)

Solution

Coverage per $1,000 = $849

Number of $1,000's in $24,000 = $24,000 ÷ $1,000
= 24

Number of $1,000's (24) **Coverage per $1,000** ($849)

Total coverage = 24 × $849
= $20,376

Exercises A

1. Determine the premium per $1,000 for each 5-year term insurance policy.

	Age	Sex	Face Value	Premium Per $1,000
a.	30	Male	$15,000	$________
b.	43	Female	18,000	________
c.	38	Female	50,000	________
d.	50	Male	35,000	________

2. Determine the premium per $1,000 for each decreasing term insurance policy.

	Age	Sex	Term	Premium Per $1,000
a.	43	Female	10-year	$________
b.	20	Male	20-year	________
c.	55	Male	10-year	________
d.	28	Female	20-year	________

3. Determine the coverage per $1,000 of initial insurance during the year shown below for each decreasing term insurance policy.

	Year	Term	Coverage Per $1,000
a.	1st	10-year	$________
b.	5th	20-year	________
c.	14th	20-year	________
d.	20th	20-year	________
e.	7th	10-year	________

B

1. Determine the premium for each 5-year term insurance policy.

	Age	Sex	Face Value	Premium Per $1,000	Premium
a.	23	Female	$12,000	$________	$________
b.	38	Female	35,000	________	________
c.	25	Male	60,000	________	________
d.	45	Male	17,000	________	________
e.	28	Female	54,000	________	________

2. Determine the premium for each decreasing term insurance policy.

	Age	Sex	Term	Initial Insurance	Premium Per $1,000	Premium
a.	35	M	10-year	$40,000	$________	$________
b.	55	M	10-year	25,000	________	________
c.	28	F	20-year	43,000	________	________
d.	58	F	10-year	75,000	________	________

3. Determine the coverage during the year shown below for each decreasing term insurance policy.

	Year	Term	Initial Insurance	Coverage Per $1,000	Total Coverage
a.	2nd	10-year	$20,000	$________	$________
b.	8th	10-year	30,000	________	________
c.	8th	20-year	40,000	________	________
d.	15th	20-year	23,000	________	________
e.	4th	10-year	38,000	________	________
f.	18th	20-year	55,000	________	________

4. Determine the coverage for the first three years for each 10-year decreasing term insurance policy.

	Initial Insurance	1st Year Coverage	2nd Year Coverage	3rd Year Coverage
a.	$21,000	$________	$________	$________
b.	34,000	________	________	________
c.	52,000	________	________	________

C

1. Ronald Baker purchased a $25,000 5-year term insurance policy.

a. How much will his beneficiary receive if he dies two years after he purchased the policy? ________

b. How much will his beneficiary receive if he dies one day after he pays his first premium? ________

c. How much will his beneficiary receive if he dies six years after he pays his first premium? ________

2. Debra Bigelow purchased a decreasing term insurance policy having an initial coverage of $30,000.

a. What is her coverage during the 5th year if the policy is for a period of 10 years? ________

b. What is her coverage during the 5th year if the policy is for a period of 20 years? ________

c. How much more coverage will she have during the 5th year on the 20-year policy than on the 10-year policy? ________

3. William Kent purchased a $40,000 20-year decreasing term policy at age 50. He died during the 19th year of the policy.

a. What were his annual premiums? ________

b. How much did he pay the insurance company over the period of 19 years? ________

c. How much did his beneficiary receive at the time of his death? ________

d. How much more did he pay the insurance company than his beneficiary received? ________

Section 3 Ordinary-Life, Limited-Payment Life, and Endowment Insurance

In Part 1 we examined the first of the four basic varieties of life insurance. At this time we are going to take a look at the remaining three. The distinguishing features of each of these policies are listed below.

1. Ordinary Life or Whole Life Insurance

Premiums paid	At time of death
↓	↓
Entire life	Beneficiary receives face value of policy.

2. Limited Payment Life Insurance

Premiums paid	At time of death whether during payment period or afterward
↓	↓
A fixed period of time or until death – whichever comes first.	Beneficiary receives face value of policy.

If premiums are paid for:	The policy is called:
20 years	20-Payment Life Insurance
30 years	30-Payment Life Insurance

3. Endowment Life Insurance

Premiums paid	If death occurs during payment period	If death does not occur during payment period
↓	↓	↓
A fixed period of time or until death – whichever comes first.	*Beneficiary* receives face value of policy.	*Insured* receives face value at end of period.

If premiums are paid for:	The policy is called:
20 years	20-Year Endowment Insurance
30 years	30-Year Endowment Insurance

Unit 1

ANNUAL LIFE INSURANCE PREMIUM RATES PER $1,000

Age		Ordinary Life		Life Paid Up At Age 65		20-Payment Life		Endowment At Age 65		20-Year Endowment	
Male	Female	$5,000-$9,999	$10,000-$19,999	$5,000-$9,999	$10,000-$19,999	$5,000-$9,999	$10,000-$19,999	$5,000-$9,999	$10,000-$19,999	$5,000-$9,999	$10,000-$19,999
20	23	$ 15.07	$ 14.32	$15.98	$15.23	$ 22.90	$ 22.15	$ 18.74	$ 17.99	$ 46.70	$ 45.95
25	28	17.08	16.33	18.41	17.66	25.43	24.68	21.85	21.10	46.93	46.18
30	33	19.63	18.88	21.65	20.90	28.42	27.67	26.00	25.25	47.36	46.61
35	38	22.94	22.19	26.09	25.34	32.02	31.27	31.71	30.96	48.19	47.44
40	43	27.25	26.50	32.39	31.64	36.40	35.65	39.88	39.13	49.66	48.91
45	48	32.85	32.10	41.72	40.97	41.72	40.97	52.01	51.26	52.01	51.26
50	53	40.23	39.48	57.21	56.46	48.33	47.58	70.72	69.97	55.76	55.01
55	58	49.93	49.18	86.46	85.71	56.64	55.89	109.55	108.80	61.64	60.89
60	63	62.47	61.72			67.60	66.85			70.59	69.84
65	68	79.43	78.68			82.77	82.02			84.27	83.52
70	73	102.25	101.50			103.94	103.19			104.53	103.78

Different life insurance companies sell slight variations of these policies. The more important of these are the following two.

1. **Whole Life Paid Up At 65** – This policy is exactly the same as the normal whole life policy except that we pay no premiums after reaching the age of 65. Our insurance, however, continues beyond that age. There are other kinds of policies such as;

   Whole Life Paid Up At 60,
   Whole Life Paid Up At 55.

2. **Endowment at Age 65** – Rather than the fixed period being 20 or 30 years as with the normal endowment policies, the period with this coverage ends at age 65 no matter what our age was at the time of purchase.

The table on page 364 gives a sample of the rates used by a life insurance company.

Example 1 Determine the annual premium on a $15,000 20-year endowment policy purchased by Fred Ellis at the age of 35.

Explanation

Age	**20-Year Endowment**
Male	**$10,000-$19,999**
35 →	47.44

Annual cost per $1,000 (47.44)

Solution

Annual cost per $1,000 = $47.44
Number of $1,000's in $15,000 = $15,000 ÷ $1,000
= 15

Number of $1,000's (15) — **Cost per $1,000** ($47.44)

Annual premium = 15 × $47.44
= $711.60

Unit 1

Many of us find it rather difficult to pay out $711.60 in a lump sum each year as Fred Ellis had to do in Example 1. In view of this, life insurance companies permit payments to be made either semiannually, quarterly, or monthly. Frequent payments, though, require the need for additional employees to record these payments. Therefore, the insurance companies charge us an additional fee to pay the salaries of these employees.

The table below is used to compute the premium if it is paid more than once each year.

Period	Premium
Semiannually	51% of annual premium
Quarterly	26% of annual premium
Monthly	9% of annual premium

Example 2 Gracie Hammond purchased a $15,000 Ordinary Life insurance policy at age 28. If premiums are paid monthly, how large will each payment be?

Solution

$$\text{Annual cost per } \$1{,}000 = \$16.33$$

$$\text{Number of } \$1{,}000\text{'s in } \$15{,}000 = \$15{,}000 \div \$1{,}000 = 15$$

$$\text{Annual premium} = 15 \times \$16.33 = \$244.95$$

$$\text{Monthly premium} = 9\% \text{ of Annual premium} = .09 \times \$244.95 = \$22.05$$

Exercises A

1. Determine the annual cost per $1,000 for each life insurance policy.

	Sex	Age		Face Value	Cost Per $1,000
a.	F	43	20-Year Endowment	$ 8,000	$______
b.	F	28	Ordinary Life	6,000	______
c.	M	50	Life-Paid-Up-At-65	12,000	______
d.	F	58	20-Payment Life	7,000	______
e.	M	35	Endowment-At-65	15,000	______

2. Determine the semiannual premium for each annual premium.

	Annual Premium	Semiannual Premium		Annual Premium	Semiannual Premium
a.	$200	$______	d.	$125.50	$______
b.	168	______	e.	146.50	______
c.	209	______	f.	269.85	______

3. Determine the quarterly premium for each annual premium.

	Annual Premium	Quarterly Premium		Annual Premium	Quarterly Premium
a.	$300	$______	d.	$240.07	$______
b.	180	______	e.	123.50	______
c.	235	______	f.	237.62	______

4. Determine the monthly premium for each annual premium.

	Annual Premium	Monthly Premium		Annual Premium	Monthly Premium
a.	$560	$______	c.	$326.40	$______
b.	475	______	d.	429.72	______

B

1. Determine the annual premium for each life insurance policy.

	Sex	Age	Policy	Face Value	Cost Per $1,000	Total Premium
a.	M	45	20-Year Endowment	$ 7,500	$______	$______
b.	M	20	Life-Paid-Up-At-65	12,500	______	______
c.	M	35	Endowment-At-65	18,500	______	______

Unit 1

2. Determine the semiannual premium for each life insurance policy.

	Sex	Age	Policy	Face Value	Cost Per $1,000	Annual Premium	Semiannual Premium
a.	F	23	Ordinary Life	$10,000	$______	$______	$______
b.	M	35	Life-Paid-Up-At-65	8,000	______	______	______
c.	F	43	20-Year Endowment	12,000	______	______	______
d.	M	50	Endowment-At-65	16,000	______	______	______

3. Determine the quarterly premium for each life insurance policy.

	Sex	Age	Policy	Face Value	Cost Per $1,000	Annual Premium	Quarterly Premium
a.	F	63	Ordinary Life	$10,000	$____	$____	$____
b.	M	55	Endowment-At-65	5,000	____	____	____
c.	M	45	Life-Paid-Up-At-65	14,000	____	____	____
d.	F	23	20-Year Endowment	17,000	____	____	____

4. Determine the monthly premium for each life insurance policy.

	Sex	Age	Policy	Face Value	Cost Per $1,000	Annual Premium	Monthly Premium
a.	F	48	20-Payment Life	$ 8,000	$____	$____	$____
b.	F	33	20-Year Endowment	12,000	____	____	____
c.	M	30	Endowment-At-65	15,000	____	____	____
d.	F	53	Ordinary Life	11,000	____	____	____

C

1. James Hayden purchased a $15,000 20-payment life insurance policy at age 25.

a. If he paid his premiums annually, what would each payment be? ________

b. If he paid his premiums quarterly, what would each payment be? ________

c. How much could he save each year by making annual payments rather than quarterly payments? ________

2. At age 23, Mary Parker purchased a $10,000 20-year-endowment policy.

a. What annual premium will she have to pay? ________

b. If she lives the entire 20 years, what will be the total of the premiums she will have paid? ________

c. How much will she receive from the company at the end of the 20-year period? ________

d. How much more will she receive from the company than she paid in premiums? ________

3. George Chapman purchased a $5,000 ordinary-life policy at age 50.

a. What annual premium will he have to pay? ________

b. According to the mortality table on page 352, indicate, to the nearest year, how many years the insurance company expects George to live. ________

c. If he should live exactly the number of years he is supposed to, what will be the total of the premiums he will have paid the insurance company? ________

d. How much more (or less) will he have paid the company than his beneficiary will receive at the time of his death? ________

D

1. Why should the rates at 45 be exactly the same for a 20-payment life policy as they are for a life-paid-up-at-65 policy? ________

2.

a. Why should the rates for an endowment-at-65 policy be less than the rates for a 20-year-endowment when a person is under 45? ______________________

b. Why should the rates for an endowment-at-65 policy be more than the rates for a 20-year-endowment when a person is over 45? ______________________

3.

a. Why should the rates for a life-paid-up-at-65 policy be less than the rates for a 20-payment life policy when the person is under 45? ______________________

b. Why should the rates for a life-paid-up-at-65 policy be more than the rates for a 20-payment-life policy when a person is over 45? ______________________

4. Why should the rates on a 20-year-endowment policy be more than the rates on a 20-payment-life policy taken out at the same age? ______________________

____

____

Unit 2 Disability Insurance

The same concern of providing for our dependents in the event of our death, drives us to find some sort of financial protection should we be disabled either through sickness or accident. Coverage that is specifically designed for these situations is called *disability insurance*.

The cost of this insurance depends upon four factors:

1. How much money do we want the insurance company to pay us each month? The more we want, the greater the cost of the insurance. However, the amount we are usually permitted to purchase is only a fraction of our earnings.
2. When do we want the insurance company to begin sending benefits to us? If payment begins as of the 30th day of our disability, then the cost will be lower than if it begins on the 15th day. And if it begins on the 60th day, the cost will be even lower.

3. For what period of time do we want the benefits to continue? Our illness may last for many years. If the policy calls for payments to be made for 1 year, then our cost will be less than if they continue for a two-year period. Payments made for a lifetime will cost far more than for 1 or 2 years.
4. What is our occupation? If we happen to be a pilot, chances are that we would be unable to buy disability insurance. The greater the risk of accident in our occupation, the higher the cost of this insurance.

In the table below, the people classified as 4A are the least likely to be involved in an accident:

accountant, lawyer, consulting engineer, classroom teacher.

Those in class 1A (rates not shown) are most likely to be involved in an accident:

truck driver, electrical lineman, practical nurse, chef.

DISABILITY INSURANCE

Annual Premium per $100 of Monthly Benefit—Male
60-Day Waiting Period

	4A			3A			2A		
Age	1-Year	2-Year	Lifetime*	1-Year	2-Year	Lifetime*	1-Year	2-Year	Lifetime*
20	$ 8.40	$10.15	$24.12	$10.59	$12.33	$30.38	$13.50	$15.24	$39.11
25	9.17	11.15	26.82	11.37	13.35	32.98	14.31	16.29	41.86
30	11.00	13.37	32.13	13.37	15.75	38.82	16.54	18.91	48.47
35	13.40	16.37	38.61	15.95	18.92	45.84	19.35	22.32	56.26
40	16.75	20.54	46.98	19.52	23.31	54.83	23.21	27.00	66.18
45	20.60	25.45	57.69	23.55	28.40	66.08	27.47	32.33	78.15
50	25.24	37.61	70.02	28.25	34.26	78.63	32.26	38.27	65.96

Maximum: $3,500 per month

*To age 65 for sickness.

Unit 2

Example 1 Determine the annual premium for disability insurance that pays $400 monthly over a period of two years. The insured is 45 years old and employed in a 3A occupation.

Explanation

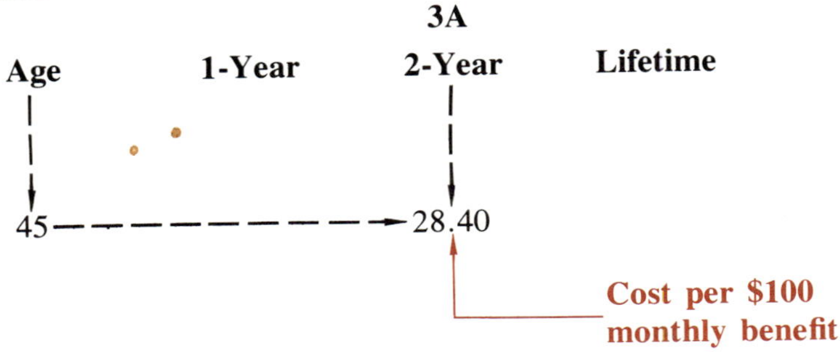

Solution

Cost per \$100 of monthly benefit = \$28.40

Number of \$100's in \$400 = \$400 ÷ \$100

= 4

Number of \$100's → 4; Cost per \$100 → \$28.40

Annual premium = 4 × \$28.40

= \$113.60

The cost of disability insurance for a woman is much higher than that for a man of the same age. A woman who is insured by the company whose rates we happen to be using will pay *150% of the man's rate*. There are a few insurance companies though, that charge women the same premium as they do men.

As in the case of life insurance, the premiums on disability insurance can be paid either semiannually or quarterly.

Semiannual premium = 51% of annual premium
Quarterly premium = 26% of annual premium

Example 2 Brenda McAdams' occupation is classified as 4A. At age 25 she purchased disability insurance that would pay her \$1,500 monthly for a period of 1 year. Determine the semiannual premium on this policy.

Solution

Computation for a man's premium

$$\text{Annual premium} = \text{Number of \$100's} \times \text{Cost per \$100}$$
$$= 15 \times \$9.17$$
$$= \$137.55$$

$$\text{Semiannual premium} = 51\% \text{ of Annual premium}$$
$$= .51 \times \$137.55$$
$$= \$70.15$$

Computation for a woman's premium

$$\text{Semiannual premium} = 150\% \text{ of a man's premium}$$
$$= 1.50 \times \$70.15$$
$$= \$105.23$$

Exercises

A

1. Annual premiums paid by males for disability insurance are shown below. Determine the annual premium a female will have to pay for the same coverage.

	Male Premium	Female Premium
a.	$ 60	$________
b.	137	________
c.	159	________
d.	$ 48.20	$________
e.	95.62	________
f.	134.87	________

2. Determine the annual premium per $100 of monthly benefit for each disability coverage. Each insured is a male.

	Age	Occupation Class	Coverage (Years)	Premium Per $100
a.	30	4A	1	$________
b.	45	2A	2	________
c.	20	3A	Lifetime	________
d.	25	4A	Lifetime	________
e.	50	3A	2	________

3. Determine the annual premium for each disability coverage. Each insured is a male.

	Age	Occupation Class	Coverage (Years)	Premium Per $100	Monthly Benefit	Annual Premium
a.	25	2A	2	$ 1629	$ 300	$
b.	40	4A	1		400	
c.	30	2A	Lifetime		800	
d.	45	3A	1		1,000	
e.	50	4A	2		1,200	

4. Determine the annual premium for each disability coverage. Each insured is a female.

	Age	Occupation Class	Coverage (Years)	Premium Per $100	Monthly Benefit	Male Premium	Female Premium
a.	20	4A	Lifetime	$	$500	$	$
b.	30	3A	1		800		
c.	45	2A	2		600		
d.	50	4A	1		1,100		
e.	25	2A	Lifetime		1,200		

5. Determine the semiannual premium for each disability coverage. Each insured is a male.

	Age	Occupation Class	Coverage (Years)	Premium Per $100	Monthly Benefit	Annual Premium	Semiannual Premium
a.	30	3A	2	$	$700	$	$
b.	25	4A	1		900		
c.	35	2A	2		800		
d.	50	3A	Lifetime		750		

6. Determine the quarterly premium for each disability coverage. Each insured is a male.

	Age	Occupation Class	Coverage (Years)	Premium Per $100	Monthly Benefit	Annual Premium	Quarterly Premium
a.	45	4A	Lifetime	$	$ 600	$	$
b.	20	3A	2		1,400		
c.	25	2A	1		1,300		
d.	40	3A	2		850		

B

1. At age 35 Paul Fleming purchased a disability-insurance policy that would pay him $1,200 per month over a period of two years in the event he was disabled. His occupation was classified as a 2A risk.

a. How much was his annual premium? ________

b. By renewing the policy year after year, he continued to pay the same premium that he paid at age 35. If he did this for a period of eight years, what did this insurance cost him? ________

c. During the eighth year he received disability benefits from the insurance company for a period of nine months. What was the total of these benefits? ________

d. How much more did he receive from the insurance company than he paid them in premiums? ________

2. An insurance company wrote a policy for a 20-year old cashier—3A rating—that would pay $500 monthly for life in the event the person was permanently disabled.

a. If the person was a male, what annual premium would he have to pay? ________

b. If the person was a female, what annual premium would she have to pay? ________

c. How much less does a male have to pay annually than a female? ________

Hospital, Surgical, and Medical Insurance Unit 3

Hospital costs have risen dramatically during the past few years. In fact, during a single year the cost rose by just under 30%. Needless to say, people are concerned that even a short stay at a hospital may devour their entire savings. Therefore, the great majority of people carry some form of hospital expense insurance.

Coverage varies widely from company to company and the greater the protection, the more we must pay. Some plans call for benefits to begin the very first day we enter the hospital, while others begin after a period of several days.

Similarly, there are plans under which we would receive benefits for a maximum hospital stay of 180 days. Under other plans, however, payments would continue for our entire life in the event we are confined to a hospital. The coverage of the two plans in the table below calls for benefits to be paid over an unlimited period of time.

Age	Plan A (Payments begin 1st day) Annual Premium Per Adult	Plan B (Payments begin 6th day) Annual Premium Per Adult
16-35	$110	$ 61
36-45	125	75
46-50	140	88
51-55	154	104
56-65	160	125
66-75	166	132
Over 75	207	163
All children	77	48

$40 PER DAY HOSPITAL PLANS

Plan C: $30 Per Day Payments	Plan D: $30 Per Day Payments
Payments begin the 1st day	Payments begin the 6th day
Premiums are 75% of those in Plan A	Premiums are 75% of those in Plan B

Semiannual premium = 51% of annual premium
Quarterly premium = 26% of annual premium

Example 1 Joe Boswell is 37 years old. How much will his semiannual premium be under Plan C?

Solution

Plan A annual premium = $125

Plan A semiannual premium = 51% of annual premium
= .51 × $125
= $63.75

Plan C premium = 75% of Plan A premium
= .75 × $63.75
= $47.81

Example 2 Edgar Miller is 41 years old; his wife is 34. They have three children. What will the quarterly premium be under Plan B?

Solution

Total annual premium = Mr. Miller + Mrs. Miller + Children

$$= \$75 + \$61 + \$48$$
$$= \$184$$

Quarterly premium = 26% of annual premium

$$= .26 \times \$184$$
$$= \$47.84$$

The $40 per day of hospital cost of Plan A falls far short of the current daily cost of a hospital room. In addition, the coverage above does not include:

1. The cost of surgery
2. The use of the operating room
3. Doctor's visits
4. Private duty nursing care

and on and on. To provide, at least to some extent, for these very large costs, people will purchase *major medical insurance*.

A sample of the nature of the coverage contained in this insurance is shown below.

MAJOR MEDICAL BENEFITS

80% of the first $2,500 of net expenses
Total excess over $2,500
Surgical Maximum: $3,000
Maximum Benefits: $100,000

Although this plan calls for a maximum of $3,000 to be paid to a surgeon, this is really not the case. Listed in the policy is a schedule of maximum benefits that will be paid for various operations:

Appendectomy	$600
Cataract removal	1,200
Gall-Bladder	900
Tonsillectomy	900

Therefore, if the surgeon's fee is \$1,000 for removing tonsils, then the insurance company will pay \$900 and we will have to pay the remaining \$100.

Of great importance is the fact that we realize that the net expense in the Benefits is the amount *not covered* by other insurance.

Example 3 A total bill for both hospital and surgical care amounted to \$9,276. Insurance plans other than the major medical plan covered \$3,941 of this cost. How much had to be paid by the insurance company under the major medical plan?

Explanation The payment under the major medical plan consists of two parts:

80% of first \$2,500 of net expense
Total of excess over \$2,500

Solution

$$\text{Net expense} = \$9{,}276 - \$3{,}941 = \$5{,}335$$

$$\begin{aligned}\text{Major medical payment} &= 80\% \text{ of } \$2{,}500 + \text{Total excess}\\ &= .80 \times \$2{,}500 + (\$5{,}335 - \$2{,}500)\\ &= \$2{,}000 + \$2{,}835\\ &= \$4{,}835\end{aligned}$$

Example 4 A person was insured under Plan A of hospital insurance and also under major medical insurance. He spent 67 days in a hospital at \$165 per day. How much of the cost of the room was covered by his major medical insurance?

Solution

Number of days → 67; Cost per day → $165

$$\text{Hospital room cost} = 67 \times \$165 = \$11{,}055$$

Number of days → 67; Coverage per day → $40

$$\text{Plan A coverage} = 67 \times \$40 = \$2{,}680$$

$$\begin{aligned}\text{Net expense} &= \text{Hospital room cost} - \text{Plan A coverage}\\ &= \$11{,}055 - \$2{,}680\\ &= \$8{,}375\end{aligned}$$

$$\begin{aligned}\text{Major medical coverage} &= 80\%\text{ of }\$2{,}500 + \text{Total excess}\\ &= .80 \times \$2{,}500 + (\$8{,}375 - \$2{,}500)\\ &= \$2{,}000 + \$5{,}875\\ &= \$7{,}875\end{aligned}$$

Exercises A

Any reference to hospital plans or major medical plans in these exercises imply the plans described in this unit.

1. Determine the annual premium for each hospital expense coverage.

	Age	Plan	Annual Premium
a.	27	A	$______
b.	36	B	______
c.	48	A	______
d.	59	A	______
e.	71	B	______

2. Determine the annual premium for each hospital expense coverage under Plan C.

	Age	Annual Premium Under Plan A	Annual Premium Under Plan C
a.	17	$______	$______
b.	29	______	______
c.	38	______	______

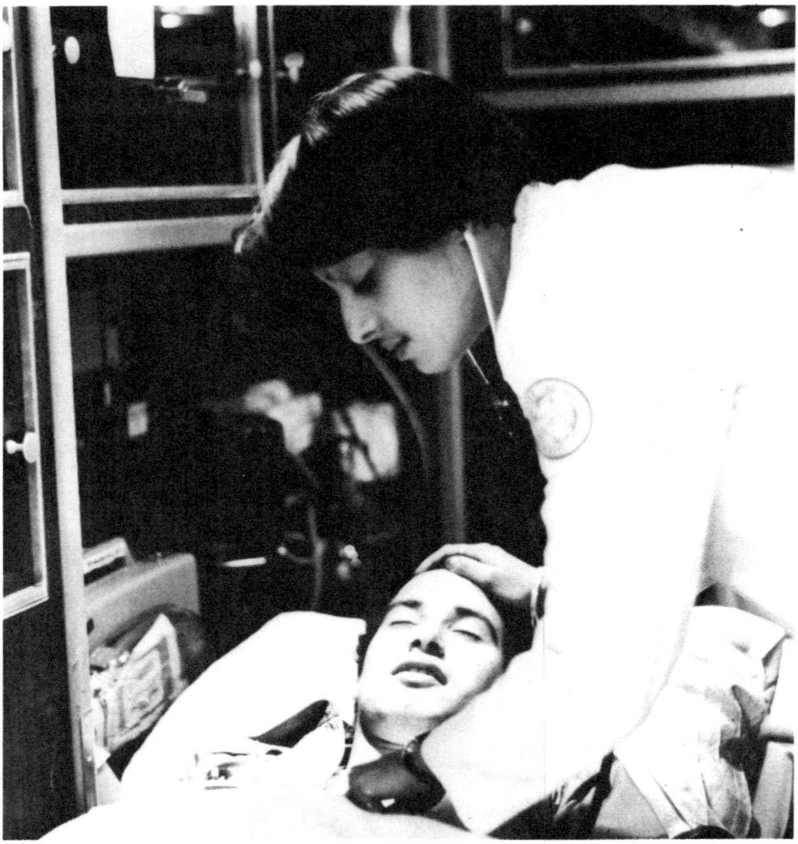

3. Determine the semiannual premium for each hospital expense coverage under Plan C.

	Age	Annual Premium Under Plan A	Annual Premium Under Plan C	Semiannual Premium Under Plan C
a.	24	$________	$________	$________
b.	47	________	________	________
c.	54	________	________	________

4. Determine the quarterly premium for each hospital expense coverage under Plan D.

	Age	Annual Premium Under Plan B	Annual Premium Under Plan D	Quarterly Premium Under Plan D
a.	19	$________	$________	$________
b.	45	________	________	________
c.	51	________	________	________

5. Determine the total annual family premium for each hospital expense coverage.

	Age		Number of		Premium			
	Husband	Wife	Children	Plan	Husband	Wife	Children	Total
a.	24	21	0	A	$____	$____	$____	$____
b.	38	34	2	A	____	____	____	____
c.	41	Deceased	3	B	____	____	____	____
d.	Deceased	29	1	B	____	____	____	____

6. Determine the total annual family premium for each coverage under Plan C.

	Age		Number of	Premium Under Plan A				Premium Under
	Husband	Wife	Children	Husband	Wife	Children	Total	Plan C
a.	27	27	1	$____	$____	$____	$____	$____
b.	52	46	2	____	____	____	____	____
c.	Deceased	31	3	____	____	____	____	____

B

1. Determine the amount paid by the insurance company for each period spent in the hospital.

	Plan	Hospital Stay—Days	Number of Days Covered	Payment Per Day	Total Payment
a.	A	15	____	$____	$____
b.	C	27	____	____	____
c.	A	3	____	____	____
d.	C	8	____	____	____
e.	B	4	____	____	____
f.	B	11	____	____	____
g.	D	5	____	____	____
h.	D	39	____	____	____

2. Determine the amount paid by major medical insurance under each condition.

	Net Expense	Payment On $2,500 or Less	Payment On Excess Over $2,500	Total Payment
a.	$ 1,600	$____	$____	$____
b.	2,300	____	____	____
c.	1,750	____	____	____
d.	3,000	____	____	____
e.	5,800	____	____	____
f.	12,650	____	____	____

3. Determine the amount paid by major medical insurance for each hospital bill.

	Hospital Costs	Am't Paid By Other Ins. Co.	Net Expense	Pay. On $2,500 Or Less	Pay. On Excess	Total Payment
a.	$ 3,500	$2,400	$______	$______	$______	$______
b.	4,600	3,100				
c.	7,800	2,300				
d.	14,200	4,700				
e.	12,860	7,530				

4. Determine the amount paid for the hospital stay by major medical coverage if the person was also covered by Plan D hospital insurance.

	Hospital Stay-Days	Hospital Charge Per Day	Total Charge	Hospital Ins. Coverage	Net Expense	Major Medical Coverage
a.	10	$120	$______	$______	$______	$______
b.	14	150				
c.	8	165				
d.	12	185				
e.	30	160				
f.	41	170				

C

1. Mr. Ellis, 45, and Mrs. Ellis, 46, have two children.

a. What is the annual premium for the family on Plan A hospital insurance? ______

b. What is the annual premium for the family on Plan C hospital insurance? ______

c. How much less will the Plan C insurance cost than the Plan A? ______

2. Stephen Barnard spent 23 days in a hospital at the time he was involved in an automobile accident. His hospital and doctor's bills showed the following expenses:

Room: $175 per day
Other hospital expenses: $1,268
Surgeon: $1,250

He was covered under Plan A hospital insurance, and major medical insurance.

a. What was the total cost for the room alone? ________

b. What was the total of both hospital and surgical expenses? ________

c. How much of the total expense was covered by the hospital insurance? ________

d. How much of the total expense was covered by the major medical insurance? ________

e. How much did Mr. Barnard himself have to pay? ________

Unemployment Insurance Unit 4

A major concern of many people is the thought of being unemployed for a long period and having no income to support their families. To soften the blow of unemployment, every state in the nation now has some form of *unemployment insurance*.

Unfortunately, state laws governing this insurance differ widely making it almost impossible to show any uniform method by which benefits are computed. The information below summarizes a few of their common features.

1. The employer pays the cost of the insurance in all but three states: Alabama, Alaska, New Jersey. In these states the employee also contributes to the cost.
2. Every state requires that we work a minimum number of weeks or earn a minimum amount of money as a requirement for being insured. A minimum number of weeks ranges from 14 to 40.
3. Twelve states pay benefits from the very first day we are unemployed. The remaining states require a waiting period of one week.
4. There are 42 states that will pay unemployment benefits for as long as 26 weeks. One state will pay for 39 weeks. The remaining states will pay between 26 and 39 weeks.
5. Only 8 states compute the benefits in terms of the number of our dependents.

6. The weekly benefit we receive is based on the amount we had earned prior to unemployment. Each state, though, has a minimum and a maximum weekly benefit. The minimum ranges from \$10 to \$30. The maximum ranges from \$64 to \$183.
7. Every state requires that we will accept reasonable employment if it can be found for us. If we do not, our benefits are stopped.

Example 1 The weekly benefit a person receives in one of the states is 1/24 of his highest quarterly earnings (hqe) during the previous year. Determine the weekly benefit if the hqe was \$1,958.

Solution

$$\begin{aligned}\text{Weekly benefit} &= \tfrac{1}{24} \text{ of hqe}\\ &= \tfrac{1}{24} \times \$1{,}958\\ &= \$81.58\end{aligned}$$

Example 2 In the state where Eve Walker lives the weekly benefit is determined as follows:

60% of 1/13 of the highest quarterly earnings.

If her highest quarterly earnings were \$2,695, how much will her weekly benefit be?

Solution

$$\begin{aligned}\text{Weekly benefit} &= 60\% \text{ of } \tfrac{1}{13} \text{ of hqe}\\ &= .60 \times \left[\tfrac{1}{13} \times \$2{,}695\right]\\ &= .60 \times \$207.31\\ &= \$124.39\end{aligned}$$

Explanation Eve Walker may not receive as large a benefit as \$124.39 if the state in which she lives does not pay so high a weekly benefit.

Exercises

1. In each exercise the weekly benefit is a fixed percent of the average weekly wage the person had received. Determine each person's weekly benefit.

	Average Weekly Wage	Percent Of Weekly Wage	Weekly Benefit
a.	$ 84	50%	$______
b.	141	50%	______
c.	170	60%	______
d.	138	60%	______
e.	217	55%	______
f.	196	55%	______

2. In each exercise the weekly benefit is a fixed percent of the annual wage the person had received. Determine each person's weekly benefit.

	Annual Wage	Percent Of Annual Wage	Weekly Benefit
a.	$ 8,000	1.1%	$______
b.	9,400	1.1%	______
c.	6,000	2.3%	______
d.	7,100	2.3%	______
e.	10,200	1.2%	______
f.	8,950	1.2%	______
g.	7,000	.8%	______
h.	11,300	.8%	______
i.	7,600	1.9%	______
j.	8,300	1.9%	______

3. In each exercise the weekly benefit is a fixed fraction of the highest quarterly earnings a person received during the previous year of employment. Determine each person's weekly benefit.

	Highest Quarterly Earnings	Fraction	Weekly Benefit
a.	$1,260	$\frac{1}{20}$	$______
b.	1,740	$\frac{1}{20}$	______
c.	1,825	$\frac{1}{25}$	______
d.	2,300	$\frac{1}{25}$	______
e.	1,978	$\frac{1}{23}$	______
f.	2,392	$\frac{1}{23}$	______
g.	2,262	$\frac{1}{26}$	______
h.	2,964	$\frac{1}{26}$	______

4. The maximum benefits a person can receive during any period of unemployment is usually computed as a fraction of his earnings during the year prior to unemployment. Determine the maximum benefits.

	Prior Year's Earnings	Fraction Permissible	Maximum Benefits
a.	$ 6,400	$\frac{1}{2}$	$________
b.	9,500	$\frac{1}{2}$	________
c.	9,600	$\frac{1}{3}$	________
d.	11,400	$\frac{1}{3}$	________
e.	11,800	$\frac{1}{4}$	________
f.	12,600	$\frac{1}{4}$	________
g.	8,200	$\frac{2}{5}$	________
h.	11,500	$\frac{3}{10}$	________

5. The maximum benefits a person can receive during any period of unemployment and the weekly benefit received are shown below. Determine the number of weeks during which each person will receive benefits.

	Weekly Benefit	Maximum Benefits	Number Of Weeks Permitted By Law: Minimum	Maximum	Number Of Weeks Benefits Will Be Paid
a.	$ 40	$ 800	10	26	________
b.	32	576	10	26	________
c.	65	1,495	12	28	________
d.	78	2,808	10	39	________
e.	105	3,255	18	26	________
f.	58	464	13	26	________

Unit 5 Homeowners Insurance

Section 1 The Cost of Homeowners Insurance

Possibly the largest investment most of us will make during our lifetime is the purchase of a house. Having done this, we immediately become concerned with the need to protect our investment. The insurance designed for this purpose is called *Homeowners Insurance*.

Fire is more likely to cause damage to our property than any of the other catastrophies listed in this policy. In view of this, we are paying largely for fire protection when we purchase homeowners insurance.

HOMEOWNERS INSURANCE		
Annual Rates per $100		
Town	House	
Class	Brick	Frame
A	.094	.145
B	.101	.153
C	.108	.159
D	.112	.169
E	.123	.174
F	.137	.183
G	.194	.245
H	.356	.385
K	.428	.456

Two features of this table stand out:

1. The rates for a frame house are higher than those for a brick house in the same community. This is so because a frame house is made of wood and can be destroyed by fire far more easily than a brick house.
2. The rates increase as we move from class A to class K. The various classes indicate the type of protection against fire that is available in the community. Town class A is very likely a large city with both water hydrants and a fire station close by. On the other hand, class K is an isolated area with no hydrants and no fire station within five miles.

Example 1 A brick house in a class D community was insured for $32,000. Determine the annual premium on this coverage.

Explanation

Town Class — **House**: **Brick**, **Frame**

D ——→ .112 ← **Cost per $100**

Solution

Number of $100's in $32,000 = $32,000 ÷ $100
= 320

Number of $100's — **Cost per $100**

Annual Premium = 320 × $.112
= $35.84

Unit 5

Most homeowners policies contain an *80% clause*. This clause is extremely important for it states that if the coverage we purchase is *less than* 80% of the value of the house itself, then we will have to share in any loss that occurs. If the coverage is *more than* 80% of the value, the insurance company will pay the entire loss.

The fraction of the loss for which the insurance company is liable is

$$\frac{\text{Coverage we purchased}}{\text{80\% of value of house.}}$$

As an example, if 80% of the value of the house is \$60,000 and we insure the house for \$50,000, then the insurance company will pay

$$\frac{50{,}000}{60{,}000} \text{ of any loss.}$$

Example 2 A house is valued at \$65,000 but is insured for only \$25,000. If the house is destroyed by fire, what fraction of the loss will be paid by the insurance company?

Solution

$$\begin{aligned} \text{80\% of the value} &= \text{80\% of \$65,000} \\ &= .80 \times \$65{,}000 \\ &= \$52{,}000 \\ \text{Fraction covered by insurance} &= \frac{25{,}000}{52{,}000} \\ &= \frac{25}{52} \end{aligned}$$

Example 3 A house is valued at \$46,000 but insured for \$39,000. If fire destroys the property to the extent of \$6,000, how much of this loss will be paid by the insurance company?

Solution

$$\begin{aligned} \text{80\% of the value} &= \text{80\% of \$46,000} \\ &= .80 \times \$46{,}000 \\ &= \$36{,}800 \end{aligned}$$

Since the house is insured for \$39,000 which is more than \$36,800, the insurance company will pay the entire \$6,000.

Example 4 A house is valued at \$62,000 but insured for only \$40,000. Damage caused by a fire amounts to \$18,000. How much of this loss will be covered by the insurance company?

Solution

$$80\% \text{ of the value} = 80\% \times \$62{,}000 = .80 \times \$62{,}000 = \$49{,}600$$

$$\text{Fraction covered by insurance} = \frac{40{,}000}{49{,}600} = \frac{400}{496}$$

$$\text{Company payment} = \tfrac{400}{496} \times \$18{,}000 = \$14{,}516.13$$

Exercises A

1. Determine the cost per \$100 for each homeowners insurance policy.

	Town Class	Structure	Cost Per \$100
a.	C	Brick	\$________
b.	F	Frame	________
c.	D	Frame	________
d.	K	Brick	________
e.	G	Frame	________

2. Determine the annual premium for each homeowners insurance policy.

	Town Class	Structure	Cost Per \$100	Amount Of Insurance	Number Of \$100's	Annual Premium
a.	E	Frame	\$________	\$30,000	________	\$________
b.	C	Frame	________	50,000	________	________
c.	A	Brick	________	35,000	________	________
d.	G	Brick	________	24,000	________	________
e.	B	Frame	________	62,000	________	________
f.	H	Brick	________	58,000	________	________
g.	D	Brick	________	37,500	________	________

B

1. The homeowners policy in these exercises contain an 80% clause. Determine the least amount of coverage that must be purchased if the insurance company is to pay the entire cost of any fire damage.

	Property Value	Amount Of Insurance		Property Value	Amount Of Insurance
a.	$30,000	$________	d.	$35,000	$________
b.	50,000	________	e.	48,000	________
c.	70,000	________	f.	67,000	________

2. Determine the fraction of the cost of any fire damage that will be paid by the insurance company.

	80% Of Property Value	Amount Of Insurance	Fraction Covered By Insurance
a.	$40,000	$20,000	________
b.	30,000	25,000	________
c.	45,000	27,000	________
d.	54,000	42,000	________
e.	64,000	56,000	________

3. Determine the extent of the fire damage that will be paid by the insurance company in each of these situations.

	80% Of Property Value	Amount Of Insurance	Fraction Covered By Insurance	Fire Loss	Company Payment
a.	$24,000	$18,000	________	$ 6,000	$________
b.	30,000	27,000	________	2,100	________
c.	42,000	35,000	________	5,400	________
d.	56,000	49,000	________	7,200	________
e.	35,000	40,000	________	9,500	________
f.	65,000	52,000	________	12,000	________

4. Determine the fraction of the cost of any fire damage that will be paid by the insurance company.

	Property Value	80% Of Property Value	Amount Of Insurance	Fraction Covered By Insurance
a.	$40,000	$________	$28,000	________
b.	35,000	________	20,000	________
c.	60,000	________	40,000	________
d.	56,000	________	32,000	________

5. Determine the extent of the fire damage that will be paid by the insurance company in each of the following situations.

	Property Value	80% Of Property Value	Amount Of Insurance	Fraction Covered By Insurance	Fire Loss	Company Payment
a.	$40,000	$________	$24,000	________	$1,600	$________
b.	50,000	________	36,000	________	2,500	________
c.	35,000	________	20,000	________	3,500	________
d.	45,000	________	40,000	________	7,300	________
e.	55,000	________	33,000	________	6,200	________
f.	75,000	________	48,000	________	4,620	________

Section 2 Extended Coverages

Coverage under homeowners insurance includes far more than simply damage by fire. Although the policies of different insurance companies may vary somewhat, usually the protection under this coverage includes the following:

1. Damage to garage, barn, toolhouse, and the like:
   10% of basic coverage
2. Expense for renting an apartment while our house is being repaired because of damage:
   20% of basic coverage
3. Damage to contents of house:
   50% of basic coverage
4. Damage that we accidently cause to the property of others:
   10% of basic coverage
5. Injury we accidently inflict on other people whether they are on our property or not:
   100% of basic coverage

The damage referred to above may be caused not only by fire but also by such things as: windstorm, rainstorm, hail, riot, aircraft, etc. In view of this, each of the concepts developed in Section 1 pertaining to damage by fire applies equally well to damage caused by any of the conditions above.

Most policies contain either a *$50-deductible* or *$100-deductible clause*. Under the $50-deductible clause, we would have to pay for the first $50 worth of damage to our home, while the insurance company would pay the balance.

Example 1 During a hailstorm a house was damaged to the extent of $375. If this property was insured under a homeowners policy having the $100-deductible clause, what is the maximum amount of this loss that the insurance company will pay?

Unit 5

Solution

Damage → $375; Deductible amount → $100

$$\text{Insurance coverage} = \$375 - \$100$$
$$= \$275$$

Example 2 Frank Yerby purchased a $40,000 homeowners policy. Determine the extent of each of the following coverages.

(1) Damage to house

(2) Damage to additional buildings.

(3) Damage to contents

(4) Living expenses during repair

(5) Accidental damage to other property

(6) Accidental injury to other people

Solution

Basic coverage: $40,000

(1) Coverage on house = 100% of basic coverage
= 1.00 × $40,000
= $40,000

(2) Coverage on additional buildings = 10% of basic coverage
= .10 × $40,000
= $4,000

(3) Coverage on contents = 50% of basic coverage
= .50 × $40,000
= $20,000

(4) Coverage for living expense = 20% of basic coverage
= .20 × $40,000
= $8,000

(5) Coverage for damage to property of other people = 10% of basic coverage
= .10 × $40,000
= $4,000

(6) Coverage for injury to others = 100% of basic coverage
= 1.00 × $40,000
= $40,000

Exercises A

1. Determine the amount of money paid to the insured under each homeowners policy. These policies contained the deductible clause shown below.

	Extent Of Damage	Deductible Amount	Amount Paid By Insurance Company
a.	$ 700	$100	$________
b.	2,500	50	________
c.	4,375	50	________
d.	75	100	________
e.	3,680	50	________

2. Determine the extent of coverage for each item under the homeowners policies below.

	Coverage	House (100%)	Additional Buildings (10%)	Living Expenses (20%)	Contents (50%)	Property Of Others (10%)	Injury To People (100%)
a.	$30,000	$________	$________	$________	$________	$________	$________
b.	50,000	________	________	________	________	________	________
c.	35,000	________	________	________	________	________	________
d.	27,000	________	________	________	________	________	________
e.	58,000	________	________	________	________	________	________
f.	76,000	________	________	________	________	________	________

B

1. A house insured under a homeowners policy for $42,000 was seriously damaged by fire. The house was insured in excess of 80% of its value and the policy did not contain a deductible clause.

a. Damage to the house amounted to $36,000. How much of this damage is covered by the insurance? ________

b. Damage to the furniture, clothing, and other valuables amounted to $23,500. How much of this damage is covered by the insurance? ________

c. During the 14 months that it took to rebuild the house, the owner lived in an apartment at a cost of $325 per month. How much of this expense is covered by the insurance? ________

2. Ralph Collins drove a golfball that struck a person in the eye. The person lost the sight of that eye and sued Ralph for $150,000. The basic coverage under Ralph's homeowners policy is for $70,000. Ralph lost the legal suit.

a. For how much of the damage did the insurance company pay? ________

b. For how much of the damage did Ralph have to pay? ________

Review of Chapter 10

Refer to the tables in this chapter when doing the computation for the problems below.

1. Albert Derio is 16 years old.

a. According to the mortality table, what is his life expectancy? ________

b. Until what age can the average 16-year old expect to live? ________

c. How many of the 16-year olds per 1,000 will probably die before reaching the age of 17? ________

2. Of a group of 700,000 people of age 60, how many will probably die before reaching the age of 61? ________

3. Eva Sandford purchased a $25,000 20-year term policy.

a. If she lives the entire 20 years, how much will she receive from the insurance company? ________

b. If she had purchased a 20-year endowment insurance policy and lives the 20 years, how much will she receive from the insurance company? ________

4. Determine the annual premium on a $25,000 10-year decreasing term insurance policy that was purchased by a man at age 45. ________

5. A woman purchased a $30,000 20-year decreasing term policy. What is her coverage during the 14th year after purchase? ________

6. At 33, Betty Higgins purchased a $14,000 20-payment life-insurance policy.

a. If she paid the premiums annually, how large would each payment be? ________

b. If she paid the premiums semiannually, how large would each payment be? ________

c. In one year, how much could she save by paying the premiums annually rather than semiannually? ________

d. In twenty years, how much could she save by paying the premiums annually rather than semiannually? ________

e. Will Betty receive the $14,000 from the company at the end of the 20 years? ________

7. Mr. Porter, who is employed in an 3A risk occupation, purchased a disability-insurance policy to cover him for two years in the event he was ill and could not work. Were this to happen, he would receive $600 per month in benefits. Mr. Porter is 45.

a. How much was his annual premium for this coverage? ________

b. If Mrs. Porter had had the same job as her husband and purchased the same insurance, what would her annual premium have been? Mrs. Porter is 45. ________

8. Jane Mallon carried the major medical insurance whose benefits are shown on page 377. She had no other hospital insurance. Her hospital and surgical bills as a result of an accident totaled $4,726.

a. How much of this bill was covered by insurance? ________

b. How much of this bill did Jane have to pay? ________

9. Thomas Johnson, age 36, insured himself, his wife, age 35, and his two children under Plan C of the hospital plans described on page 376.

a. What total annual premium will he have to pay? ________

b. If the premiums are paid quarterly, how large will they be? ________

10. Grace Bigelow lives in a state where the weekly unemployment compensation benefits are 1.9% of the annual wage the person had received when employed. Determine her weekly benefit if her annual wage had been $8,000. ________

11. Determine the annual premium for a brick house in a class F community that was insured for $47,000 under a homeowners policy. ________

12. A house is covered by a homeowners policy that contains an 80% clause but no deductible clause. The value of the house is $50,000. It is insured, however, for only $36,000. Fire damages the house to the extent of $4,200.

a. What fraction of the damage will be paid by the insurance company? ________

b. How much of the damage will be paid by the insurance company? ________

c. How much of the damage will have to be paid by the owner of the house? ________

Review

U. S. BUREAU OF THE CENSUS
U.S. CENSUS

Chapter 11

Retirement Income

It may strike us as being somewhat premature that at this point in our life we would be involved in a study of retirement income. However, as we have already discovered, whether we choose to or not, we begin to pay for our social security retirement income with our very first paycheck. Since this process has already begun for some of us and will soon affect all of us, it would seem wise that we take a look at just what benefits our payroll deductions are purchasing.

Social Security Benefits Unit 1

Section 1 The Primary Insurance Amount

To become insured under the federal social security program we must earn a required number of *quarters of coverage*. During any single year the social security agency credits us with 1 quarter of coverage for each $250 of our earnings. Thus, a person whose annual salary is $250 will earn

1 quarter of coverage: $\$250 \div \$250 = 1$.

Similarly, if the annual salary is $500, the person will earn

2 quarters of coverage: $\$500 \div \$250 = 2$.

However, we can not earn more than 4 quarters of coverage during a single year. Therefore, if our annual salary is $1,000 or more, we will earn the maximum of 4 quarters of coverage for the year.

$$\$1{,}000 \div \$250 = 4.$$

The number of quarters of coverage we accumulate determines whether we are currently insured or fully insured.

1. Currently Insured—Generally speaking, to be currently insured requires that we earn *6 quarters of coverage* during the preceding *3 years*.

The important advantage of being currently insured is that in the event of our death a number of our dependents will receive monthly survivors' benefits. These benefits will be examined shortly.

2. Fully Insured—Under this arrangement we must earn 40 quarters of coverage. This is the equivalent of having paid the social security tax over a ten-year period. Once we have reached the point of being fully insured we are protected for the duration of our life whether we are employed or not.

As fully insured people we are entitled to receive monthly social security benefits when we reach the age of 65 – or 62, if we prefer. The size of these monthly benefits depends on what our *average monthly earnings* had been during the years from age 22 to age 65. However, before computing this average, the social security agency does two things to increase the size of that number:

1. The lowest five years of our earnings are eliminated from the computation.
2. Our salary for any year is adjusted to what it would have been had that money been earned around the time of our retirement.

As an example, because of inflation, *earnings of $10,000 in 1970* are equivalent to *earnings of $19,000 in 1980*. Therefore, for the person who retires in 1980, the inflated earnings will be used in computing the average rather than the original earnings.

Having determined the average monthly earnings, called the Average Indexed Monthly Earnings, the social security agency then uses that number to compute the *primary insurance amount* (PIA). This is the amount of money upon which all monthly benefits are based — the payments we receive and the payments our dependents receive.

The table below is used when determining the primary insurance amount.

If the average monthly earnings are:		The PIA is :
Greater than	At least	
$ 0	$180	90% of average
180	905	$162 + 32% of excess over $180
905		$394 + 15% of excess over $905

Example 1 Determine the PIA for the average monthly earnings of $154.

Explanation

Greater than	At least	
$0	$180	90% of average

$154 is between these numbers (Greater than / At least)

This is the computation for the PIA. (90% of average)

Solution

$$\text{PIA} = 90\% \text{ of Average monthly earnings}$$
$$= .90 \times \$154$$
$$= \$138.60$$

Example 2 Determine the PIA for average monthly earnings of $745.

Explanation

Greater than	**At least**	
180	905	$162 + 32% of excess over $180

$745 is between these numbers (180 and 905)

Base payment ($162)

Payment on excess (32% of excess over $180)

Solution

$$\text{Base payment} = \$162$$

$$\begin{aligned}\text{Excess over } \$180 &= \$745 - \$180 \\ &= \$565\end{aligned}$$

$$\begin{aligned}\text{Payment on excess} &= 32\% \text{ of Excess} \\ &= .32 \times \$565 \\ &= \$180.80\end{aligned}$$

$$\begin{aligned}\text{PIA} &= \text{Base payment} + \text{Payment on excess} \\ &= \$162 + \$180.80 \\ &= \$342.80\end{aligned}$$

Example 3 Determine the PIA for average monthly earnings of $1,344.

Solution

$$\text{Base payment} = \$394$$

$$\begin{aligned}\text{Excess over } \$905 &= \$1{,}344 - \$905 \\ &= \$439\end{aligned}$$

$$\begin{aligned}\text{Payment on excess} &= 15\% \text{ of Excess} \\ &= .15 \times \$439 \\ &= \$65.85\end{aligned}$$

$$\begin{aligned}\text{PIA} &= \text{Base payment} + \text{Payment on excess} \\ &= \$394 + \$65.85 \\ &= \$459.85 \text{ or } \$459.90\end{aligned}$$

Explanation Notice that the PIA was rounded from $459.85 to $459.90. *Computation for the PIA and all other benefits are always rounded to the next higher 10¢ amount.*

Exercises

1. Round each to the next higher 10¢ amount.

a. $123.67	$________	**d.** $205.21	$________	**g.** $456.94	$________
b. 156.89	________	**e.** 317.02	________	**h.** 129.98	________
c. 137.52	________	**f.** 238.92	________	**i.** 199.91	________

2. Determine the PIA for each average monthly earning.

	Av. Monthly Earnings	PIA		Av. Monthly Earnings	PIA
a.	$140	$______	d.	$145	$______
b.	160	______	e.	162	______
c.	110	______	f.	178	______

3. Determine the PIA for each average monthly earning.

	Av. Monthly Earnings	PIA		Av. Monthly Earnings	PIA
a.	$260	$______	d.	$675	$______
b.	480	______	e.	849	______
c.	590	______	f.	903	______

4. Determine the PIA for each average monthly earning.

	Av. Monthly Earnings	PIA		Av. Monthly Earnings	PIA
a.	$ 940	$______	d.	$1,180	$______
b.	1,100	______	e.	1,246	______
c.	1,060	______	f.	1,358	______

5. Determine the PIA for each average monthly earning.

	Av. Monthly Earnings	PIA		Av. Monthly Earnings	PIA
a.	$738	$______	c.	$1,267	$______
b.	159	______	d.	689	______

6. Determine the number of quarters of coverage each person earned during a year in which their total earnings were the amounts shown below.

	Earnings For Year	Quarters Of Coverage		Earnings For Year	Quarters Of Coverage
a.	$ 400	______	e.	$20,000	______
b.	600	______	f.	175	______
c.	850	______	g.	499	______
d.	2,000	______	h.	500	______

Section 2 Retirement Benefits

We are now prepared to examine the actual monthly retirement benefits we will receive under the social security law. To simplify the explanation, we will assume that it was the husband that is covered and not the wife. Similar benefits are paid if the situation is reversed.

Should the husband retire at age 65 – or older – his monthly benefit will be the entire PIA. He can, though, retire as early as age 62. However, should he retire before age 65, for each month prior to that age his benefit will be reduced by 1/180 of the PIA.

As soon as the husband begins to receive benefits the wife also becomes qualified to receive them – assuming that she is at least 62 years of age. If she is 65 or over at that time, her monthly check will be 50% of the PIA.

Should she elect to receive her benefits as early as age 62, her monthly payment will be reduced accordingly.

The more important benefits received at the time of retirement are summarized in the table below.

FAMILY RETIREMENT BENEFITS

Family Member	Percent of PIA
Husband	
65 or older	100%
64	93.33%
63	86.67%
62	80%
Wife	
65 or older	50%
64	45.83%
63	41.67%
62	37.5%
Wife with child under 18	50%
Child under 18	50%
Child under 22 attending school	50%
Family maximum benefit	175%

Example 1 Mr. Swenson retired at age 66. His wife was 65 at the time. The social security agency computed his PIA at $384.60. What monthly benefit will both he and his wife receive?

Solution

$$\begin{aligned}\text{Husband's benefit} &= 100\% \text{ of } \quad \text{PIA}\\ &= 1.00 \quad \times \$384.60\\ &= \$384.60\end{aligned}$$

$$\begin{aligned}\text{Wife's benefit} &= 50\% \text{ of } \quad \text{PIA}\\ &= .50 \quad \times \$384.60\\ &= \$192.30\end{aligned}$$

$$\begin{aligned}\text{Total benefits} &= \$384.60 + \$192.30\\ &= \$576.90\end{aligned}$$

Example 2 Mr. Williams retired on his 63rd birthday. Mrs. Williams was 55 and they had a child under the age of 18. What was the total family benefit if the PIA was $467.80?

Solution

$$\begin{aligned}\text{Husband's benefit} &= 86.67\% \text{ of } \quad \text{PIA}\\ &= .8667 \quad \times \$467.80\\ &= \$405.44 \text{ or } \$405.50\end{aligned}$$

Because child is under 18

$$\begin{aligned}\text{Wife's benefit} &= 50\% \text{ of } \quad \text{PIA}\\ &= .50 \quad \times \$467.80\\ &= \$233.90\end{aligned}$$

$$\begin{aligned}\text{Child's benefit} &= 50\% \text{ of } \quad \text{PIA}\\ &= .50 \quad \times \$467.80\\ &= \$233.90\end{aligned}$$

$$\begin{aligned}\text{Total} &= \$405.50 + \$233.90 - \$233.90\\ &= \$873.30\end{aligned}$$

$$\begin{aligned}\text{Family maximum benefit} &= 175\% \text{ of } \quad \text{PIA}\\ &= 1.75 \quad \times \$467.80\\ &= \$818.65 \text{ or } \$818.70\end{aligned}$$

Monthly benefit family receives = $818.70

Explanation

Total of the three benefits → $873.30

Family maximum benefit → $818.70

The family receives this amount because this is the smaller of the two.

In a family of more than two people, it is best to determine the family maximum benefit. As in this example, it may be smaller than the sum of all benefits.

Example 3 Determine the monthly benefits for both a husband and wife who are over 65 years of age. The average monthly earnings of the husband is $864.

Solution

Computation for PIA →

Base payment = $162

Payment on excess = 32% of excess over $180 (**$864 − $180**)
= .32 × $684
= $218.88

PIA = $162 + $218.88
= $380.88 or $380.90

Husband's benefit = 100% of PIA
= 1.00 × $380.90
= $380.90

Wife's benefit = 50% of PIA
= .50 × $380.90
= $190.45 or $190.50

Family benefit = $380.90 + $190.50
= $571.40

Exercises A

1. In each exercise the husband receives his first monthly benefit on the birthday shown. Determine his monthly benefit.

	Birthday	PIA	Percent of PIA	Monthly Benefit
a.	65th	$375.40	______%	$______
b.	65th	416.70	______%	______
c.	62nd	275	______%	______
d.	62nd	358	______%	______
e.	63rd	210	______%	______
f.	64th	360	______%	______

2. In each exercise the wife receives her first monthly benefit on the birthday shown. Determine her monthly benefit.

	Birthday	PIA	Percent of PIA	Monthly Benefit
a.	65th	$450	______%	$______
b.	65th	385	______%	______
c.	62nd	340	______%	______
d.	63rd	510	______%	______
e.	64th	472	______%	______

3. In each exercise the husband receives his first monthly benefit on the birthday shown. Determine the monthly benefit.

	Birthday	Av. Monthly Earnings	PIA	Percent Of PIA	Monthly Benefit
a.	65th	$ 650	$______	______%	$______
b.	65th	800	______	______%	______
c.	62nd	960	______	______%	______
d.	62nd	1,200	______	______%	______

4. In each exercise the wife receives her first monthly benefit on the birthday shown. Determine her monthly benefit.

	Birthday	Av. Monthly Earnings	PIA	Percent Of PIA	Monthly Benefit
a.	65th	$230	$______	______%	$______
b.	62nd	500	______	______%	______
c.	63rd	950	______	______%	______

5. Determine the family maximum benefit in each exercise.

	PIA	Family Max Benefit		PIA	Family Max Benefit
a.	$420	$______	c.	$326.50	$______
b.	537	______	d.	487.30	______

6. Determine the family maximum benefit in each exercise.

	Av. Monthly Earnings	PIA	Family Max Benefit
a.	$ 730	$________	$________
b.	680	________	________
c.	985	________	________
d.	1,063	________	________

7. In each exercise the ages of both husband and wife are shown. None of the children in these families is under the age of 22. Determine the total family benefit in each situation.

	PIA	Husband's Age	Wife's Age	Husband's Benefit	Wife's Benefit	Family Benefit
a.	$340	65	57	$________	$________	$________
b.	420	66	60	________	________	________
c.	280	67	66	________	________	________
d.	410	62	65	________	________	________
e.	500	67	63	________	________	________
f.	400	64	62	________	________	________

8. In each exercise the ages of the husband, wife and children are shown. Determine the total family benefit in each situation. Remember to compare the total with the family maximum benefit.

	PIA	Husband's Age	Husband's Benefit	Wife's Age	Wife's Benefit	Child's Age	Child's Benefit	Total Benefit
a.	$375	65	$________	57	$________	24	$________	$________
b.	427	65	________	65	________	31	________	________
c.	480	65	________	52	________	16	________	________
d.	250	69	________	55	________	17	________	________
e.	310	62	________	55	________	16	________	________
f.	520	64	________	53	________	16	________	________

B

1. Mr. Burns started to collect his social security checks at age 65. His wife was 50 at the time and they had two children—15 and 17 years of age. His average monthly earnings were $820.

a. Determine his PIA. ________

b. What was his monthly benefit? ________

c. What was his wife's monthly benefit? ________

d. What was the monthly benefit of each child? ________

e. What total monthly benefit did the entire family receive? ________

f. What total monthly benefit did the entire family receive after the older child reached the age of 18? This child was not attending school. ________

2. Mr. Campara retired at age 65 at which time he received his first social security check. His average monthly earnings were $680.

a. Determine his PIA. ________

b. What was his monthly benefit? ________

c. What would his monthly benefit have been had he retired at age 62? ________

d. How much more does he receive monthly by having retired at age 65 rather than at age 62? ________

3. The social security agency found that Mr. Flanagan's average monthly earnings were $820 at the time he retired at age 66.

a. Determine his PIA. ________

b. What was his monthly benefit? ________

c. According to the mortality table on page 352, how many years, to the nearest year, can Mr. Flanagan expect to live? ________

d. How much should the social security agency plan on paying Mr. Flanagan over his expected lifetime? ________

Section 3 Survivors' Benefits

Normally we think of social security only in terms of retirement benefits. Equally as important, though, are the benefits the family receives in the event the father dies. The family is qualified for these benefits only if the father had been currently or fully insured at the time of his death.

Upon the death of the father, children under the age of 22, if attending school, automatically qualify for monthly benefits The widow who is under the age of 60, however, qualifies for benefits only under either of the following conditions:

1. She is caring for a child under the age of 18 who is attending school.
2. She is caring for a disabled child who had been disabled before reaching age 22.

Unit 1

The monthly benefit of a widow who is 65 or older is the entire primary insurance amount. She can, though, elect to receive her benefits as early as age 60. But should she do so, the benefits will be reduced accordingly.

At the time of the husband's death, in addition to the monthly payments, the family receives a one time death benefit of $255.

The more important points made above are summarized in the following table.

SURVIVORS' BENEFITS

Family Member	Percent of PIA
Widow	
65 or older	100%
64	94.3%
63	88.6%
62	82.9%
61	77.2%
60	71.5%
Widow any age caring for either a child under 18 or a disabled child	75%
Child under 18	75%
Child under 22 attending school	75%
Child disabled before 22	75%
Lump-sum death benefit	$255

Example 1 Mrs. Bailey had just had her 63rd birthday at the time of her husband's death. His average monthly earnings were $734.

(a) How large was the lump-sum death benefit?

(b) What will her monthly benefit be?

Solution

(a) Lump-sum death benefit = \$255

(b) Base payment = \$162

Computation for PIA

\$734 − \$180

Payment on excess = 32% of excess over \$180
= .32 × \$554
= \$177.28

PIA = \$162 + \$177.28
= \$339.28 or \$339.30

Percent at age 63

Mrs. Bailey's monthly benefit = 88.6% of PIA
= .886 × \$339.30
= \$300.62 or \$300.70

Example 2 Mr. Newman died leaving a wife of 40 and two children aged 12 and 14. The social security agency computed his PIA at \$490. Determine the total monthly benefit the family will receive.

Solution

Mrs. Newman's benefit = 75% of PIA
= .75 × \$490
= \$367.50

Benefit of child of 12 = 75% of PIA
= \$367.50

Benefit of child of 14 = 75% of PIA
= \$367.50

Total = \$367.50 + \$367.50 + \$367.50
= \$1,102.50

Family maximum benefit = 175% of PIA
= 1.75 × \$490
= \$857.50

Monthly benefit family receives = \$857.50

Explanation

Total of the three benefits — \$1,102.50

Family maximum benefit — \$857.50

The family receives this amount because it is the smaller of the two.

Exercises A

1. Determine the widow's benefit if she is under 60 years of age and caring for a child under the age of 18.

	PIA	Widow's Benefit		PIA	Widow's Benefit
a.	$354	$________	c.	$268.50	$________
b.	279	________	d.	403.70	________

2. Determine the child's benefit if under 18 years of age and the father is deceased.

	Average Monthly Earnings	PIA	Child's Benefit
a.	$ 640	$________	$________
b.	527	________	________
c.	176	________	________
d.	1,042	________	________

3. Determine the widow's monthly benefit if she elects to receive her first check on the birthday shown below.

	PIA	Birthday	Percent of PIA	Widow's Benefit
a.	$320	60th	________%	$________
b.	258	61st	________%	________
c.	425.40	62nd	________%	________
d.	467.30	64th	________%	________

4. Determine the widow's monthly benefit if she elects to receive her first check on the birthday shown below.

	Average Monthly Earnings	PIA	Birthday	Percent of PIA	Widow's Benefit
a.	$155	$________	60th	________%	$________
b.	456	________	61st	________%	________
c.	695	________	62nd	________%	________
d.	972	________	63rd	________%	________

5. The father in each family has died. Determine the monthly benefit his family will receive. Remember to compare the total with the family maximum benefit.

	PIA	Widow's Age	Widow's Benefit	1st Child's Age	1st Child's Benefit	2nd Child's Age	2nd Child's Benefit	Total Benefit
a.	$310	37	$______	10	$______	—	$______	$______
b.	406	29	______	6	______	—	______	______
c.	275	56	______	32	______	29	______	______
d.	428	64	______	40	______	—	______	______
e.	512	24	______	3	______	1	______	______
f.	540	43	______	17	______	14	______	______

B

1. When Mr. Pitt died the social security agency found that his average monthly earnings were $941. At the time of his death his wife was 48 years old and his disabled son was 25 years old. The young man had been permanently disabled at age 14.

a. What was the lump-sum death benefit? ______

b. What was the PIA? ______

c. What monthly benefit did Mrs. Pitt receive? ______

d. What monthly benefit did the son receive? ______

2. When Mr. Waring died his wife was 42 years of age and his daughter who was attending college was 20 years of age. His average monthly income was computed to be $1,215.

a. What was the lump-sum death benefit? ______

b. What was the PIA? ______

c. What monthly benefit did Mrs. Waring receive? ______

d. What monthly benefit did the daughter receive? ______

Section 4 The Medicare Program

As part of the social security insurance we receive coverage under an arrangement called the Medicare program. To be eligible for this insurance, we must be at least 65 years of age. The program itself consists of two parts:

1. Medicare Hospital Insurance
2. Medicare Medical Insurance

Let us take a look at each of these coverages separately.

Medicare Hospital Insurance

1. Included as part of the social security tax deduction taken from our earnings during the years we work is a small premium that pays for the cost of this coverage. Therefore, there are no premiums to be paid after retirement.

2. In the event we are confined to a hospital, the amount we are required to pay is shown in the table below. The remainder is paid under the Medicare program.

HOSPITAL CARE

Number of Days	Payment
First 60	$160
Next 30	$ 40 per day
Next 60 (lifetime reserve)	$ 80 per day

3. The lifetime reserve of 60 days can be used only once in our life. As an example, if we were confined to a hospital for 100 days, the the last 10 days will be paid from the lifetime reserve of 60 days. This implies that we will have 50 days remaining in this reserve to be used during any future long illness.

$$60 - 10 = 50 \text{ days.}$$

4. The first 90 days for a single period of illness can be used over and over again. As long as 60 days elapse between the first period of illness and the second, then these illnesses are considered as two distinct periods of illness.

Example 1 Mr. Sherman spent 75 days in a hospital. His total bill for hospital services amounted to $11,550. How much of this bill was covered by Medicare insurance?

Solution

Computation for Mr. Sherman's cost →

First 60 days = $160
Days in excess of 60 = 75 − 60
= 15

Number of days ↓ ↓ **Cost per day**

Cost for excess days = 15 × $40
= $600

Mr. Sherman's cost = $160 + $600
= $760

Medicare cost = $11,550 − $760
= $10,790

Example 2 Because of a serious illness, Mrs. Williamson was confined to a hospital for 96 days. If she was covered by Medicare insurance, how much did this stay cost her?

Solution

First 60 days = $160

Number of days ↓ ↓ **Cost per day**

Next 30 days = 30 × $40
= $1,200

Number of days ↓ ↓ **Cost per day**

Remaining 6 days = 6 × $80
= $480

Total cost = $160 + $1,200 + $480
= $1,840

Unit 1

Another phase of the Medicare Hospital Insurance program is concerned with the cost of care in a nursing home.

1. In the event we are confined to a nursing home, the amount we are required to pay is shown in the table below. The remainder is paid under the Medicare program.

NURSING-HOME CARE	
Number of Days	**Payment**
First 20	Nothing
Next 80	$20 per day

2. Nursing-home coverage applies only if the following conditions are met:

(a) We must have been in a hospital for a period of at least 3 days before being admitted into the nursing home.

(b) No more than 14 days can elapse between the day we leave the hospital and the day we enter the nursing home.

(c) We must require skilled nursing care at the nursing home.

Example 3 Mrs. Fitzsimmons spent 74 days at a nursing home that charged $42 per day. How much of this cost will be covered by Medicare Hospital Insurance?

Solution

Computation of Mrs. Fitzsimmon's cost →

First 20 days = $0

Days in excess of 20 = 74 − 20
= 54

Number of days → 54; **Cost per day** → $20

Cost for excess days = 54 × $20
= $1,080

Mrs. Fitzsimmon's cost = $0 + $1,080
= $1,080

Number of days → 74; **Cost per day** → $42

Nursing-home bill = 74 × $42
= $3,108

Medicare payment = $3,108 − $1,080
= $2,028

Now let us take a look at the second coverage under the Medicare program.

Medicare Medical Insurance

1. The coverage under the medical program is not automatic. We are given the option of deciding whether or not we want to purchase this insurance. If we agree to purchase it, then the premiums are deducted from our monthly social security benefits.
2. Through this coverage the greater part of our doctor and medical bills will be paid. During any one year the total we are required to pay is

$60 + 20% of the excess over $60.

The remainder is paid by the Medicare program.

Example 4 Mr. Harper had surgery performed on him during the past year. The total of the doctor and medical bills amounted to $3,526. How much of this amount was covered by Medicare Medical Insurance?

Solution

Computation for Mr. Harper's cost

Mr. Harper's base cost = $60

Excess over $60 = $3,526 − $60
= $3,466

Mr. Harper's excess cost = 20% of excess
= .20 × $3,466
= $693.20

Base Cost **Cost on excess**

Mr. Harper's total cost = $60 + $693.20
= $753.20

Medicare cost = $3,526 − $753.20
= $3,772.80

Unit 1

Exercises A

1. Determine the hospital cost that will have to be paid by each person who is covered by Medicare Hospital Insurance.

	Hospital Stay (Days)	Insured's Cost		Hospital Stay (Days)	Insured's Cost
a.	15	$______	d.	72	$______
b.	53	______	e.	81	______
c.	65	______	f.	87	______

2. Determine the nursing home cost that will have to be paid by each person who is covered by Medicare Hospital Insurance.

	Nursing Home Stay (Days)	Insured's Cost		Nursing Home Stay (Days)	Insured's Cost
a.	8	$______	d.	25	$______
b.	14	______	e.	78	______
c.	47	______	f.	93	______

3. Determine the hospital cost that will have to be paid by each person who is covered by Medicare Hospital Insurance. None of them had used any of their lifetime reserve days.

	Hospital Stay (Days)	Insured's Cost		Hospital Stay (Days)	Insured's Cost
a.	100	$______	c.	135	$______
b.	126	______	d.	148	______

4. Determine the Medicare payment for each hospital charge.

	Hospital Bill	Hospital Stay (Days)	Insured's Cost	Medicare Payment
a.	$ 1,675	10	$______	$______
b.	7,538	45	______	______
c.	8,912	53	______	______
d.	11,836	65	______	______
e.	10,952	74	______	______
f.	15,007	82	______	______

5. Determine the Medicare payment for each nursing home charge.

	Nursing Home Cost	Nursing Home Stay (Days)	Insured's Cost	Medicare Payment
a.	$ 273	7	$______	$______
b.	756	18	______	______
c.	1,378	32	______	______
d.	2,016	56	______	______
e.	3,696	84	______	______

6. Determine the Medicare payment for each nursing home confinement.

	Nursing Home Stay (Days)	Cost per Day	Total Cost	Insured's Cost	Medicare Cost
a.	14	$37	$______	$______	$______
b.	23	38	______	______	______
c.	31	42	______	______	______
d.	56	40	______	______	______
e.	68	43	______	______	______

B

1. The following medical costs were incurred over the period of one year. Each of these people is covered by Medicare Medical Insurance. Determine how much of this cost was paid by the insured.

	Medical Cost	Base Amount	Excess Over $60	20% Of Excess	Insured's Cost
a.	$ 150	$______	$______	$______	$______
b.	280	______	______	______	______
c.	40	______	______	______	______
d.	52	______	______	______	______
e.	856	______	______	______	______
f.	1,050	______	______	______	______
g.	1,762	______	______	______	______
h.	4,374	______	______	______	______

2. The following medical costs were incurred over the period of one year. Determine the Medicare payment on each of these charges.

	Medical Cost	Base Amount	Excess Over $60	20% Of Excess	Insured's Cost	Medicare Cost
a.	$ 170	$______	$______	$______	$______	$______
b.	45	______	______	______	______	______
c.	782	______	______	______	______	______
d.	2,340	______	______	______	______	______
e.	3,945	______	______	______	______	______

C

1. Mrs. Yarby was covered by the medicare program at the time she was taken ill and had to go to the hospital. She spent 27 days there, for which the cost was $4,723, and another 16 days in a nursing home at a cost of $736.

a. How much of the hospital bill did Mrs. Yarby have to pay? ______

Unit 1

b. How much of the nursing-home bill did she have to pay? ________

c. What is the total amount that Mrs. Yarby had to pay to both institutions? ________

2. Mr. Darden, who is covered by medicare for both hospital and medical insurance, had a serious operation for which the surgeon's charge was $1,400. He spent 70 days in the hospital and another 38 days in a nursing home before returning to his own home. The hospital costs came to $11,457, while the nursing home charged him $1,678.

a. How much of the hospital bill did Mr. Darden have to pay? ________

b. How much of the nursing-home bill did Mr. Darden have to pay? ________

c. How much of the surgeon's fee did Mr. Darden have to pay? ________

d. How much of the entire bill did the Medicare coverage have to pay? ________

Unit 2 Company Retirement Plans

A great many corporations and their employees have long felt that social security benefits alone are not quite sufficient for most people. As a result, retirement funds have been designed that require joint contributions by both the company and the employee. Upon retirement the employee receives monthly benefits until death.

Although each plan differs somewhat from every other plan, the features described below are common to many of them.

1. The amount deducted from an employee's salary depends upon the age at which the person began to work for the company.
2. Since women on the average live longer than men, a woman will contribute more than a man of the same age.
3. The benefits an employee receives are based on average earnings over the last few years of employment.

As an illustration we are using the retirement plan of the Price-Davis Corporation. The rate that each employee must contribute is shown in the table below.

PRICE-DAVIS CORPORATION

Employee Contribution Rate

Entrance Age	Percent of Contribution		Entrance Age	Percent of Contribution	
	Men	Women		Men	Women
20	4.80	5.27	40	6.04	6.81
21	4.80	5.31	41	6.14	6.92
22	4.81	5.36	42	6.23	7.04
23	4.83	5.40	43	6.34	7.15
24	4.87	5.45	44	6.44	7.27
25	4.91	5.51	45	6.55	7.39
26	4.96	5.57	46	6.65	7.50
27	5.01	5.64	47	6.77	7.62
28	5.06	5.71	48	6.88	7.75
29	5.12	5.78	49	7.00	7.88
30	5.19	5.85	50	7.12	8.02
31	5.26	5.93	51	7.25	8.18
32	5.34	6.02	52	7.39	8.34
33	5.43	6.11	53	7.54	8.51
34	5.51	6.21	54	7.69	8.67
35	5.59	6.30	55	7.84	8.83
36	5.67	6.39	56	7.99	9.00
37	5.76	6.50	57	8.14	9.16
38	5.86	6.60	58	8.30	9.33
39	5.95	6.71	59	8.45	9.51

Example 1 Ben Nelson was employed by the Price-Davis Corporation at age 24. His weekly earnings are now \$385. Determine his weekly contribution to the retirement fund.

Explanation

Percent of Contribution

Entrance Age → Men Women

24 ——————→ 4.87

Percent of salary contributed

Solution

Weekly contribution = 4.87% of Earnings

= .0487 × \$385

= \$18.75

Unit 2

Perhaps equally as important to us as the amount we contribute to the retirement fund is the answer to the question: *How much will I receive in return for the money I contribute?* For the employees of the Price-Davis Corporation, their annual benefits are found by using the formula below.

$$\text{Annual benefit} = \frac{\text{Number of years employed}}{60} \times \begin{array}{c}\text{Average annual}\\ \text{salary during}\\ \text{last 3 years}\end{array}$$

Example 2 For the three years before retirement Clara Lee's average annual income was \$14,672. If she worked for the Price-Davis Corporation for 28 years, how large was her annual retirement benefit?

Solution

$$\begin{aligned}\text{Annual benefit} &= \frac{\text{Number of years employed}}{60} \times \begin{array}{c}\text{Average annual}\\ \text{salary during}\\ \text{last 3 years}\end{array}\\ &= \frac{28}{60} \times \$14{,}672\\ &= \frac{\$410{,}816}{60}\\ &= \$6{,}846.93\end{aligned}$$

Example 3 Joseph LaCombe worked for the Price-Davis Corporation for a period of 37 years. During the last three years his annual earnings were \$19,628, \$21,416, and \$22,351 respectively. What was his annual benefit upon retirement?

Solution

$$\begin{aligned}\text{Average of last 3 years} &= \frac{\$19{,}628 + \$21{,}416 + \$22{,}351}{3}\\ &= \frac{\$63{,}395}{3}\\ &= \$21{,}131.67\end{aligned}$$

$$\begin{aligned}\text{Annual benefit} &= \frac{\text{Number of years employed}}{60} \times \begin{array}{c}\text{Average annual}\\ \text{salary during}\\ \text{last 3 years}\end{array}\\ &= \frac{37}{60} \times \$21{,}131.67\\ &= \frac{\$781{,}871.79}{60}\\ &= \$13{,}031.20\end{aligned}$$

Exercises

A

The following exercises are based on the retirement plan of the Price-Davis Corporation.

1. Rename each percent numeral as a decimal numeral.

a. 4.93% ________	c. 5.40% ________	e. 7.03% ________
b. 5.02% ________	d. 6.39% ________	f. 8.00% ________

2. Determine the weekly contribution each employee will make to the retirement fund.

	Weekly Earnings	Entrance Age	Sex	Contribution Rate	Weekly Contribution
a.	$150	49	Male	________%	$________
b.	185	35	Female	________%	________
c.	217	41	Male	________%	________
d.	268	23	Male	________%	________
e.	257	28	Female	________%	________
f.	324	34	Female	________%	________

3. Determine the contribution each employee will make to the retirement fund over the period of one year.

	Weekly Earnings	Entrance Age	Sex	Contribution Rate	Weekly Contribution	Contribution For the Year
a.	$210	24	Male	________%	$________	$________
b.	250	21	Female	________%	________	________
c.	385	47	Female	________%	________	________
d.	346	51	Male	________%	________	________

B

1. Determine the annual benefit received by each retired employee.

	Number of Years Of Employment	Average Annual Earnings During Last 3 Years	Annual Benefit
a.	30	$16,000	$________
b.	20	18,000	________
c.	15	14,624	________
d.	45	12,576	________
e.	36	14,380	________

2. The annual earnings during the three years prior to retirement of each employee is shown below. Determine the annual retirement benefit of each of these employees.

	Number of Years of Employment	Last Three Years Annual Earnings	Average	Annual Benefit
a.	40	$14,000, $15,000, $16,000	$______	$______
b.	24	$12,000, $12,500, $13,000	______	______
c.	18	$17,400, $18,200, $19,000	______	______
d.	42	$21,600, $22,300, $24,200	______	______

C

1. Marion Torres began working for the Price-Davis Corporation at age 22 and retired at 64. Her average annual salary during the last three years before retirement was $16,560.

a. What was her annual retirement benefit? ______

b. If she received her benefits in monthly payments, how large was each check? ______

2. Jack Freeman retired from Price-Davis Corporation at age 65. His employment with the company began at age 31. During the three years prior to retirement his average annual earnings were $15,480.

a. What was his annual retirement benefit? ______

b. According to the mortality table on page 352, to the nearest whole number of years, how many years is Mr. Freeman expected to live after retirement? ______

c. If he lives the number of years predicted, what total amount of money will he collect under the retirement plan? ______

Unit 3 Individual Retirement Plans

Section 1 Annuities Purchased from Insurance Companies

Even the total of both the social security and company retirement benefits is less than the income many of us feel we will need. Realizing this, we often purchase an *annuity* from a life insurance company.

An annuity is simply a contract between an individual and an insurance company. On the one hand, we agree to give the company a certain amount of money. In return, the insurance company agrees to provide us with a monthly income for the duration of our life. The size of the monthly income will depend on:

1. The amount of money paid to the insurance company.
2. The age at which the annuity is purchased.
3. The age at which the monthly payments are to begin.

There are two varieties of annuities:

Single Premium	Annual Premium
One lump-sum premium at date of purchase. No other premiums paid.	Annual premiums are paid until date of maturity.

Under both of these arrangements benefits begin on the maturity date.

In some instances, people receive their first few benefits and then pass away. Normally, the insurance company is under no obligation to make any further payments. In view of this, many people prefer to accept a slightly smaller monthly benefit in return for a guarantee that someone will receive benefits for a fixed period of time. As an example, under a

10 Year Certain Annuity

If we live more than 10 years	If we live less than 10 years
We receive monthly benefits for our entire life.	After our death the person whom we name as beneficiary will receive monthly benefits until the end of the 10th year after the maturity date.

Unit 3

Example 1 At age 23, George Turner purchased a $15,000 single-premium annuity. If his monthly benefits are to begin at age 65, how large will each of these benefits be? (See table, page 424.)

SINGLE-PREMIUM ANNUITY							ANNUAL-PREMIUM ANNUITY						
Monthly Life Income Purchased by $1,000 Single Premium							Monthly Life Income (10 Year Certain) Purchased by $100 Annual Premium				Monthly Life Income (10 Year Certain) Purchased by $100 Annual Premium		
MALE				FEMALE			MALE				FEMALE		
Age at Date Of Maturity			Age	Age at Date Of Maturity			Age at Date Of Maturity			Age	Age at Date Of Maturity		
60	65	70		60	65	70	60	65	70		60	65	70
$15.11	$19.43	$24.98	**20**	$13.49	$17.31	$22.26	$35.74	$48.98	$66.33	**20**	$31.92	$43.62	$59.11
14.71	18.91	24.31	**21**	13.13	16.84	21.66	34.29	47.12	63.93	**21**	30.63	41.96	56.97
14.31	18.41	23.66	**22**	12.78	16.39	21.08	32.88	45.30	61.60	**22**	29.37	40.34	54.89
13.93	17.92	23.03	**23**	12.44	15.95	20.52	31.51	43.53	59.33	**23**	28.14	38.77	52.87
13.56	17.44	22.41	**24**	12.11	15.53	19.97	30.17	41.81	57.12	**24**	26.94	37.23	50.90
13.19	16.97	21.81	**25**	11.78	15.11	19.43	28.87	40.14	54.97	**25**	25.78	35.74	48.98
12.84	16.51	21.23	**26**	11.47	14.71	18.91	27.60	38.51	52.87	**26**	24.65	34.29	47.12
12.50	16.07	20.66	**27**	11.16	14.31	18.41	26.37	36.93	50.84	**27**	23.55	32.88	45.30
12.16	15.64	20.10	**28**	10.86	13.93	17.92	25.17	35.38	48.85	**28**	22.48	31.51	43.53
11.84	15.22	19.57	**29**	10.57	13.56	17.44	24.00	33.88	46.92	**29**	21.44	30.17	41.81
11.52	14.82	19.04	**30**	10.29	13.19	16.97	22.87	32.42	45.05	**30**	20.42	28.87	40.14
11.21	14.42	18.53	**31**	10.01	12.84	16.51	21.76	31.00	43.22	**31**	19.44	27.60	38.51
10.91	14.03	18.04	**32**	9.74	12.50	16.07	20.69	29.62	41.44	**32**	18.47	26.37	36.93
10.62	13.66	17.55	**33**	9.48	12.16	15.64	19.64	28.27	39.71	**33**	17.54	25.17	35.38
10.34	13.29	17.08	**34**	9.23	11.84	15.22	18.62	26.96	38.02	**34**	16.63	24.00	33.88
10.06	12.94	16.63	**35**	8.98	11.52	14.82	17.63	25.68	36.38	**35**	15.74	22.87	32.42
9.79	12.59	16.18	**36**	8.74	11.21	14.42	16.66	24.44	34.79	**36**	14.88	21.76	31.00
9.53	12.25	15.75	**37**	8.51	10.91	14.03	15.72	23.23	33.23	**37**	14.04	20.69	29.62
9.27	11.93	15.33	**38**	8.28	10.62	13.66	14.81	22.06	31.72	**38**	13.22	19.64	28.27
9.02	11.61	14.92	**39**	8.06	10.34	13.29	13.92	20.91	30.25	**39**	12.43	18.62	26.96
8.78	11.30	14.52	**40**	7.84	10.06	12.94	13.05	19.80	28.82	**40**	11.66	17.63	25.68
8.55	10.99	14.13	**41**	7.63	9.79	12.59	12.21	18.71	27.43	**41**	10.90	16.66	24.44
8.32	10.70	13.75	**42**	7.43	9.53	12.25	11.39	17.66	26.07	**42**	10.17	15.72	23.23
8.10	10.41	13.38	**43**	7.23	9.27	11.93	10.59	16.63	24.75	**43**	9.46	14.81	22.06
7.88	10.13	13.03	**44**	7.04	9.02	11.61	9.81	15.63	23.47	**44**	8.76	13.92	20.91
7.67	9.86	12.68	**45**	6.85	8.78	11.30	9.06	14.66	22.22	**45**	8.09	13.05	19.80
7.46	9.60	12.34	**46**	6.67	8.55	10.99	8.32	13.71	21.00	**46**	7.43	12.21	18.71
7.26	9.34	12.01	**47**	6.49	8.32	10.70	7.60	12.79	19.81	**47**	6.79	11.39	17.66
7.07	9.09	11.69	**48**	6.31	8.10	10.41	6.91	11.89	18.66	**48**	6.17	10.59	16.63
6.88	8.85	11.37	**49**	6.14	7.88	10.13	6.23	11.02	17.54	**49**	5.56	9.81	15.63
6.70	8.61	11.07	**50**	5.98	7.67	9.86	5.57	10.17	16.45	**50**	4.97	9.06	14.66
6.52	8.38	10.77	**51**	5.82	7.46	9.60	4.93	9.34	15.39	**51**	4.40	8.32	13.71
6.34	8.16	10.48	**52**	5.66	7.26	9.34	4.30	8.54	14.35	**52**	3.84	7.60	12.79
6.17	7.94	10.20	**53**	5.51	7.07	9.09	3.69	7.76	13.35	**53**	3.30	6.91	11.89
6.01	7.73	9.93	**54**	5.37	6.88	8.85	3.10	6.99	12.37	**54**	2.77	6.23	11.02
5.85	7.52	9.66	**55**	5.22	6.70	8.61	2.52	6.25	11.41	**55**	2.25	5.57	10.17
	7.32	9.41	**56**		6.52	8.38		5.53	10.49	**56**		4.93	9.34
	7.12	9.15	**57**		6.34	8.16		4.83	9.58	**57**		4.30	8.54
	6.93	8.91	**58**		6.17	7.94		4.15	8.70	**58**		3.69	7.76
	6.75	8.67	**59**		6.01	7.73		3.48	7.85	**59**		3.10	6.99
	6.57	8.44	**60**		5.85	7.52		2.83	7.02	**60**		2.52	6.25
		8.21	**61**			7.32			6.21	**61**			5.53
		7.99	**62**			7.12			5.42	**62**			4.83
		7.57	**63**			6.93			4.65	**63**			4.15
		7.37	**64**			6.75			3.91	**64**			3.48
			65			6.57			3.18	**65**			2.83

Explanation

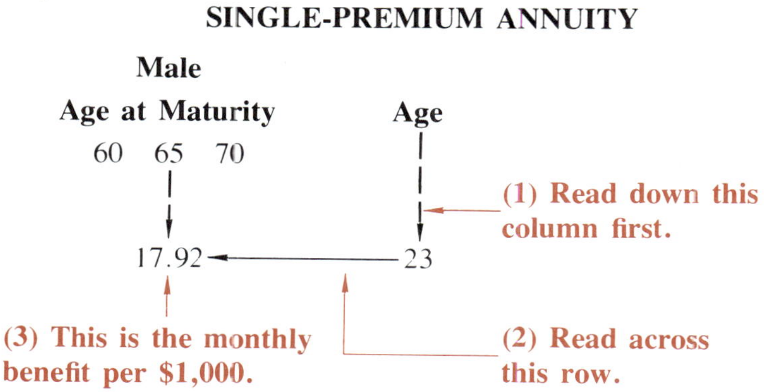

Solution

Number of $1,000's — Benefit per $1,000

Monthly benefit = 15 × $17.92
= $268.80

Example 2 At age 37, Grace Becker purchased an annual-premium annuity by agreeing to make annual payments of $1,800 until age 60. How large will her monthly benefits be when she reaches 60?

Explanation

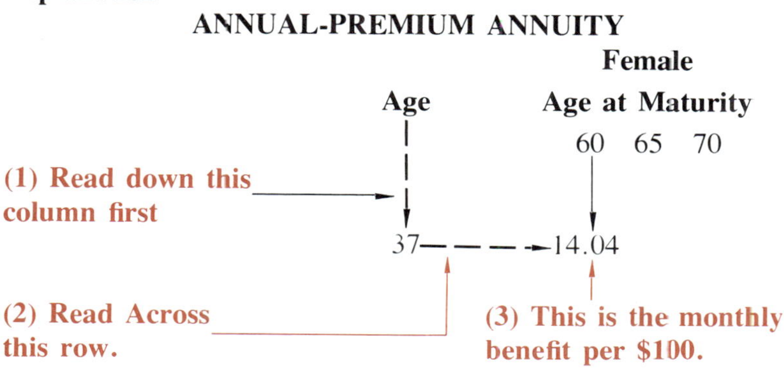

Solution

Number of $100's — Benefit per $100

Monthly benefit = 18 × $14.04
= $252.72

Unit 3

Exercises A

1. Determine the monthly retirement income that can be purchased by these $1,000 single-premium annuities.

	Age at Purchase	Age at Maturity	Sex	Monthly Income
a.	25	60	Male	$______
b.	27	60	Female	______
c.	34	65	Female	______
d.	41	65	Male	______
e.	54	70	Male	______
f.	49	70	Female	______

2. Determine the monthly retirement income that can be purchased by these $100 annual-premium annuities.

	Age at Purchase	Age at Maturity	Sex	Monthly Income
a.	22	60	Female	$______
b.	32	60	Male	______
c.	27	65	Female	______
d.	31	70	Female	______
e.	43	65	Male	______

3. Determine the monthly retirement income that can be purchased by these single-premium annuities.

	Lump-sum Payment	Age at Purchase	Age at Maturity	Sex	Monthly Income Per $1,000	Monthly Income
a.	$10,000	27	65	Male	$______	$______
b.	15,000	25	60	Male	______	______
c.	25,000	34	70	Female	______	______
d.	40,000	40	65	Male	______	______
e.	65,000	43	60	Female	______	______
f.	85,000	45	65	Female	______	______

4. Determine the monthly income that can be purchased by these annual-premium annuities.

	Annual Premium	Age at Purchase	Age at Maturity	Sex	Monthly Income Per $100	Monthly Income
a.	$ 500	20	65	Female	$______	$______
b.	1,100	29	60	Male	______	______
c.	1,100	29	60	Female	______	______
d.	1,400	45	70	Male	______	______
e.	2,600	32	60	Female	______	______

B

1. Charles Samuels purchased a $25,000 single-premium annuity. His monthly benefits will begin when he reaches the age of 60.

a. How large will the benefit be if the annuity was purchased at age 20? ________

b. How large will the benefit be if the annuity was purchase as age 40? ________

c. How much more will each of the benefits be if the purchase is made at age 20 rather than age 40? ________

2. Evelyn Noyes purchased a $30,000 single-premium annuity at age 36.

a. How large will her monthly benefits be if they begin at age 60? ________

b. How large will her monthly benefits be if they begin at age 70? ________

c. How much more will she receive each month if the benefits are to begin at age 70 rather than age 60? ________

3. Mr. Hawkins purchased a $1,000 annual-premium annuity at age 30. His monthly benefits are to begin when he reaches the age of 65.

a. How large will his monthly benefit be? ________

b. If Mrs. Hawkins had purchased the same annuity at the same age, how large would her monthly benefit be? ________

c. How much will Mr. Hawkins receive during his life expectancy of 15 years (Federal government mortality table)? ________

d. How much will Mrs. Hawkins receive during her life expectancy of 18 years 2 months (Federal government mortality table)? ________

e. How much more will one receive than the other during their life expectancy periods? ________________

4. John Blanchard purchased a $30,000 single-premium annuity at age 40. His monthly benefits are to begin at age 65.

a. How large will his monthly benefit be? ________

b. His life expectancy at age 65 is 15 years 1 month. If he lives this period, what will the total of his benefits be? ________

c. If he had deposited the $30,000 in a bank at a 6% rate compounded annually, to what amount would the $30,000 grow in the 25-year period? (See compound interest table, page 232.)

____

d. How much more - or less - will his total benefits be, than the value of his money had he kept it in a bank?

____

Section 2 The Keogh Plan and The Individual Retirement Account

For a number of years the Federal government has encouraged many of us to create our own retirement fund. This encouragement comes by way of permitting us to defer the payment of income tax on any money deposited in the fund. Unfortunately, everyone does not qualify for either of the programs designed by the government;

The Keogh Plan
The Individual Retirement Account (IRA)

These plans have certain similar provisions:

1. Money deposited in the retirement fund can not be withdrawn until age 59 1/2.
2. We must begin to withdraw money by age 70 1/2.
3. The money must be withdrawn during our life-expectancy period.
4. The money we withdraw is taxed during the year it is withdrawn.

They differ, though, in the following ways:

Keogh Plan

1. Designed primarily for people who operate their own business – doctors, lawyers, shopowners, artists . . . .
2. The maximum that can be deposited in the fund each year is $7,500 or 15% of our annual income, whichever of these two is the smaller amount.

Individual Retirement Account

1. Designed primarily for people who are employed by companies that have no retirement plan.
2. The maximum that can be deposited in the fund each year is $1,500 or 15% of our annual income, whichever of the two is the smaller amount.

The people who choose to invest in these plans are, of course, interested in delaying payment of their income tax until the time of retirement. However, they are also interested in the answers to these two questions:

1. How much money will there be in the account at the time I retire?
2. How much income will I have from this fund when I retire?

Actually, the first question was answered some time ago when we studied the growth of annual deposits. The factor table below is similar to the one we used at that time. The factors in this one, however, are a result of compounding the interest daily as is common practice.

ANNUITY FACTOR TABLE
(Interest Compounded Daily)

	Interest Rate					
Years	5%	5½%	6%	6½%	7%	7½%
15	23.0449	24.1185	25.2537	26.4545	27.7250	29.0698
20	35.5288	37.8023	40.2536	42.8979	45.7515	48.8329
25	51.6139	55.8861	60.5854	65.7582	71.4560	77.7370
30	72.3390	79.7845	88.1443	97.5398	108.1088	120.0101
35	99.0428	111.3672	125.4993	141.7242	160.3733	181.8359
40	133.4496	153.1051	176.1325	203.1516	234.8989	272.2578

In each of the examples and exercises in this section we will assume that the interest is compounded daily.

Example 1 Sarah Billings deposited $5,200 annually in her Keogh Plan account over a period of 25 years. The interest rate paid on this account was 6 1/2%. How large was her account at the end of the 25 years?

Explanation

Years → 25 ——————————→ 65.7582 ← **Interest Rate $6\frac{1}{2}\%$**

This is the factor on 6 1/2% after 25 years.

Solution

Total deposited = Annual deposit × Annuity Factor
= $5,200 × 65.7582
= $341,942.64

Explanation Continued:

The annuity factor in this example can also be interpreted as follows:

> If $1 is deposited each year over a period of 25 years, the amount on deposit will be $65.7582 if the interest rate is 6 1/2% compounded daily.

The answer to the second question raised earlier,

> "How much income will I receive when I retire?"

involves nothing more that what we already know about the payment of loans on the installment plan. Now, however, it is the bank that is returning our money to us in monthly installment payments. A table of factors for these monthly payments is shown below.

MONTHLY BENEFIT FACTORS FOR A KEOGH OR IRA PLAN
(Interest Compounded Daily)

	Interest Rate					
Years	5%	$5\frac{1}{2}\%$	6%	$6\frac{1}{2}\%$	7%	$7\frac{1}{2}\%$
11	.009904	.010159	.010417	.010680	.010946	.011216
12	.009289	.009547	.009809	.010076	.010347	.010621
13	.008771	.009033	.009299	.009570	.009845	.010124
14	.008330	.008594	.008865	.009139	.009419	.009703
15	.007950	.008218	.008492	.008770	.009054	.009343
16	.007618	.007891	.008168	.008451	.008739	.009032
17	.007329	.007605	.007886	.008172	.008464	.008762
18	.007073	.007352	.007637	.007928	.008224	.008526
19	.006846	.007129	.007417	.007711	.008012	.008318
20	.006644	.006929	.007221	.007519	.007824	.008134

Example 2 At age 61 Jonathan Byers decided he would withdraw his Keogh Plan retirement fund monthly over a period of 12 years. The fund showed a balance of $75,000. If the annual rate of interest on the account is 7%, how large will each monthly benefit be?

Explanation

Years → 12 - - - - - - → .010347 ← **Interest Rate 7%**

Solution

Size of fund → $75,000; **Monthly factor** → .010347

$$\text{Monthly benefit} = \$75{,}000 \times .010347$$
$$= \$776.03$$

Exercises A

1. Determine the annuity factor for each rate over the period shown.

	Interest Rate	Years	Annuity Factor
a.	5%	30	________
b.	6%	20	________
c.	$7\frac{1}{2}$%	25	________
d.	$5\frac{1}{2}$%	35	________

2. One dollar is deposited each year over the period shown in each exercise. How large will the account be at the close of the period?

	Interest Rate	Years	Size of Account
a.	7%	25	$________
b.	$6\frac{1}{2}$%	40	________
c.	5%	30	________
d.	$7\frac{1}{2}$%	15	________
e.	6%	35	________

3. Determine the monthly benefit factor for each rate over the period shown.

	Interest Rate	Years	Monthly Benefit Factor
a.	5%	14	______
b.	7%	16	______
c.	6%	12	______
d.	$5\frac{1}{2}$%	18	______
e.	$6\frac{1}{2}$%	13	______

B

1. Determine the size of the IRA fund if the following annual deposits are made over the period shown.

	Annual Deposits	Interest Rate	Years	Annuity Factor	Size of Fund
a.	$ 100	5%	30	______	$______
b.	100	$7\frac{1}{2}$%	30	______	______
c.	400	6%	15	______	______
d.	1,200	7%	35	______	______
e.	1,500	$5\frac{1}{2}$%	40	______	______

2. Determine the size of the Keogh Fund if the following annual deposits are made over the period shown.

	Annual Deposits	Interest Rate	Years	Annuity Factor	Size of Fund
a.	$3,000	6%	40	______	$______
b.	5,000	5%	25	______	______
c.	5,000	$7\frac{1}{2}$%	25	______	______
d.	2,500	$6\frac{1}{2}$%	30	______	______
e.	7,500	7%	15	______	______

3. The size of each of the following IRA or Keogh Fund at the time of retirement is shown below. Determine the monthly benefit over the period of years shown.

	Size of Fund	Interest Rate	Years	Monthly Benefit Factor	Monthly Benefit
a.	$ 20,000	5%	12	______	$______
b.	50,000	6%	15	______	______
c.	70,000	7%	18	______	______
d.	45,000	$5\frac{1}{2}$%	16	______	______
e.	125,000	$6\frac{1}{2}$%	11	______	______

C

1. Annual deposits of $6,000 are made into a Keogh Fund for a period of 20 years. The fund pays an interest rate of 7 1/2%.

a. How large will the fund be at the end of the period? (Round your answer to the nearest dollar.) ________

b. Monthly benefits are withdrawn from this fund over the period of 12 years. How large is each monthly benefit? ________

2. At age 40, Matthew Kelly begins to make annual payments of $1,200 into an IRA. His plans are to withdraw the money over a 14-year period starting at age 65. The annual rate paid on this fund is 6%.

a. For how many years did he make contributions to the fund? ________

b. How large will the fund be at age 65? ________

c. How large will the monthly benefits be during retirement? ________

3. At age 37 Madeline Korn began to make annual payments of $4,000 into a Keogh Fund that paid a 6 1/2% rate of interest. At age 62 she retired and began to withdraw money monthly. The amount withdrawn was based on her life expectancy.

a. For how many years did she make contributions to the fund? ________

b. How large will the fund be at age 62? ________

c. According to the mortality table on page 352, to the nearest year, what is her life expectancy at age 62? ________

d. How large will her monthly benefits be during the life expectancy period? ________

Review of Chapter 11

When doing the computation for the exercises that follow, it will be necessary to refer to the tables in this chapter.

1. Determine the social security primary insurance amount for each average monthly earning.

	Av. Monthly Earnings	PIA		Av. Monthly Earnings	PIA
a.	$160	$________	d.	$ 675	$________
b.	175	________	e.	980	________
c.	350	________	f.	1,140	________

Review

2. In each exercise the husband receives his first monthly social security benefit on the birthday shown. Determine his monthly benefit.

	Birthday	PIA	Percent of PIA	Monthly Benefit
a.	65	$269	______%	$______
b.	62	350	______%	______
c.	63	420	______%	______

3. In each exercise the wife receives her first monthly social security benefit on the birthday shown. Determine her monthly benefit.

	Birthday	PIA	Percent of PIA	Monthly Benefit
a.	65	$370	______%	$______
b.	62	431	______%	______

4. Determine the family maximum benefit on a primary insurance amount of $396. ______

5. Mr. Perkins was 66 at the time he began to collect social security benefits on average monthly earnings of $792. His wife was 59 at the time, and they have no children.

a. What is the PIA on these earnings? ______
b. How large was Mr. Perkins' monthly benefit? ______
c. How large was Mrs. Perkins' monthly benefit? ______

6. On her 61st birthday, Mrs. Bethea began to receive social security benefits based on her deceased husband's average monthly earnings of $410.

a. What was the PIA on this amount? ______
b. What was Mrs. Bethea's monthly benefit? ______

7. Mr. Akorn died leaving a wife, 27, and a daughter, 5. His PIA was computed as $520.

a. What was Mrs. Akorn's monthly benefit? ______
b. What was the daughter's monthly benefit? ______
c. What is the family maximum benefit for this PIA? ______
d. Had there been a second child in the family, what would the monthly benefit have been for this child? ______

8. Determine the hospital cost that will have to be paid by each person who is covered by Medicare Hospital Insurance.

	Hospital Stay (Days)	Insured's Cost		Hospital Stay (Days)	Insured's Cost
a.	23	$________	c.	67	$________
b.	48	________	d.	81	________

9. Determine the nursing home cost that will have to be paid by each person who is covered by Medicare Hospital Insurance.

	Nursing Home Stay (Days)	Insured's Cost		Nursing Home Stay (Days)	Insured's Cost
a.	16	$________	c.	43	$________
b.	27	________	d.	78	________

10. During the past year Mrs. Chester had medical bills amounting to 12,360. If she is covered by Medicare Medical Insurance, how much of this amount did she have to pay? ________

11. Under the Price-Davis Corporation retirement plan described in this chapter, how much will Robert Riley contribute to the retirement fund during a year in which he earned $14,630? He started to work for the company at age 26. ________

12. At the time Frank Daley retired he had been working for the Price-Davis Corporation for 28 years. His average annual earnings during the last three years of this period were $15,240. What was Mr. Daley's annual retirement benefit under the company plan?

____

13. When Grace Johnson was 27 she purchased a $20,000 single-premium annuity. She plans to begin collecting monthly benefits at age 60. How large will these monthly benefits be?

____

14. A person made annual payments of $1,100 into an IRA over a period of 35 years. The interest rate paid on this plan was 7%. How large was the fund at the end of 35 years? ________

15. Hazel Jorgenson had accumulated $100,000 in a Keogh fund by the time she retired. Her life expectancy at retirement is 17 years. If the interest rate paid on the fund is 7%, how large will her monthly benefits be during the life expectancy period?

____

COLONIAL AVE
FAIRFAX AVE

Chapter 12

The Cost of Housing

Where and how we choose to live very often relate to the things we consider important to us. Some people who rent an apartment may do so because they do not want to be burdened with the chores connected with the upkeep of a home. By renting an apartment, they leave these cares to others.

On the other hand, the people who prefer a home of their own see things quite differently. They look with pride upon their home and the condition in which they have kept it. The payments they make month after month and year after year do not lead to a drawer full of rent receipts but rather to the eventual ownership of a home for themselves and their family.

In this chapter we are going to examine some of the major expenses that face us as either the apartment dweller or the homeowner.

The Cost of Utilities Unit 1

Section 1 The Cost of Electricity
Part 1

As in the case of practically everything else, the cost of electricity has increased immensely over recent years. Perhaps if we knew the method by which the cost for the use of this energy is computed we would be in a better position to consider ways to reduce that cost.

The flow of electricity is measured in terms of a unit called the *kilowatt-hour*. To obtain some understanding of this unit we must first realize that every household appliance requiring electrical energy for operation has stamped on it a number such as, 650 watts. As an example, printed on every light bulb used in the home we will very likely find one of the terms below:

10 watts, 25 watts, 40 watts, 60 watts, 75 watts, 100 watts.

In the event a 40 watt bulb is used for the period of 1 hour, the bulb is said to have consumed

40 watt-hours of electricity. (40×1)

Similarly, if the 40 watt bulb is used for a period of 2 hours, it will have consumed

80 watt-hours of electricity. (40×2)

In general, the number of watt-hours of electricity consumed by an appliance is determined by multiplying *the number of watts required to run the appliance* by *the number of hours the appliance had been in operation*. Therefore,

$$\text{Watt-hours} = \text{Number of watts} \times \text{Number of hours.}$$

Example 1 An electric heating pad operates on 210 watts of electricity. In order to ease the pain of a strained back, a person used the heating pad for a period of 14 hours. How many watt-hours of electricity were consumed during this period?

Solution

$$\begin{aligned}\text{Number of watt-hours} &= \text{Number of watts} \times \text{Number of hours}\\ &= 210 \times 14\\ &= 2{,}940\end{aligned}$$

In an average home, thousands of watt-hours of electricity are consumed each day. Hence, measuring the consumption of electricity in watt-hours is comparable to measuring the distance from San Francisco to New York in meters – the unit is far too small.

To overcome this difficulty, a much larger unit was developed; it is called the *kilowatt-hour* (kwh).

A kilowatt-hour is equivalent to 1,000 watt-hours.

In view of this,

2,000 watt-hours are equivalent to 2 kwh (2,000 ÷ 1,000)
3,000 watt-hours are equivalent to 3 kwh (3,000 ÷ 1,000)

and in general:

The number of kilowatt-hours of electricity is equivalent to the number of watt-hours divided by 1,000.

Example 2 Determine the number of kilowatt-hours of electricity in 8,350 watt-hours.

Solution

$$\begin{aligned}\text{Number of kwh} &= \text{Number of watt-hours} \div 1{,}000\\ &= 8{,}350 \div 1{,}000\\ &= 8.35\end{aligned}$$

Example 3 Lillian Turner spent 3 1/2 hours ironing her clothes. If her iron requires 1,050 watts of electricity, how many kilowatt-hours of electricity were consumed during this period of time?

Solution

$$\begin{aligned}\text{Number of watt-hours} &= \text{Number of watts} \times \text{Number of hours}\\ &= 1{,}050 \times 3\tfrac{1}{2}\\ &= 3{,}675\end{aligned}$$

$$\begin{aligned}\text{Number of kwh} &= \text{Number of watt-hours} \div 1{,}000\\ &= 3{,}675 \div 1{,}000\\ &= 3.675\end{aligned}$$

Unit 1

Exercises

A

1. Rename these watt-hours as kilowatt-hours.

	Watt-Hours	Kilowatt-Hours		Watt-Hours	Kilowatt-Hours
a.	4,000	______	e.	6,750	______
b.	7,000	______	f.	29,324	______
c.	56,000	______	g.	700	______
d.	124,000	______	h.	225	______

2. Determine the number of watt-hours consumed by each appliance.

	Number Of Watts	Hours Operated	Watt-Hours Consumed
a.	345	3	______
b.	1,100	4	______
c.	180	7	______
d.	2,500	$5\frac{1}{2}$	______
e.	645	$4\frac{1}{3}$	______
f.	240	$3\frac{3}{4}$	______
g.	645	$6\frac{2}{3}$	______

3. Determine the number of kilowatt-hours consumed by each appliance.

	Number Of Watts	Hours Operated	Watt-Hours Consumed	Kilowatt-Hours Consumed
a.	400	15	______	______
b.	800	9	______	______
c.	350	6	______	______
d.	190	4	______	______
e.	2,400	7	______	______
f.	1,500	8	______	______

B

1. The markings on the motor of Fred Evans' electric table saw show that it operates on 1,150 watts of electricity. When Fred remodeled his home, he used the saw for a period of 75 hours. How many kilowatt-hours of electricity were consumed by the operation of the saw? ______

2. To cool off her bedroom during the summer evenings before going to bed, Doris Templer ran a portable fan for a period of 2 1/2 hours. The fan operates on 395 watts of electricity. If Doris found it necessary to operate the fan for 60 days, how many kilowatt-hours of electricity did she use? ________

3. William Carlton's electric coffeepot runs for approximately 40 minutes each time he makes coffee. If the information on the appliance shows that it uses 600 watts of electricity, how many kilowatt-hours are used each time William brews coffee?

____

4. During a recent evening, Mr. Speer watched television for 4 1/2 hours. During the same evening, Mrs. Speer ironed for 2 hours and 20 minutes and listened to their stereo set at the same time. Each used light bulbs of 30 and 150 watts respectively. In addition, the TV set operated on 180 watts, the stereo on 150 watts, and the iron on 1,200 watts. What was the total number of kilowatt-hours of electricity used that evening? ________

Part 2

The amount of money we pay for electricity depends upon the number of kilowatt-hours consumed by our electrical appliances. As the current enters our home it must first pass through an electric meter before it reaches the appliances. There are usually five dials on the face of the meter. The current moves the arrow on each dial as the appliances consume electricity.

Kilowatt-Hours

Unit 1

Each month an employee of the electric utility company reads the number of kilowatt-hours registered on our meter. This is done by examining each arrow on each dial. The number recorded is the *smaller of the two numbers* between which the arrowhead falls. Thus, the recording for the dial below is the number 6.

(1) The arrowhead is between 6 and 7.

(2) The number recorded is 6; it is the smaller number.

There is only one exception to the situation described above and that occurs when the arrowhead falls between 9 and 0. When this occurs, the number recorded is the number 9.

The arrowhead is between the 0 and 9.

The number 9 is recorded for both readings above.

Example 1 Determine the reading on the meter below.

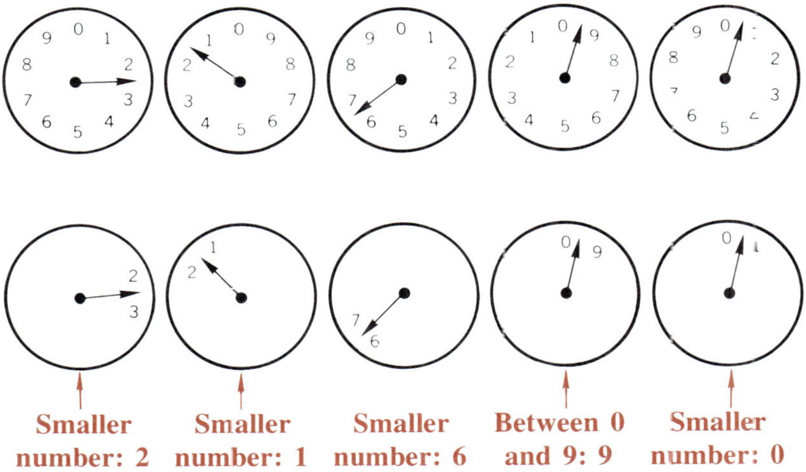

Solution

Meter reading = 21,690 kwh

Explanation

The electric meter is used in the same manner as we would use the odometer (mileage indicator) on our car. If the odometer registers 2,000 miles at the start of a trip and registers 7,000 miles at the end of the trip, then we know that the distance traveled during the trip is 5,000 miles.

Reading at end — **Reading at outset** — **Distance traveled**

7,000 − 2,000 = 5,000

Similarly, since the meter is read at both the outset and the end of a monthly period, the difference between the two readings is the number of kilowatt-hours of electricity consumed during the month.

Unit 1

Example 2 The electric meter reading on June 28 was 34,726. The previous reading taken on May 27 was 34,631. Determine the number of kilowatt-hours of electricity consumed during this period.

Solution

Reading at end ↓ ↓ **Reading at outset**

$$\text{KWH consumed} = 34{,}726 - 34{,}631$$
$$= 95$$

After determining the number of kilowatt-hours of electricity we have consumed during the month, the utility company then refers to a table such as the one below to compute our electric bill for the month.

CAPE ELECTRIC AND GAS COMPANY

Residential Rate per Month – Electric

KWH Consumed	Cost per KWH
First 10	29.3¢
Next 60	7.1¢
Next 230	6.3¢
Over 300	4.8¢

Example 3 On August 25 an electric meter reading was recorded as 35,682. On September 24, when it was again checked, it registered 36,054. Use the rate table of the Cape Electric and Gas Company to determine the cost of electricity for this period.

Solution

Reading at end → 36,054; Reading at outset → 35,682

$$\text{KWH consumed} = 36{,}054 - 35{,}682 = 372$$

Cost for:

First 10	Next 60	Next 230	Over 300
10 × 29.3¢	60 × 7.1¢	230 × 6.3¢	72 × 4.8¢
= 10 × $.293	= 60 × $.071	= 230 × $.063	= 72 × $.048
= $2.93	= $4.26	= $14.49	= $3.456

Total cost = $2.93 + $4.26 + $14.49 + $3.456
= $25.136 or $25.14

Exercises

A

1. Determine the reading for each meter.

a. 

______

d. 

______

b. 

______

e. 

______

c. 

______

f. 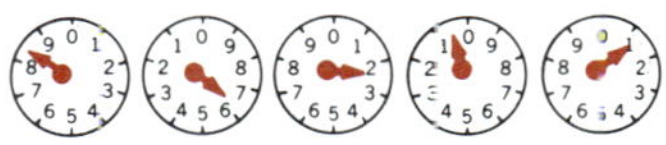

______

2. Determine the number of kilowatt-hours of electricity consumed during the month in each exercise.

	Meter At Beginning Of Month	Meter At End Of Month	
a.			______
b.			______
c.			______

B

The rate table of the Cape Electric and Gas Company is to be used when doing the exercises below.

1. Determine the electric bill for the month for each customer.

	KWH Consumed	Cost for: First 10 KWH	Cost for: Over 10 KWH	Total Cost
a.	25	$______	$______	$______
b.	46	______	______	______
c.	65	______	______	______

2. Determine the electric bill for the month for each customer.

	KWH Consumed	Cost for: First 10 KWH	Cost for: Next 60 KWH	Cost for: Over 70 KWH	Total Cost
a.	80	$______	$______	$______	$______
b.	150	______	______	______	______
c.	206	______	______	______	______
d.	254	______	______	______	______

3. Determine the electric bill for the month for each customer.

	KWH Consumed	Cost for: First 10 KWH	Cost for: Next 60 KWH	Cost for: Next 230 KWH	Cost for: Over 300 KWH	Total Cost
a.	320	$______	$______	$______	$______	$______
b.	375	______	______	______	______	______
c.	412	______	______	______	______	______
d.	634	______	______	______	______	______

4. Based in the meter readings shown below, determine the electric bill for the month.

	Reading At Beginning Of Month	Reading At End Of Month		Reading At Beginning Of Month	Reading At End Of Month
a.	37,065	37,113	c.	16,974	17,341
b.	21,576	21,725	d.	60,284	60,952

	KWH Consumed	Cost for: First 10 KWH Or Less	Cost for: Next 60 KWH Or Less	Cost for: Next 230 KWH Or Less	Cost for: Over 300 KWH	Total Cost
a.	______	$______	$______	$______	$______	$______
b.	______	______	______	______	______	______
c.	______	______	______	______	______	______
d.	______	______	______	______	______	______

C

Utility companies are permitted to add an adjustment charge to their bill. The adjustment charge depends on the amount of money the utility company had to pay to produce electricity that month.

1. The adjustment charge for the month of June was $.0341 per kwh. Determine the total adjustment charge each customer had to pay.

	KWH Consumed	Adjustment Charge		KWH Consumed	Adjustment Charge
a.	62	$________	c.	378	$________
b.	204	________	d.	725	________

2. The adjustment charge for the month of September for customers of the Cape Electric and Gas Company was $.02741 per kwh. Determine the total bill each customer received.

			Cost for:				
	KWH Consumed	Adjustment Charge	First 10 KWH Or Less	Next 60 KWH Or Less	Next 230 KWH Or Less	Over 300 KWH	Total Cost
a.	51	$_____	$_____	$_____	$_____	$_____	$_____
b.	130	_____	_____	_____	_____	_____	_____
c.	368	_____	_____	_____	_____	_____	_____
d.	741	_____	_____	_____	_____	_____	_____

D

1. The information on the back of an electric clock shows that it operates on 2 watts of electricity. What is the total cost of operating this clock for an entire month of 31 days if the cost of each kilowatt-hour is 6.5¢? ________

2. A 65-gallon hot-water heater requires 2,500 watts of electricity to operate. In order to heat the water, electric current must flow through the appliance approximately 6 hours per day. During a 30-day month, what will be the cost of heating water for this home if the price of each kilowatt-hour is 5.8¢? ________

Unit 1

3. An electric clothes dryer operates on 4,200 watts of electricity. Mrs. Armond uses the dryer for approximately 2 hours and 30 minutes every Monday. How much will it cost Mrs. Armond to dry her clothes during a month in which there are four Mondays if the cost of each kilowatt-hour is 6.71¢? ________

4. Each weekday evening after dinner, Milton and Ruth Riker sit down to a routine of 4 1/2 hours with their TV set. Their son George goes to his room to study and to listen to the radio for approximately 3 hours before going to bed. The lamp he uses requires a 150-watt bulb, while the light burning in the room where his parents sit is only 50 watts. The TV set operates on 210 watts, and the radio on 32 watts. During a month of 22 weekdays, what will the cost of electricity be for just this use alone? Use the Cape Electric and Gas Company rate table. ________

Section 2 The Cost of Gas

We purchase gas for cooking or heating by the *cubic foot*. Perhaps it would be easier for us to visualize a cubic foot of gas if we think in terms of a cubic foot of water.

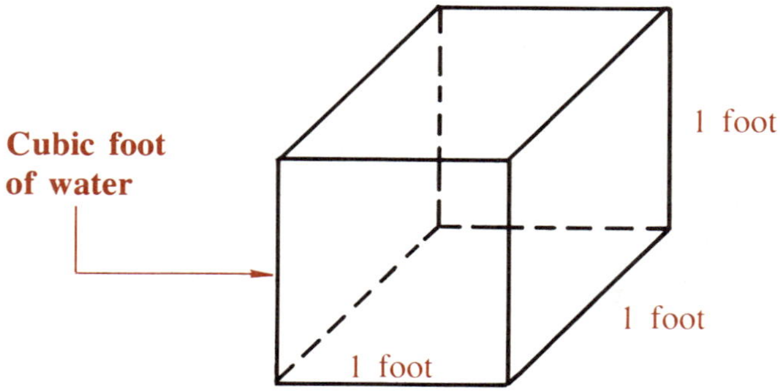

Now let us imagine that the water is drained out of this container and into a long plastic hose. Although the containers do not have the same shape, the volume of water (1 cubic foot) remains the same whether the container is a box or a hose.

As the gas enters our home it passes through a gas meter similar to the one pictured here.

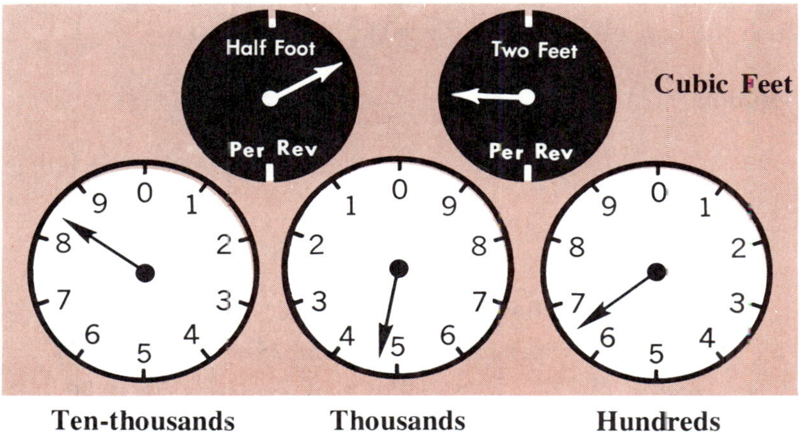

Other than for the minor points described below, a reading is taken from this meter in exactly the same manner as from the electric meter.

1. The two small dials at the top of the meter are completely ignored.
2. The reading from the remaining three dials registers *hundreds of cubic feet*.

Therefore, the reading of the meter above is 846 *hundred* cubic feet.

The gas rate table below is designed in the same manner as found for the electric rate table. The computation for the cost of gas follows the same pattern as for the cost of electricity.

CAPE ELECTRIC AND GAS COMPANY

Residential Rate per Month—Gas

Hundreds of Cubic Feet Consumed	Cost Per Hundred Cubic Feet
First 3	65¢
Next 22	40.1¢
Next 25	36.3¢
Over 50	30.4¢

Unit 1

Example On March 17, a gas meter reading was recorded as 628. On April 18, when it was again checked, it registered 732. Use the rate table of the Cape Electric and Gas Company to determine the cost of gas for this period.

Solution

Reading at end → 732 **Reading at outset** → 628

Hundreds of cu ft consumed = 732 − 628
= 104

Cost for hundreds of cubic feet:

First 3	**Next 22**	**Next 25**	**Over 50**
3 × 65¢	22 × 40.1¢	25 × 36.3¢	54 × 30.4¢
= 3 × $.65	= 22 × $.401	= 25 × $.363	= 54 × $.304
= $1.95	= $8.822	= $9.075	= $16.416

Total cost = $1.95 + $8.822 + $9.075 + $16.416
= $36.263 or $36.26

Exercises A

1. Determine the reading for each meter.

a. ______

d. ______

b. 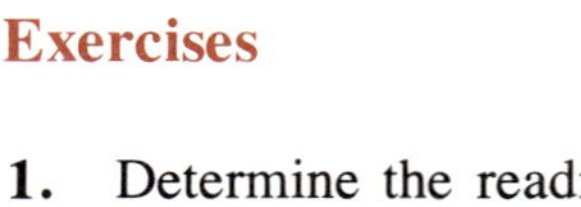 ______

e.  ______

c.  ______

f.  ______

2. Determine the number of hundreds of cubic feet of gas consumed during the month in each exercise.

	Meter At Beginning Of Month	Meter At End Of Month	
a.			________
b.			________
c.			________

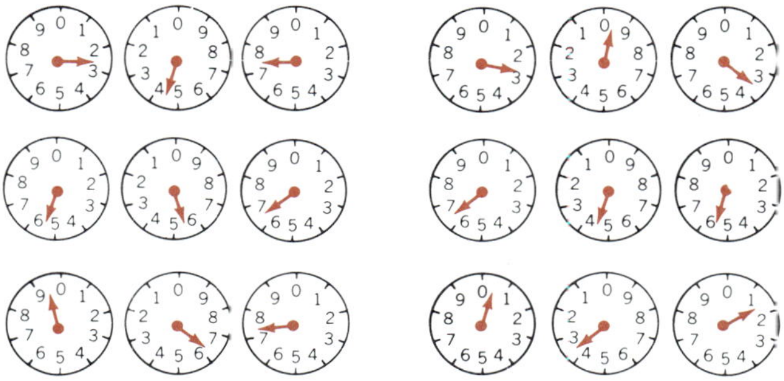

B

The rate table for the Cape Electric and Gas Company is to be used when doing the exercises below.

1. Determine the gas bill for the month for each customer.

	Hundreds Of Cu Ft Consumed	Cost for hundreds of cu ft: First 3	Cost for hundreds of cu ft: Over 3	Total Cost
a.	9	$________	$________	$________
b.	15	________	________	________
c.	21	________	________	________

2. Determine the gas bill for the month for each customer.

	Cu Ft Consumed	Cost for hundreds of cu ft: First 3	Next 22	Over 25	Total Cost
a.	30	$________	$________	$________	$________
b.	36	________	________	________	________
c.	42	________	________	________	________
d.	47	________	________	________	________

3. Determine the gas bill for the month for each customer.

	Cu Ft Consumed	Cost for hundreds of cu ft: First 3	Next 22	Next 25	Over 50	Total Cost
a.	56	$________	$________	$________	$________	$________
b.	67	________	________	________	________	________
c.	103	________	________	________	________	________
d.	165	________	________	________	________	________

Unit 1

4. Based on the meter readings shown below, determine the gas bill for the month.

	Reading At Beginning Of Month	Reading At End Of Month		Reading At Beginning Of Month	Reading At End Of Month
a.	385	398	c.	888	957
b.	714	740	d.	643	801

	Cu Ft Consumed	Cost for hundreds of cu ft: First 3 Or Less	Next 22 Or Less	Next 25 Or Less	Over 50	Total Cost
a.	________	$________	$________	$________	$________	$________
b.	________	________	________	________	________	________
c.	________	________	________	________	________	________
d.	________	________	________	________	________	________

C

The bills for the Cape Electric and Gas Company resemble the forms below. Use the rate tables of that company to determine the total cost for both gas and electricity for each customer.

1.

CAPE ELECTRIC AND GAS COMPANY

Servant of a Great State

Service Period: July 7 August 5

Next Meter Reading Date: September 6

Service	Meter Readings: From	To			Amount
Electricity	6842	6878	________	KWH	$________
Gas	165	193	________	Hundred cubic feet	$________
				Total	$________

2.

Service Period: May 15 June 16

Next Meter Reading Date: July 15

Service	Meter Readings: From	To			Amount
Electricity	3591	3725	________		$________
Gas	847	889	________		$________
				Total	$________

3.

	Service Period			Next Meter Reading Date	
	October 22	November 24		December 21	
	Meter Readings				Amount
Service	From	To			
Electricity	23684	23859	______		$______
Gas	913	024	______		$______
				Total	$______

4.

	Service Period			Next Meter Reading Date	
	February 16	March 15		April 16	
	Meter Readings				Amount
Service	From	To			
Electricity	61652	61982	______		$______
Gas	416	604	______		$______
				Total	$______

Section 3 The Cost of Water

In the event we rent an apartment we will probably not be called upon to pay for the water we consume. However, as a homeowner we are not quite so fortunate unless we have our own well and are permitted to use it.

Water is piped into our home and passes through a water meter that measures the amount we consume. Meters resemble the one below and are usually read once every three months (quarterly).

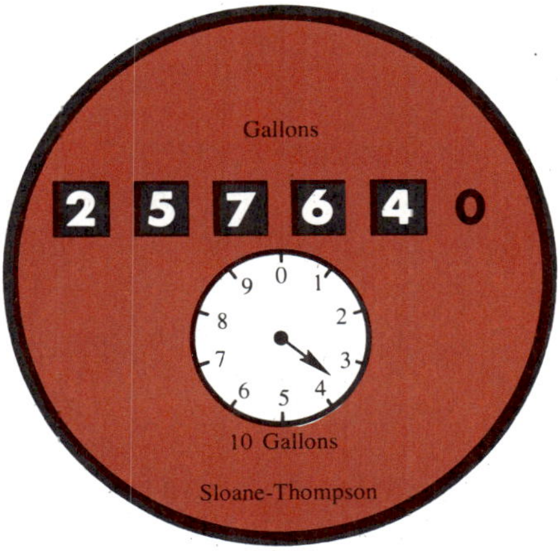

Although the meter above is designed to measure the consumption of water by the gallon, others measure the consumption by the cubic foot. The water is sold in either 1,000 gallon units or 1,000

Unit 1

cubic foot units. Both the single cubic foot and the single gallon are much too small a quantity by which to price the cost of water.

Having examined both electric and gas rate tables, we should find the water rate table quite familiar.

BLOOMINGTON WATER COMPANY

Quarterly Water Rates

Number of Thousands Of Gallons Consumed	Cost Per 1,000 Gallons
First 5	$1.80
Next 20	1.45
Over 25	1.05

Minimum Charge: $5.00

Example On March 27 a water meter reading was recorded as 136,240. On June 29 when it was again checked, it registered 195,260. Use the rate table of the Bloomington Water Company to determine the cost of water for the quarterly period.

Solution

Reading at end → 195,260; **Reading at outset** → 136,240

Gallons of water consumed = 195,260 − 136,240
= 59,020

Number of 1,000's of gallons consumed = 59,020 ÷ 1,000
= 59.02

Cost for thousands of gallons

First 5	Next 20	Over 25
5 × $1.80	20 × $1.45	(59.02 − 25) × $1.05
= $9.00	= $29.00	= 34.02 × $1.05
		= $35.721

Total cost = $9.00 + $29.00 + $35.721
= $73.721 or $73.72

Exercises A

The rate table of the Bloomington Water Company is to be used when doing the exercises below.

1. Determine the quarterly water bill for each period.

	Thousands of Gallons Consumed	Cost for Thousands of Gallons: First 5	Cost for Thousands of Gallons: Over 5	Total Cost
a.	6.7	$_____	$_____	$_____
b.	12.4	_____	_____	_____
c.	16.21	_____	_____	_____
d.	21.34	_____	_____	_____

2. Determine the quarterly water bill for each period.

	Thousands of Gallons Consumed	Cost for Thousands of Gallons: First 5	Cost for Thousands of Gallons: Next 20	Cost for Thousands of Gallons: Over 25	Total Cost
a.	27.2	$_____	$_____	$_____	$_____
b.	31.4	_____	_____	_____	_____
c.	42.21	_____	_____	_____	_____
d.	53.04	_____	_____	_____	_____
e.	62.15	_____	_____	_____	_____

3. Determine the number of thousands of gallons of water consumed during each quarterly period.

	Reading At Beginning Of Quarter	Reading At End Of Quarter	Number of Gallons Consumed	Thousands of Gallons Consumed
a.	005820	007510	_____	_____
b.	069510	074370	_____	_____
c.	369570	382430	_____	_____
d.	927650	986790	_____	_____
e.	998140	027370*	_____	_____

*This meter has passed the 1,000,000 gallon point and has begun recording from 0.

4. Determine the quarterly water bill for each period.

	Reading At Beginning Of Quarter	Reading At End Of Quarter	Number of Gallons Consumed	Thousands of Gallons Consumed	Total Cost
a.	007430	012570	_____	_____	$_____
b.	243750	258960	_____	_____	_____
c.	582910	614720	_____	_____	_____
d.	738260	805960	_____	_____	_____

B

The Livingston Township water rates shown below are to be used when doing the exercises that follow.

LIVINGSTON TOWNSHIP

Quarterly Water Rates

Minimum Charge $9.50

$4.75 per 1,000 cubic feet for first 2,000 cubic feet
3.85 per 1,000 cubic feet in excess of 2,000 cubic feet

1. Determine the quarterly water bill for each period.

	Thousands of Cubic Feet Consumed	Cost for Thousands of cu ft: First 2	Over 2	Total Cost
a.	3.6	$______	$______	$______
b.	5.07	______	______	______
c.	4.28	______	______	______
d.	10.14	______	______	______

2. Determine the quarterly water bill for each period.

	Reading At Beginning Of Quarter	Reading At End Of Quarter	Number of Cubic Feet Consumed	Thousands of Cubic Feet Consumed	Total Cost
a.	004520	007730	______	______	$______
b.	023710	027840	______	______	______
c.	059460	068520	______	______	______
d.	743180	765060	______	______	______

C

1. How many cubic feet of water will have to be consumed by a resident of Livingston Township in order to cover the minimum charge? (See table for the exercises of group B.) ______

2. A water softener requires regenerating twice each week. During each regeneration, 90 gallons of water are consumed. The cost of water is $1.35 per 1,000 gallons.

a. How many times during the year is the water softener regenerated? ________

b. What is the annual cost for water for regenerating this water softener? ________

3. A cubic foot of water is equivalent to approximately 7.5 gallons of water. Approximately how many gallons of water are there in 5,000 cubic feet of water? ________

4. Donald Bates owns a home in Livingston Township. During a quarterly period he used 6,000 cubic feet of water.

a. Determine the cost of water for this period. ________

b. How many gallons of water were consumed during this period? ________

c. If Mr. Bates had been serviced by the Bloomington Water Company, what would his bill have been for this period? ________

d. How much more – or less – would his water bill have been had he purchased his water from the Bloomington Water Company? ________

5. During a particularly dry summer Maria Kern had to water her lawn approximately 6 hours each day for a period of 47 days. Water flowed through the hose at the rate of 2 1/2 gallons per minute.

a. How many gallons of water were consumed each hour? ________

b. How many gallons of water were consumed each day? ________

c. How many gallons of water were consumed during the 47-day period? ________

d. If the cost of water was $1.25 per 1,000 gallons, what was the cost for watering the lawn for this period? ________

Section 4 The Cost of Telephone Calls

We have become so dependent upon the telephone that having one in our apartment is no longer considered a luxury but rather a necessity. Telephone rates vary from state to state and even within a single state the rate will depend upon the service we purchase. A 4-party line may be as low as $4.25; a private line may run as high as $7.47.

However, whichever service we choose, the basic cost usually permits us to make an unlimited number of calls at no charge within a region called our *local area*. This region is simply the communities surrounding the one in which we live.

The table on page 459 is typical of most telephone rate tables. It is designed for calls made to points within a particular state but outside the local area. Notice that the cost of a telephone call depends primarily on two factors:

1. The length of the call
2. The distance from the person making the call to the person receiving the call

Notice that the heading of the table contains the words *Dial-Direct Calls*. These are telephone calls made without the assistance of the operator.

Example 1 A telephone call was made from a home phone on Thursday at 7:05 PM. The distance between the person who made the call and the person who received the call was 84 miles. If the conversation lasted 15 minutes, what was the charge for this call?

Explanation

Miles From	Miles Through	Evening: Initial 2 Min	Evening: Each Add'l Min
81	96 ——→	.52	.17
84 is between these numbers		**Base cost for first 2 minutes**	**Cost for each minute in excess of 2**

RATES FOR DIAL-DIRECT CALLS WITHIN THE STATE

		Home Phones (Calls outside "local area")						Coin Phones	
Distance -Miles-		Day 8AM-5PM Mon-Fri		Evening 5PM-11PM Mon-Fri 8AM-11PM Sat-Sun		Night 11PM-8AM Everyday		Everyday All Hours	
From	Through	Initial 2 Min.	Each Add'l Min.	Initial 2 Min.	Each Add'l Min.	Initial 1 Min.	Each Add'l Min.	Initial 3 Min.	Each Add'l Min.
0	10	$.10	$.03	$.10	$.03	$.10	$.03	$.10	$.05
11	15	.16	.06	.16	.05	.10	.03	.30	.06
16	20	.22	.07	.21	.07	.10	.05	.39	.07
21	25	.27	.09	.26	.07	.15	.06	.46	.09
26	32	.32	.10	.26	.08	.15	.07	.52	.10
33	48	.37	.12	.27	.08	.15	.07	.59	.12
49	64	.42	.13	.27	.09	.15	.08	.65	.13
65	80	.47	.16	.27	.11	.15	.09	.73	.16
81	96	.52	.17	.27	.11	.15	.09	.79	.17
97	112	.57	.18	.27	.11	.15	.09	.85	.18
113	Up	.62	.19	.27	.11	.15	.09	.91	.19

Coin Phone Calls: Cost rounded to nearest 5¢

Solution

$$\text{Base cost} = \$.52$$

$$\text{Excess minutes} = 15 - 2 = 13$$

Cost per minute → $.17; Excess minutes → 13

$$\text{Excess cost} = \$.17 \times 13 = \$2.21$$

$$\text{Total cost} = \text{Base cost} + \text{Excess cost} = \$.52 + \$2.21 = \$2.73$$

Example 2 A call was made from a phone booth to a person who was 38 miles away. The conversation lasted 17 minutes. What was the cost of this call?

Explanation

Miles From	Miles Through	Coin Phones: Initial 3 Min.	Coin Phones: Each Add'l Min.
33	48	.59	.12

38 is between these numbers

Base cost for first 3 minutes

Cost for each minute in excess of 3

Solution

$$\text{Base cost} = \$.59$$

$$\text{Excess minutes} = 17 - 3 = 14$$

Cost per minute → $.12; Excess minutes → 14

$$\text{Excess cost} = \$.12 \times 14 = \$1.68$$

$$\text{Total cost} = \text{Base cost} + \text{Excess cost} = \$.59 + \$1.68 = \$2.27 \text{ rounded to } \$2.25$$

Notice that the cost, $2.27, is rounded to $2.25. A coin phone call is always rounded to the nearest 5¢ amount. A coin phone does not accept pennies.

Exercises A

Home-phone calls in these exercises were made to points outside the local area but within the state.

1. Determine the base cost for each home-phone call.

	Distance -Miles-	Time Of Day	Base Cost
a.	8	Night	$____
b.	27	Day	____
c.	35	Day	____
d.	49	Evening	____
e.	84	Night	____

2. Determine the total cost for each home-phone Day call.

	Distance -Miles-	Base Cost	Minutes	Excess Minutes	Excess Cost	Total Cost
a.	7	$____	2	____	$____	$____
b.	15	____	3	____	____	____
c.	29	____	5	____	____	____
d.	51	____	9	____	____	____
e.	72	____	4	____	____	____

3. Determine the total cost for each home-phone call made during a weekday.

	Distance -Miles-	Time Of Day	Base Cost	Minutes	Excess Minutes	Excess Minutes	Total Cost
a.	19	Night	$____	4	____	$____	$____
b.	12	Night	____	6	____	____	____
c.	17	Day	____	14	____	____	____
d.	46	Evening	____	21	____	____	____
e.	92	Night	____	26	____	____	____

4. Determine the cost for each coin-phone call.

	Distance -Miles-	Minutes	Base Cost	Excess Cost	Total Cost	Rounded Cost
a.	7	23	$____	$____	$____	$____
b.	53	6	____	____	____	____
c.	69	11	____	____	____	____
d.	18	17	____	____	____	____
e.	99	34	____	____	____	____

5. Determine the cost for each home-phone call.

	Distance -Miles-	Hour When Call Was Made	Base Cost	Minutes	Excess Cost	Total Cost
a.	41	Mon-2:00PM	$______	9	$______	$______
b.	27	Thurs-10:00PM	______	4	______	______
c.	9	Tues-8:10PM	______	7	______	______
d.	18	Fri-8:00AM	______	16	______	______
e.	98	Sat-9:20PM	______	29	______	______
f.	86	Wed-4:30PM	______	12	______	______
g.	41	Sun-1:05AM	______	32	______	______

B

Each of the calls in these exercises was made to points within the same state.

1. Dorothy used a coin phone when she called her mother from college. The college is located 145 miles from home. The call lasted 5 minutes and was made at 2:30 on a Wednesday afternoon.

a. What was the cost of the call? ______

b. What would the cost of the call have been had it been made from a home phone? ______

c. How much money could have been saved had the call been made from a home phone? ______

2. Judy Carlton is now married and lives in a town that is 125 miles from her parent's home. Each week she phones them and the conversation lasts about 15 minutes.

a. How much would each call cost if it is made on Friday at 2:00PM? ________

b. How much would each call cost if it is made on Saturday at 2:00PM? ________

c. How much can be saved by making the call on Saturday rather than Friday? ________

3. The federal tax rate on phone bills is 4%. What federal tax will have to be paid on each of the following monthly phone bills?

	Phone Bill	Federal Tax		Phone Bill	Federal Tax
a.	$12.65	$________	c.	$23.42	$________
b.	16.73	________	d.	27.61	________

4. The basic charge for Roger Denton's private line is $7.63 per month. During the past month he made the following daytime calls during weekdays to points not within the local area.

9	7-mile calls for less than 2 minutes each
4	15-mile calls for less than 2 minutes each
1	23-mile call for 8 minutes
1	51-mile call for 12 minutes

a. What was the total cost of the calls that were not in the local area? ________

b. Excluding the federal tax, what was Roger's phone bill for the month? ________

c. Including the federal tax, what was Roger's phone bill for the month? ________

C

The partial table below gives the rates for long distance dial-direct interstate home-phone calls. Use this table when doing the exercises below.

Distance -Miles-		8AM - 5PM Mon - Fri		5PM - 11PM Sun - Fri		11PM - 8AM Daily Plus all day Sat to 5PM Sun	
From	Through	First Min.	Each Add'l Min.	First Min.	Each Add'l Min.	First Min.	Each Add'l Min.
71	124	$.43	$.29	$.27	$.19	$.17	$.12
125	196	.44	.30	.28	.20	.17	.12
197	292	.46	.32	.29	.21	.18	.13
293	430	.48	.34	.31	.23	.19	.14
431	925	.50	.34	.32	.23	.20	.14
926	1910	.52	.36	.33	.24	.20	.15
1911	3000	.54	.38	.35	.25	.21	.16

1. Determine the cost of each interstate dial-direct call that was made from a home phone.

	Distance -Miles-	Hour When Call Was Made	Base Cost	Minutes	Excess Cost	Total Cost
a.	152	Tues-9:00AM	$____	6	$____	$____
b.	204	Thurs-6:20PM	____	7	____	____
c.	623	Wed-11:35PM	____	12	____	____
d.	1,427	Sat-2:10PM	____	16	____	____
e.	954	Sun-6:15PM	____	23	____	____
f.	368	Mon-7:25AM	____	9	____	____
g.	2,143	Fri-10:19PM	____	14	____	____

2. An interstate telephone call was made from a home phone to a point 1,706 miles away. The conversation lasted 18 minutes.

a. What would the cost of this call be if it was made on Wednesday at 4:45PM? ________

b. What would the cost of this call be if it was made on Wednesday at 5:15PM? ________

c. How much could have been saved by waiting the half-hour to make the call? ________

3. An interstate phone call lasting 3 minutes was made on a Monday to a point 75 miles away.

a. If the call was made at 1:00PM, what was its cost? ________

b. If the call was made at 1:00AM, what was its cost? ______

c. What was the discount for making the call at 1:00AM rather than 1:00PM? ______

d. What was the percent of discount for making the call at 1:00AM rather than 1:00PM? (See discount rate in Chapter 3.) ______

Home Ownership Unit 2

Section 1 The Mortgage
Part 1

As mentioned earlier, the largest investment most of us will make during our lifetime is the purchase of a house. The cost of the average one-family house is approximately $65,000. Since very few of us have that much cash, we will very likely make a small down payment. The remainder of the money will be borrowed from either a bank or a savings and loan association.

Loans for the purchase of a home are much the same as the installment loans with which we are familiar. However, the monthly payments are made over a 25 or 30-year period rather than for a year or two. When loans are made for the purchase of a building, or land, or both, they are called *mortgage loans*. In the event we fail to repay the debt, the bank has the right to sell the property to regain its money.

Interest rates on mortgage loans are controlled by law in many states. These rates range from a low of 8 3/4% to a high of around 12 1/2%. The table below shows the monthly payments per $1,000 of debt at various rates for different periods of time.

MORTGAGE LOAN SCHEDULE
(Monthly Payment Per $1,000)

	Annual Interest Rate							
Years	9%	9½%	10%	10½%	11%	11½%	12%	12½%
15	$10.14	$10.44	$10.75	$11.05	$11.37	$11.68	$12.00	$12.32
20	9.00	9.32	9.65	9.98	10.32	10.66	11.01	11.36
25	8.39	8.74	9.09	9.44	9.80	10.16	10.53	10.90
30	8.05	8.41	8.78	9.15	9.52	9.90	10.29	10.67
35	7.84	8.22	8.60	8.98	9.37	9.76	10.16	10.55
40	7.71	8.10	8.49	8.89	9.28	9.68	10.09	10.49

Unit 2

Example 1 Clifford Temple was granted a $42,500 mortgage on the purchase of a new home. The interest rate was 9 1/2%. If the debt is to be paid off over a 25-year period, what will the monthly payments be?

Explanation

Annual Interest Rate

Years → 25 ———————→ $9\frac{1}{2}\%$ → $8.74

Monthly payment per $1,000 of debt

Solution

Number of $1,000's in $42,500 = $42,500 ÷ $1,000
= 42.5

Number of $1,000's → 42.5; Payment per $1,000 → $8.74

Monthly payment = 42.5 × $8.74
= $371.45

In many states the lending agency is permitted to charge *points* on a mortgage loan. As an example, a charge of *2 points* is another way of saying that the borrower must pay the lending agency 2% *of the amount borrowed* as a fee for obtaining the loan.

Similarly, a charge of *5 points* implies that the borrower will have to pay *5% of the amount borrowed* as a fee for simply obtaining the loan.

Example 2 Julius Spencer purchased a $70,000 home by making a down payment of 20%. For the balance he obtained a mortgage loan for which he had to pay 5 1/2 points. The lawyer whom Mr. Spencer hired to help him with the legal problems involved in buying a home charged him $450. What was the total cash Mr. Spencer had to have on the day he took possession of the house?

Solution

$$\text{Down payment} = 20\% \text{ of purchase price} = .20 \times \$70{,}000 = \$14{,}000$$

Purchase price → Down payment →

$$\text{Mortgage loan} = \$70{,}000 - \$14{,}000 = \$56{,}000$$

Loan →

$$\text{Point fee} = \$56{,}000 \times 5\tfrac{1}{2} \text{ points} = \$56{,}000 \times 5\tfrac{1}{2}\% = \$56{,}000 \times .055 = \$3{,}080$$

$$\text{Legal fee} = \$450$$

$$\text{Total cash needed} = \text{Down payment} + \text{Point fee} + \text{Legal fee} = \$14{,}000 + \$3{,}080 + \$450 = \$17{,}530$$

Within recent years many people have borrowed money under an arrangement called a *secondary mortgage loan*. The period of these loans is usually from 10 to 20 years. However, the interest rates are quite high and many run as much as 15% or 16% per year.

These loans are also guaranteed by our property. However, the company holding the first mortgage has the first claim to the property should we fail to repay the loans. Since the company holding the secondary mortgage stands a greater risk of losing its money, it charges a greater rate of interest.

SECONDARY MORTGAGE LOAN SCHEDULE
(Monthly Payment Per $1,000)

	Annual Interest Rate				
Years	14%	$14\frac{1}{2}$%	15%	$15\frac{1}{2}$%	16%
10	$15.53	$15.83	$16.13	$16.44	$16.75
15	13.06	13.66	14.00	14.34	14.69
20	11.10	12.80	13.17	13.54	13.91

Unit 2

Example 3 A secondary mortgage loan for \$12,000 was obtained at an interest rate of 14 1/2%. The debt is to be paid off monthly over a 15-year period. How large will each monthly payment be?

Explanation

Annual Interest Rate

Years → 15 ———————— 13.66 ← $14\frac{1}{2}\%$

Monthly payment per \$1,000 (13.66)

Solution

Number of \$1,000's in \$12,000 = \$12,000 ÷ \$1,000
= 12

Number of \$1,000's (12) **Payment per \$1,000** (\$13.66)

Monthly payment = 12 × \$13.66
= \$163.92

Exercises

A

1. To obtain each mortgage the borrower had to pay the number of points indicated. What was the cost for obtaining each loan?

	Mortgage	Points	Cost for Obtaining Mortgage
a.	\$20,000	1	\$________
b.	34,000	3	________
c.	39,000	4	________
d.	41,000	$2\frac{1}{2}$	________
e.	58,000	$5\frac{1}{2}$	________

2. In each exercise the purchaser made the down payment shown. To obtain a mortgage for the remainder of the money needed, the purchaser had to pay the number of points indicated. How large was the point fee in each situation?

	Purchase Price Of Property	Down Payment (Percent)	Down Payment	Mortgage	Points	Point Fee
a.	\$40,000	10%	\$________	\$________	2	\$________
b.	36,000	5%	________	________	4	________
c.	57,000	30%	________	________	5	________
d.	63,000	20%	________	________	$1\frac{1}{2}$	________
e.	74,000	35%	________	________	$4\frac{1}{2}$	________

B

1. Determine the monthly payment for each mortgage loan.

	Mortgage	Interest Rate	Mortgage Period (Years)	Payment per $1,000	Monthly Payment
a.	$12,000	9%	20	$______	$______
b.	23,000	$10\frac{1}{2}$%	25	______	______
c.	27,500	11%	35	______	______
d.	38,400	$9\frac{1}{2}$%	30	______	______
e.	69,200	$11\frac{1}{2}$%	35	______	______
f.	74,800	$12\frac{1}{2}$%	40	______	______

2. Determine the monthly payment for each secondary mortgage loan.

	Mortgage	Interest Rate	Mortgage Period (Years)	Payment per $1,000	Monthly Payment
a.	$ 7,000	14%	10	$______	$______
b.	16,100	15%	15	______	______
c.	12,300	$14\frac{1}{2}$%	20	______	______
d.	21,600	$15\frac{1}{2}$%	15	______	______

C

1. The borrower of mortgage money is often required to purchase title insurance. This insurance is designed to cover the possibility that the person who sold the property had no legal right to do so and therefore the person who purchased the property does not really own it. Determine the cost of the title insurance for each mortgage.

	Mortgage	Insurance Rate	Insurance Cost
a.	$30,000	$5.00 per $1,000	$______
b.	45,000	6.25 per $1,000	______
c.	57,400	6.75 per $1,000	______

2. Henry Stevens purchased a house for $48,700. He made a down payment of 45% and was granted a mortgage for the remainder. The title insurance he purchased at $5.50 per $1,000 covered only the amount of the mortgage. How much did this cost him? ______

3. Doris Clarke received a 9% mortgage for $30,500 when she purchased her house. She agreed to pay off this debt over a 20-year period.

a. What were her monthly payments? ________

b. How much did this amount to annually? ________

c. What was the total amount that she paid back over the 20-year period? ________

4. When purchasing his house, Carl Denton put down a 20% deposit on the $54,000 he had to pay. The remainder he borrowed at 10%, to be paid off over a period of 25 years.

a. What were his monthly payments? ________

b. What was the total he paid back over the 25-year period? ________

c. How much more did he pay back than he borrowed? ________

5. At the time that Charles Martin purchased his house, he obtained a $38,000 mortgage from his bank at a 10 1/2% interest rate.

a. If he agrees to amortize — that is, to pay off — the debt in monthly payments over 20 years, what will be the total amount that he will return to the bank? ________

b. If he agrees to amortize the debt over a 30-year period, what will be the total amount he will return to the bank? ________

c. How much will Mr. Martin save by amortizing the mortgage over a 20-year period rather than a 30-year period? ________

6. Miguel Botta was granted a mortgage of $46,000 on his house. He arranged to amortize the loan over a 25-year period.

a. What will be the total amount he will return if the interest rate is 9%? ________

b. What will be the total amount he will return if the interest rate is 11%? ________

c. How much will Mr. Botta save by being able to borrow the money at 9% rather than 11%? ________

7. Gloria Kempton purchased a house for $68,300 for which she planned to give a 40% down payment. The rest of the money she borrowed from a savings and loan association, which charged her 3 points for obtaining the mortgage.

a. How much did she have to pay just for the privilege of borrowing the money? ________

b. The savings and loan association charged her $75 for expenses it incurred in checking to see if she should receive the mortgage. In addition, she had to pay George Stacey, her lawyer, $385 for legal fees connected with his work relative to the purchase of this house. What was the total of these fees? ________

c. Gloria was required to take out title insurance to the extent of the mortgage. How much did this cost her if the rate was $5.25 per $1,000? ________

d. What was the total amount of money that she had to have in cash on the day she took possession of the house? Exclude the down payment. ________

Part 2

The monthly mortgage payment consists of two parts:

1. The interest for one month on the unpaid debt.
2. A partial payment to reduce the unpaid debt.

As an illustration let us examine the very first payment on a *\$36,000 mortgage for 25 years at 11%*. Based on our work in Part 1, we can easily determine that the monthly payment will be \$352.80.

Of this \$352.80, part will be the interest owed on \$36,000 for the first month while the remainder will be used to pay off the \$36,000.

$$\begin{aligned}\text{Interest for first month} &= \$36{,}000 \times 11\% \times 1 \text{ month} \\ &= \$36{,}000 \times .11 \times \tfrac{1}{12} \\ &= \$330\end{aligned}$$

Since \$330 of the \$352.80 is interest on the debt, the remainder is a partial payment on the debt.

$$\begin{aligned}\text{Remainder} &= \$352.80 - \$330 \\ &= \$22.80\end{aligned}$$

Amount used to decrease the debt (↑ \$22.80)

Therefore, the debt during the second month is \$22.80 less than it was during the first month.

$$\begin{aligned}\text{Debt during second month} &= \$36{,}000 - \$22.80 \\ &= \$35{,}977.20\end{aligned}$$

To determine the amount of interest paid the second month we repeat the process above — however, now the debt is \$35,977.20 rather than \$36,000.

$$\begin{aligned}\text{Interest for the second month} &= \$35{,}977.20 \times 11\% \times 1 \text{ month} \\ &= \$329.79\end{aligned}$$

We can obtain a much clearer picture of the above if this information is arranged in tabular form.

Original Debt: $36,000				
Payment Number	Monthly Payment	Interest For Month	Amount Used To Decrease Debt	Balance Of Debt
1	$352.80	$330.00	$22.80	$35,977.20
2	352.80	329.79	23.01	35,954.19

This table reveals the following:

1. The monthly payment *remains the same*.
2. The monthly interest charge *decreases* with each succeeding month.
3. The amount used to pay off the debt *increases* with each succeeding month.

Actually, there will be very little interest on the last payment of $352.80.

Example 1 Alvin Jenkins had reduced his 9% mortgage to the point where he owed only $5,724. His monthly payments were $107.56.

(a) How much of the very next payment will be interest on the debt?

(b) How much of this same payment will be used to decrease the debt?

Solution

(a) Interest for month $= \$5,724 \times 9\% \times 1$ month

$$= \$5,724 \times .09 \times \tfrac{1}{12}$$

$$= \$42.93$$

Monthly payment → **Interest**

(b) Amount used to decrease debt $= \$107.56 - \$42.93$

$$= \$64.63$$

Many people who borrow money under mortgage loans are interested in repaying the debt as quickly as possible. Therefore, should they somehow acquire a large sum of money, their first thought is to eliminate their mortgage. On the other hand, under most circumstances, the bank or savings and loan from which the money was borrowed prefers that the debt not be repaid before it is due – any early repayment means a loss of interest.

Unit 2

To discourage an early repayment, mortgage contracts frequently include a table of penalty charges similar to the one below.

THE GIBRALTER SAVINGS BANK			
If Mortgage Is Paid Off During	Penalty	If Mortgage Is Paid Off During	Penalty
1st Year	Not Permitted	6th Year	$1\frac{1}{4}$% of Original Debt
2nd Year	Not Permitted	7th Year	1% of Original Debt
3rd Year	2% of Original Debt	8th Year	$\frac{3}{4}$% of Original Debt
4th Year	$1\frac{3}{4}$% of Original Debt	9th Year	$\frac{1}{2}$% of Original Debt
5th Year	$1\frac{1}{2}$% of Original Debt	10th Year	$\frac{1}{4}$% of Original Debt

Example 2 Mr. Thorn borrowed \$32,500 from the Gibralter Savings Bank at the time he purchased his home. After making payments for a period of 3 years 2 months he found that he was financially able to pay off the balance of the mortgage. What penalty fee did he have to pay because of this early repayment?

Solution

Repayment year: 4th year

Penalty fee = $1\frac{3}{4}$% of Original Debt

$$= 1.75\% \times \$32{,}500$$
$$= .0175 \times \$32{,}500$$
$$= \$568.75$$

Exercises A

Do you recall how to do the computation in the following exercises? This computation is the same as you will be doing when finding the penalty fee on the early repayment of a mortgage.

1. Rename each percent numeral as a decimal.

a.	$1\frac{1}{2}$%	________	**d.**	$\frac{1}{2}$%	________
b.	$1\frac{1}{4}$%	________	**e.**	$\frac{1}{4}$%	________
c.	$1\frac{3}{4}$%	________	**f.**	$\frac{3}{4}$%	________

2. Determine the product in each of the following exercises.

a. $1\frac{1}{2}\% \times \$5{,}000$ $________ **c.** $\frac{1}{2}\% \times \$46{,}000$ $________

b. $1\frac{3}{4}\% \times \$32{,}000$ ________ **d.** $\frac{1}{4}\% \times \$37{,}000$ ________

B

1. Determine the penalty fee if each mortgage is repaid during the year shown.

	Original Mortgage	Repayment Year	Penalty Fee
a.	$16,000	3rd	$________
b.	27,000	7th	________
c.	23,500	5th	________
d.	36,200	8th	________
e.	41,600	4th	________

2. Determine the penalty fee if each mortgage is repaid at the end of the period shown.

	Original Mortgage	Repayment Period	Repayment Year	Penalty Fee
a.	$20,000	6 years 4 months	________	$________
b.	31,000	2 years 9 months	________	________
c.	46,000	4 years 1 month	________	________
d.	53,000	9 years 6 months	________	________

C

1. Determine the interest payment for the month on each unpaid mortgage balance.

	Unpaid Mortgage Balance	Interest Rate	Interest
a.	$12,000	6%	$________
b.	30,000	9%	________
c.	42,000	11%	________
d.	7,476	10%	________
e.	28,836	7%	________

2. How much of each monthly mortgage payment will be used to decrease the mortgage debt?

	Unpaid Mortgage Balance	Monthly Payment	Interest Rate	Interest	Amount to Decrease Debt
a.	$ 2,760	$ 85.60	8%	$______	$______
b.	16,800	197.40	10%	______	______
c.	5,160	106.24	11%	______	______
d.	37,200	361.35	9%	______	______
e.	6,144	137.46	15%	______	______

D

1. The balance of Paul Tracey's secondary mortgage is $4,944. His monthly payment on this debt is $146.34 and the annual interest rate is 14%.

a. How much interest will be included in the next monthly payment? ______

b. How much of the next payment will be used to reduce the mortgage? ______

c. What will the balance of his mortgage be after the next payment? ______

2. Arthur Jefferson was given a $30,000 mortgage to be paid off over a period of 25 years at an annual interest rate of 10%.

a. Use the Mortgage Loan Schedule to determine his monthly mortgage payment. ______

b. Complete the table below in terms of his first and second monthly mortgage payments.

Payment Number	Monthly Payment	Interest For Month	Amount Used To Decrease Debt	Balance Of Debt
1	$______	$______	$______	$______
2	______	______	______	______

Section 2 Real Estate Tax

Part 1

Ranking either equal to or perhaps even greater than the burden of the mortgage payments is the real estate tax with which every homeowner must contend. The communities in which we live provide us with services such as

1. a police department
2. a fire department
3. a sanitation department
4. schools
5. libraries
6. museums

To operate each of these services requires money and that money is most often raised by taxing the property owner.

Each property owner's share of the cost of these services is judged in terms of the value of his property. As an illustration, let us examine the very simple situation of a community in which there are only two residents – Chris and Lee.

Value of Chris' Property	Value of Lee's Property
↓	↓
\$3,000	\$2,000

The two of them agree to spend a total of \$10,000 to improve the value of both properties. To determine what each one's share of the cost will be, they find the total value of both properties:

$$\$3,000 + \$2,000 = \$5,000.$$

They soon realize that if each of them contributes \$2 for each \$1 in the value of his property, they will have the \$10,000 they need.

Money needed → \$10,000 ÷ \$5,000 ← **Total value of property**

$$\$10,000 \div \$5,000 = \$2$$

Therefore,

Chris contributes:	Lee contributes:
$3,000 \times \$2 = \$6,000$	$2,000 \times \$2 = \$4,000$

$$\text{Total contributions} = \$6,000 + \$4,000 = \$10,000$$

Unit 2

Every community, no matter what its size may be, determines the real estate tax each property owner will contribute in much the same manner as just described.

The money needed by the town ÷ **The total value of the property in the town** = **The amount each homeowner pays per $1**

Example 1 The town of Triton needs $840,000 to meet its expenses for the coming year. If the total value of all the property in the town is $24,000,000, determine the amount of money per $1 each homeowner will have to pay.

Solution

Money needed → $840,000; **Total value of property** → $24,000,000

$$\text{Amount per } \$1 = \$840{,}000 \div \$24{,}000{,}000$$
$$= \$.035$$

The rate of

$.035 per $1

found in Example 1 above is called the *tax rate*. Different towns use different ways in which to express the tax rate. As an illustration, if both numbers in the rate above are multiplied by 100, we will obtain a new name for this tax rate. Therefore, the tax rate, $.035 per $1 can also be named as

$3.50 per $100

$.035 × 100 → $3.50; **$1 × 100** → $100

Similarly, if both numbers in a tax rate such as

$4.60 per $100

are divided by 10, its new name will be

$.46 per $10

$4.60 ÷ 10 → $.46; **$100 ÷ 10** → $10

Example 2 Rename the tax rate of 6.2¢ per $1 to one quoted in dollars per $1,000.

Solution

6.2¢ renamed as $.062

6.2¢ per $1 = $.062 per $1
= $62.00 per $1,000

$.062 × 1,000 **$1 × 1,000**

The list below includes some of the more common names for the tax rate of $.054 per $1.

54 mills per $1
5.4¢ per $1
$.054 per $1
$.54 per $10
$5.40 per $100
$54.00 per $1,000

The first name above may seem a bit strange for there is no coin in the United States called a *mill*. This unit, however, is used by a number of communities to express their tax rate. When this is done it is understood that

$1 is the equivalent of 1,000 mills.

Therefore,

To rename dollars as mills	**To rename mills as dollars**
↓	↓
Multiply by 1,000	Divide by 1,000

Example 3 Rename $.067 in terms of mills.

Explanation

Multiply .067 by 1,000

.067 dollars = 67 mills

Solution

$.067 = 67 mills

Unit 2

Example 4 Rename the tax rate 75 mills per \$1 to one quoted in dollars per \$1.

Explanation

Divide 75 by 1,000

75 mills = .075 dollars

Solution

75 mills per \$1 = \$.075 per \$1

Exercises A

1. Rename each in terms of dollars.

a.	67 mills	\$________	**d.**	254 mills	\$________
b.	84 mills	________	**e.**	6 mills	________
c.	10 mills	________	**f.**	490 mills	________

2. Rename each in terms of mills.

a.	\$.026	________mills	**d.**	\$.07	________mills
b.	\$.057	________mills	**e.**	\$.13	________mills
c.	\$.128	________mills	**f.**	\$.68	________mills

3. Rename each in terms of cents.

a.	\$.052	________¢	**d.**	\$.06	________¢
b.	\$.074	________¢	**e.**	\$.09	________¢
c.	\$.156	________¢	**f.**	\$.13	________¢

4. Rename each in terms of dollars.

a.	6.4¢	\$________	**d.**	5¢	\$________
b.	7.3¢	________	**e.**	8¢	________
c.	12.8¢	________	**f.**	12¢	________

B

1. Rename each tax rate from its name at the left to the name at the right.

a.	$2.67 per $100	=	$________	per $1,000
b.	$ .47 per $10	=	$________	per $100
c.	$ .52 per $10	=	$________	per $1,000
d.	$ .061 per $1	=	$________	per $100
e.	$ .057 per $1	=	$________	per $1,000

2. Rename each tax rate from its name at the left to the name at the right.

a.	$36.40 per $1,000	=	$________	per $100
b.	$ 7.24 per $100	=	$________	per $10
c.	$56.80 per $1,000	=	$________	per $1
d.	$ 3.25 per $100	=	$________	per $1
e.	$ .248 per $10	=	$________	per $1

3. Rename each tax rate from its name at the left to the name at the right.

a.	$.026 per $1	=	________	¢ per $1
b.	$.052 per $1	=	________	mills per $1
c.	4.3¢ per $1	=	$________	per $1
d.	52 mills per $1	=	$________	per $1
e.	142 mills per $1	=	$________	per $1

4. Rename each tax rate as indicated.

a.	43 mills per $1	=	$________	per $1	=	$________	per $100
b.	59 mills per $1	=	$________	per $1	=	$________	per $1,000
c.	17 mills per $1	=	$________	per $1	=	$________	per $10
d.	$2.58 per $100	=	$________	per $1	=	________	mills per $1
e.	$56.10 per $1,000	=	$________	per $1	=	________	mills per $1
f.	$.83 per $10	=	$________	per $1	=	________	mills per $1

Unit 2

C

1. Determine the tax rate per $1 for each community. Continue the division until the remainder is zero.

	Total Expenses For the Year	Total Property Value	Tax Rate Per $1
a.	$ 450,000	$ 9,000,000	$______
b.	1,380,000	23,000,000	______
c.	1,035,000	45,000,000	______
d.	2,315,500	84,200,000	______

2. Determine the tax rate in the form shown in the last column. Continue the division until the remainder is zero.

	Total Expenses For the Year	Total Property Value	Tax Rate Per $1	Tax Rate
a.	$ 280,000	$ 7,000,000	$______	$______ per $10
b.	1,054,000	31,000,000	______	$______ per $100
c.	2,444,000	47,000,000	______	______ mills per $1
d.	1,463,800	56,300,000	______	______¢ per $1

D

1. Brighton Township is one of four communities that contribute to the support of a regional high school. Brighton's share of the cost amounted to $492,000 last year. If the total value of the property in the township is $41,000,000, what tax rate in dollars per $100 was needed to raise the money to operate the high school? ______

2. In its budget for the coming year the town of Lakeside listed its contribution to the cost of maintaining the county government as $1,344,000. What is the tax rate per $1,000 that is needed for this expenditure if the total property value in Lakeside is $64,000,000? ______

3. In order to meet its expenses for the coming year the town of West Camden will have to raise $1,040,000. Through fines for traffic and building violations the town will collect $20,100. If the total value of property in the town is $16,450,000, what will the tax rate have to be in mills per $1? ______

4. During the past year Lansing Township paid \$1,292,000 to support its school system. For the coming year it will have to raise \$1,596,000 to meet this cost. The total value of property in the township is \$38,000,000. How much more will the tax rate be in dollars per \$100 for the coming year than it had been for the past year? __________

Part 2

It would be well if we clarified a term used in Part 1 of this section. At that time we learned that the tax rate is determined in terms of the total value of the property in a town. The value to which we refer is the *assessed value* of the property and not its actual value.

Each community employs either one or more persons called *assessors*. It is their job to judge the value of every property in the community. The price they assign to each property is called the *assessed value* of that property.

There is frequently very little relation between the assessed value of property and the actual value for which the property can be sold. In a single state we will often find that the assessed value of property in one community can be as low as 8% of its actual value while in another, the assessed value is as high as 104% of its actual value!

Whether the assessed value is either a small or large percent of the actual value makes very little difference. The important point, however, is the fact that all properties within the same community must be assessed at exactly the same percent of its actual value. If this is not so, then some of the property owners will be paying more than their fair share for the community services.

Example 1 In the town of West Milford the tax rate is \$4.12 per \$100. The property in this town is assessed at 65% of its actual value. What will the tax be on a house that can be sold for \$58,000?

Solution

$$\text{Assessed value} = 65\% \text{ of Actual value}$$
$$= .65 \times \$58{,}000$$
$$= \$37{,}700$$
$$\text{Number of \$100's in \$37,700} = \$37{,}700 \div \$100$$
$$= 377$$

Number of \$100's → 377; Tax per \$100 → \$4.12

$$\text{Tax} = 377 \times \$4.12$$
$$= \$1{,}553.24$$

Example 2 The actual value of a home is \$64,700. However it is assessed for 72% of its actual value. If the tax rate in the community is \$24.83 per \$1,000, what is the tax on this property?

Solution

$$\text{Assessed value} = 72\% \text{ of Actual value}$$
$$= .72 \times \$64{,}700$$
$$= \$46{,}584$$
$$\text{Number of \$1,000's in \$46,584} = \$46{,}584 \div \$1{,}000$$
$$= 46.584$$

Number of \$1,000's → 46.584; Tax per \$1,000 → \$24.83

$$\text{Tax} = 46.584 \times \$24.83$$
$$= \$1{,}156.6807 \text{ or } \$1{,}156.68$$

Example 3 Doris Sanford can sell her home for \$71,500. Property in the community in which she lives is assessed at 36% of its actual value. If the tax rate is 26.3 mills per \$1, how much is her real estate tax on this property?

Explanation

Divide 26.3 by 1,000

26.3 mills = .0263 dollars

Tax rate renamed:

26.3 mills per \$1 = \$.0263 per \$1

Solution

Assessed value = 36% of Actual value

$$= .36 \times \$71{,}500$$
$$= \$25{,}740$$

Tax rate = $.0263 per $1

Number of $1's → Tax = 25,740 × $.0263 ← Tax per $1

$$\text{Tax} = 25{,}740 \times \$.0263$$
$$= \$676.962 \text{ or } \$676.96$$

Exercises A

1. Determine the tax on each property.

	Assessed Value	Tax Rate	Tax
a.	$12,400	$ 2.30 per $100	$________
b.	31,500	3.72 per $100	________
c.	46,700	5.84 per $100	________
d.	39,000	34.20 per $1,000	________
e.	53,000	48.36 per $1,000	________
f.	62,400	29.51 per $1,000	________
g.	42,600	.56 per $10	________
h.	29,300	.71 per $10	________

2. Determine the tax on each property.

	Assessed Value	Tax Rate	Tax
a.	$22,000	$.041 per $1	$________
b.	36,000	$.054 per $1	________
c.	62,100	$.067 per $1	________
d.	34,000	5.1¢ per $1	________
e.	54,600	8.2¢ per $1	________

3. Determine the tax on each property.

	Assessed Value	Tax Rate	Tax
a.	$29,000	48 mills per $1	$________
b.	35,000	39 mills per $1	________
c.	21,000	41.2 mills per $1	________
d.	43,000	54.2 mills per $1	________

4. Determine the tax on each property.

	Actual Value	Assessed At the Following Percent	Assessed Value	Tax Rate	Tax
a.	$41,000	20%	$______	$2.59 per $100	$______
b.	46,000	90%	______	$7.82 per $100	______
c.	64,000	45%	______	$43.76 per $1,000	______
d.	72,000	65%	______	$52.03 per $1,000	______
e.	83,000	34%	______	3.6¢ per $1	______
f.	54,000	71%	______	4.71¢ per $1	______
g.	61,000	24%	______	53 mills per $1	______
h.	68,200	55%	______	64.2 mills per $1	______

B

1. The tax rate in Orangeburg is 89¢ per $10. How much tax will have to be paid on a property whose actual value is $53,000 but is assessed at 40% of this value? ______

2. Which of the following tax rates is the greater?

a. $5.37 per $100 or $48.26 per $1,000 ______

b. 5.4¢ per $1 or $6.27 per $100 ______

c. 46 mills per $1 or 6¢ per $1 ______

3. The budget for Elmsport showed that its tax rate of $3.52 per $100 was to be spent as follows:

Borough Tax	$ .19
School Tax	1.18
Regional School Tax	1.20
County Tax	.95
Total	$3.52

Ronald DeFalco owns a home in Elmsport that is valued at $50,000 but is assessed at 90% of its value.

a. What is the total tax that Mr. DeFalco will have to pay? ______

b. How much of Mr. DeFalco's tax will go to each of the expenses listed in the budget? ______

4. When Mildred Cramer built an addition to her home, her assessed value was increased from $26,400 to $28,600. If the tax rate last year was $4.81 per $100, how much more tax did she have to pay because of the enlargement of her home? ________

5. A house in Great Meadows is worth $60,000 but is assessed at 80% of its value. A house in Cedarville is also worth $60,000 but is assessed at 40% of its value.

a. If the tax rate in Great Meadows is $3.24 per $100, what is the tax on the Great Meadows home? ________

b. If the tax rate in Cedarville is $7.02 per $100, what is the tax on the Cedarville home? ________

c. On which home is the assessed value the greater? ________

d. On which home is the tax the greater? ________

The Family Budget Unit 3

Having devoted so much of our attention to the ways in which we both spend our money and save our money, we should take a brief look at how we can keep an accounting of just what happens to all this money. The guides that are used to suggest the amount of money we must set aside for each of our needs is called a *budget*.

Budgets are usually drawn up by people who have made a study of the amount of money that the average family spends for such things as:

Food	Clothing	Saving and Investment
Shelter	Transportation	Personal

The "average" family, however, may not have the same likes or dislikes as our own. Therefore, although it may be wise to use a standard budget as a guide at the outset, after a period of testing it should be adjusted to our personal requirements. No budget should be considered a permanent guide. Our income changes, our interest changes, our needs change, and so too should our money guide change. The need for monthly payments on a car at one period in our life gives way to monthly payments on a motorboat during another period.

Example 1 The total take-home pay in a family is $950 per month. The monthly budget provides for the following expenditures:

Food–25%	Clothing–13%	Savings–12%
Shelter–30%	Transportation–11%	Personal–9%

Determine the amount of money assigned for each item.

Solution

Food

25% of take-home pay
= .25 × $950
= $237.50

Shelter

30% of take-home pay
= .30 × $950
= $285

Clothing

13% of take-home pay
= .13 × $950
= $123.50

Transportation

11% of take-home pay
= .11 × $950
= $104.50

Savings

12% of take-home pay
= .12 × $950
= $114

Personal

9% of take-home pay
= .09 × $950
= $85.50

We now want to reverse the direction we have been taking and inquire, "What *percent* of our take-home pay is spent on shelter?" And, of course, the same question can be asked about each of the other items of a budget.

To determine the answers to these questions, let us examine the shelter item in Example 1 above.

Percent for shelter **Take-home pay** **Shelter**

30% × $950 = $285

First number **Second number** **Product**

Several times earlier in our work we encountered this very same situation. On each occasion we are able to show that

First number = Product ÷ Second number

Therefore, in this illustration it is possible to conclude

Percent for shelter = Shelter ÷ Take-home pay.

By simply replacing the word **Shelter** with the word **Food**, this same formula can be used to determine the percent of take-home pay spent on food. Similar replacements will enable us to find the percent of the budget assigned to other items.

Example 2 A take-home pay of $1,100 per month for a family of 3 calls for the following expenditures.

Food – $214	Clothing – $80	Savings – $120
Shelter – $350	Transportation – $95	Personal – $105
	House operation – $136	

What percent of the take-home pay is spent for Food and for Shelter?

Solution

(1) Percent for food = Food ÷ Take-home pay

$$= \$214 \div \$1{,}100$$
$$= .19$$
$$= 19\%$$

(2) Percent for shelter = Shelter ÷ Take-home pay

$$= \$350 \div \$1{,}100$$
$$= .32$$
$$= 32\%$$

Exercises A

1. As the monthly take-home pay increases the percent of the budget assigned for food decreases. Determine the amount assigned for food in each budget for a family of two.

	Monthly Take-home Pay	Percent Assigned For Food	Amount Assigned For Food
a.	$ 800	27%	$________
b.	900	25%	________
c.	1,000	24%	________
d.	1,100	22.5%	________
e.	1,200	21.6%	________

2. As the size of a family increases, the percent of the budget assigned for clothing increases. Each of the following families has a monthly take-home pay of $1,100 per month. Determine the amount assigned for clothing in each budget.

		Percent Assigned For Clothing	Amount Assigned For Clothing
a.	Family of 2	11.1%	$________
b.	Family of 3	12.2%	________
c.	Family of 4	12.7%	________
d.	Family of 5	13.3%	________

3. As the monthly take-home pay increases, so too does the amount assigned for savings increase. Determine the amount assigned for savings in each budget for a family of three.

	Monthly Take-home Pay	Percent Assigned For Savings	Amount Assigned For Savings
a.	$ 800	4.8%	$________
b.	1,000	6.6%	________
c.	1,400	9.2%	________
d.	1,600	13.5%	________
e.	2,000	16.4%	________

B

1. Determine the percent of the take-home pay assigned for shelter in each budget for a family of four.

	Monthly Take-home Pay	Amount Budgeted For Shelter	Percent Budgeted For Shelter
a.	$ 900	$243	________%
b.	1,200	372	________%
c.	1,800	468	________%

2. Determine the percent of the take-home pay assigned for clothing in each budget for a family of three.

	Monthly Take-home Pay	Amount Budgeted For Clothes	Percent Budgeted For Clothes
a.	$ 800	$ 75	________%
b.	1,000	104	________%
c.	1,500	145	________%

C

1. The following budget was prepared for a family of four having a take-home pay of $900 each month.

Food – $258	Home Utilities – $110
Clothing – $95	Transportation – $68
Rent – $285	Savings – $45
Personal – $39	

a. How much money is budgeted for food for the year? ________

b. How much money is budgeted for clothing for the year? ________

c. During the year the monthly rent is increased by 8%. How much will the new budget monthly allotment for rent have to be? ________

d. If the increase in rent has to be deducted from the amount budgeted for savings, how much will the new budget monthly allotment for savings have to be? ________

2. The following is a suggested budget for a family of three having a take-home pay of $1,200 per month. Determine the percent of the take-home pay assigned to each item.

Food – $245	Transportation – $125
Clothing – $135	Savings – $210
Rent – $327	Personal – $158

Food	Clothing	Rent	Transportation	Savings	Personal
________%	________%	________%	________%	________%	________%

3. James Fenway's annual earnings are $12,950. He saves no money. If the cost of living – this includes all items on the normal budget other than savings – rises 9% during the year, how much will his salary have to increase next year for him to be able to live in the manner he is currently living? ________

Review of Chapter 12

1. Rename the following watt-hours as kilowatt-hours.

	Watt-Hours	Kilowatt-Hours		Watt-Hours	Kilowatt-Hours
a.	6,000	________	c.	4,650	________
b.	19,000	________	d.	375	________

Review

2. Determine the number of kilowatt-hours of electricity consumed by each appliance.

	Number Of Watts	Hours Operated	Watt-Hours Consumed	Kilowatt-Hours Consumed
a.	500	14	________	________
b.	2,100	$2\frac{1}{2}$	________	________

3. The electric meter at the beginning and end of a month appear below. Determine the number of kilowatt-hours of electricity consumed during the month. ________

Meter At Beginning Of Month

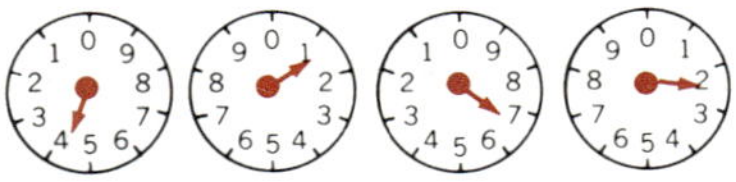

Meter At End Of Month

4. Determine the cost of electricity for a month in which 142 kilowatt-hours were consumed. Use the rate table of the Cape Electric and Gas Company on page 444. ________

5. Nora Stanley roasted a turkey in an electric oven for a period of 7 hours 30 minutes. The oven operates on 2,200 watts of current. If the cost of each kilowatt-hour is 6.7¢, what was the total cost for roasting the turkey? ________

6. During the period of one month a gas range and gas oven consumed 89 hundred cubic feet of gas. Determine the cost of this gas. Use the rate table of the Cape Electric and Gas Company on page 449. ________

7. The readings of a water meter at the beginning and end of a quarterly period are shown below.

Beginning of Period	End of Period
468,470	481,630

If this home was serviced by the Bloomington Water Company, page 454, determine the cost of water for this period. ________

8. Determine the cost of each home-phone call made to points outside the local calling area.

	Distance -Miles-	Hour When Call Was Made	Base Cost	Minutes	Excess Cost	Total Cost
a.	16	Wed-3:00PM	$________	8	$________	$________
b.	75	Mon-11:04AM	________	14	________	________
c.	93	Sun-4:00PM	________	17	________	________
d.	158	Tue-1:30AM	________	12	________	________

9. Alan Green was granted a $34,000 mortgage by a savings and loan association but he had to pay 4 1/2 points to obtain the loan. How much did this cost him? ________

10. When the Rymers had an addition built on their home they borrowed $12,500 on a secondary mortgage. This loan was to be paid off monthly at an interest rate of 14% over a period of 15 years.

a. How large were the monthly payments? ________

b. How much did the Rymers pay the mortgage company over the 15-year period? ________

c. How much more did the Rymers repay the mortgage company than they received? ________

11. The balance on the Engel's mortgage is $8,320. Their monthly payments are $143.62 while the annual interest rate on the loan is 9%.

a. How much of the next payment is interest on the debt? ________

b. How much of that payment will be used to decrease the debt? ________

12. Rename each tax rate from its name at the left to the name at the right.

a. $4.75 per $100 = $________ per $1,000

b. $35.60 per $1,000 = $________ per $10

c. $5.40 per $100 = ________ mills per $1

13. The Durkee's property is worth $67,000 but is assessed at only 30% of its value. The tax rate in the community in which the property is located is $5.95 per $100.

a. What is the assessed value of the property? ________

b. What is the real estate tax on this property? ________

14. The budget for an $850 per month take-home pay for a family of two calls for a 28% allotment for housing.

a. How much money will be spent monthly for this item? ________

b. How much money will be spent annually for this item? ________

15. On a $1,400 monthly take-home pay a budget calls for an expenditure of $336 for food per month for a family of four. What percent of the take-home pay is budgeted for food? ________

Appendix

The Fundamental Operations and Percent

This appendix is designed as a review of the fundamental operations of arithmetic on whole numbers, decimals, and fractions. You will find that everything included in this appendix has been taught to you at one time or another during your first six or seven years of school. The purpose of including it here is to help refresh your memory in the event that the mechanics of some of the operations may have slipped your mind. In view of this, the explanations that are included are designed primarily to tell you how to perform the necessary steps in arriving at your answer rather than why these steps are necessary.

Addition Unit 1

Section 1 Addition of Whole Numbers

Exercises A

Find the sums.

1.	1 5	2.	7 2	3.	4 0	4.	5 2	5.	4 3
6.	2 5	7.	4 5	8.	8 0	9.	1 4	10.	3 5
11.	6 6	12.	1 0	13.	5 3	14.	2 6	15.	7 4

16.	1	17.	5	18.	0	19.	7	20.	4
	9		7		5		7		6
21.	2	22.	1	23.	8	24.	3	25.	6
	7		6		4		6		7
26.	6	27.	0	28.	7	29.	8	30.	6
	3		8		1		7		2
31.	2	32.	5	33.	1	34.	6	35.	0
	8		8		7		8		6
36.	6	37.	3	38.	9	39.	8	40.	4
	5		7		4		6		7
41.	0	42.	9	43.	6	44.	9	45.	7
	9		3		1		2		6
46.	7	47.	2	48.	5	49.	7	50.	4
	5		0		9		3		8
51.	8	52.	1	53.	0	54.	8	55.	7
	2		8		7		5		1
56.	8	57.	2	58.	9	59.	3	60.	3
	3		9		6		8		0
61.	9	62.	8	63.	9	64.	7	65.	8
	8		7		5		0		1
66.	6	67.	3	68.	8	69.	4	70.	7
	0		9		8		9		9
71.	9	72.	9	73.	5	74.	6	75.	8
	1		7		0		9		9

B

Find the sums.

1.	2.	3.	4.	5.
4 8 5	2 7 6	6 5 6	1 7 8	4 4 9

6.	7.	8.	9.	10.
7 4 8	3 7 7	5 9 6	8 1 8	9 3 6

11.	12.	13.	14.	15.
6 5 8	8 7 7	7 9 4	6 8 7	9 5 8

16.	17.	18.	19.	20.
8 8 7	6 8 9	9 7 8	9 8 8	9 9 7

C

Find the sums.

1.	2.	3.	4.
13 34	17 10	46 12	62 27

5.	6.	7.	8.
53 46	21 58	72 16	68 21

9.	10.	11.	12.
65 27	34 44	46 27	58 63

13.	14.	15.	16.
64 57	45 76	96 24	47 65

17.	18.	19.	20.
65 66	78 74	87 75	98 56

D

Find the sums.

1.	56 23 35	**2.**	62 27 57	**3.**	19 84 36	**4.**	28 63 16
5.	85 39 41	**6.**	76 27 32	**7.**	19 86 54	**8.**	36 45 87
9.	72 43 88	**10.**	57 65 98	**11.**	76 59 87	**12.**	98 69 87

E

Find the sums.

1.	624 372 246	**2.**	175 653 365	**3.**	319 584 623	**4.**	581 367 419
5.	657 743 917	**6.**	585 693 778	**7.**	908 675 247	**8.**	488 874 565
9.	675 788 696	**10.**	897 689 758	**11.**	986 869 788	**12.**	876 987 969

Section 2 Addition of Decimals

Example 1 Find the sum of the following numbers.

$4.26, $3, $8.04, $27

Explanation

$3. → Insert decimal point to right of last digit ← $27.

Solution

Align decimal points

```
 $4.26
  3.00  ← Insert two zeros
  8.04
+27.00  ← Insert two zeros
------
$42.30
```

Example 2 Find the sum of the following numbers

$3.45, $2.58, 76¢, 8¢

Explanation

Rename 76¢ as $.76
Rename 8¢ as $.08

Solution

```
 $3.45
  2.58
   .76
+  .08
------
 $6.87
```

Exercises A

Find the sums.

1.	2.	3.	4.
$8.22 4.69	$7.90 5.36	$6.78 8.45	$9.27 6.99

5.	6.	7.	8.
$14.67 37.09	$27.42 53.65	$49.75 13.86	$16.49 27.19

9.	$159.95 42.05	10.	$105.50 96.49	11.	$299.98 68.75	12.	$697.19 489.95

B

Find the sums.

1.	$6.37 2.41 1.16	2.	$1.85 4.23 6.70	3.	$9.42 3.27 5.86	4.	$7.29 8.56 3.35
5.	$24.09 32.16 65.42	6.	$11.41 13.28 7.64	7.	54.23 8.64 10.17	8.	$41.75 29.25 12.19
9.	$247.65 109.16 693.27	10.	$352.29 847.63 285.07	11.	$679.95 325.09 437.65	12.	$937.49 895.78 876.88

C

In each of the following exercises, rearrange the numerals vertically before adding.

1. $2.41 + $3.76 + $4.45 = ________
2. $2.56 + $4.07 + $3.98 = ________
3. $5.48 + $6.29 + $8.05 = ________
4. $3.75 + $7.34 + $9.70 = ________
5. $12.03 + $6.74 + $14.39 + $2.49 = ________
6. $36.24 + $18.56 + $2.27 + $12.50 = ________
7. $4.36 + $43.17 + $29.25 + $3.45 = ________
8. $17.95 + $15 + $42.50 + $8 = ________
9. $54.80 + $21.55 + $39 + $2.04 = ________
10. $123.60 + $43 + $15 + $2.73 = ________
11. $144.75 + $5.65 + $86 + $7 + $14.36 = ________
12. $46.30 + 54¢ + 85¢ + $3.95 + $8.50 = ________
13. 67¢ + $12.85 + 14¢ + $58 + $16.20 = ________
14. $125 + 75¢ + $40 + $18.12 + 56¢ = ________
15. $57 + 98¢ + 64¢ + $176.58 + $39 = ________
16. $258.42 + $43 + $125 + 72¢ + 134¢ = ________

Section 3 Addition of Fractions

Example 1 Find the sum of 5/8 and 2/8.

Solution

$$\begin{array}{r} \frac{5}{8} \\ \frac{2}{8} \\ \hline \frac{7}{8} \end{array} \qquad 5 + 2$$

Denominators remain the same

Add numerators

Example 2 Find the sum of 2/3 and 3/4.

Solution

$$\begin{array}{rcll} \frac{2}{3} & = & \frac{8}{12} & \longleftarrow \frac{2\times4}{3\times4} \\ +\ \frac{3}{4} & = & \frac{9}{12} & \longleftarrow \frac{3\times3}{4\times3} \\ \hline & & \frac{17}{12} & = 1\frac{5}{12} \end{array}$$

Rename fractions as 12ths

Example 3 Find the sum of 2 3/4 and 15 5/6.

Solution

$$\begin{array}{rcll} 2\frac{3}{4} & = & 2\frac{9}{12} & \longleftarrow \frac{3\times3}{4\times3} \\ +\ 15\frac{5}{6} & = & 15\frac{10}{12} & \longleftarrow \frac{5\times2}{6\times2} \\ \hline & & 17\frac{19}{12} & = 17 + 1\frac{7}{12} = 18\frac{7}{12} \end{array}$$

Rename fractions as 12ths

Add whole numbers

Add fractions

Exercises

A

Find the sums.

1. $\begin{array}{r}\frac{3}{5}\\ \frac{1}{5}\\ \hline\end{array}$ 2. $\begin{array}{r}\frac{2}{7}\\ \frac{3}{7}\\ \hline\end{array}$ 3. $\begin{array}{r}\frac{1}{3}\\ \frac{1}{3}\\ \hline\end{array}$ 4. $\begin{array}{r}\frac{3}{8}\\ \frac{2}{8}\\ \hline\end{array}$ 5. $\begin{array}{r}\frac{4}{9}\\ \frac{3}{9}\\ \hline\end{array}$

6. $\begin{array}{r}\frac{12}{11}\\ \frac{13}{11}\\ \hline\end{array}$ 7. $\begin{array}{r}\frac{15}{12}\\ \frac{16}{12}\\ \hline\end{array}$ 8. $\begin{array}{r}\frac{15}{14}\\ \frac{14}{14}\\ \hline\end{array}$ 9. $\begin{array}{r}\frac{2}{6}\\ \frac{3}{6}\\ \hline\end{array}$ 10. $\begin{array}{r}\frac{13}{15}\\ \frac{17}{15}\\ \hline\end{array}$

11. $\begin{array}{r}\frac{3}{4}\\\frac{3}{4}\\\hline\end{array}$	12. $\begin{array}{r}\frac{1}{2}\\\frac{1}{2}\\\hline\end{array}$	13. $\begin{array}{r}\frac{1}{4}\\\frac{3}{4}\\\hline\end{array}$	14. $\begin{array}{r}\frac{5}{6}\\\frac{1}{6}\\\hline\end{array}$	15. $\begin{array}{r}\frac{4}{6}\\\frac{5}{6}\\\hline\end{array}$
16. $\begin{array}{r}\frac{5}{12}\\\frac{7}{12}\\\hline\end{array}$	17. $\begin{array}{r}\frac{7}{12}\\\frac{7}{12}\\\hline\end{array}$	18. $\begin{array}{r}\frac{11}{12}\\\frac{5}{12}\\\hline\end{array}$	19. $\begin{array}{r}\frac{11}{12}\\\frac{10}{12}\\\hline\end{array}$	20. $\begin{array}{r}\frac{8}{12}\\\frac{10}{12}\\\hline\end{array}$
21. $\begin{array}{r}\frac{11}{16}\\\frac{5}{16}\\\hline\end{array}$	22. $\begin{array}{r}\frac{15}{16}\\\frac{5}{16}\\\hline\end{array}$	23. $\begin{array}{r}\frac{5}{16}\\\frac{13}{16}\\\hline\end{array}$	24. $\begin{array}{r}\frac{7}{16}\\\frac{11}{16}\\\hline\end{array}$	25. $\begin{array}{r}\frac{10}{16}\\\frac{15}{16}\\\hline\end{array}$
26. $\begin{array}{r}\frac{1}{3}\\\frac{1}{3}\\\frac{2}{3}\\\hline\end{array}$	27. $\begin{array}{r}\frac{3}{4}\\\frac{1}{4}\\\frac{1}{4}\\\hline\end{array}$	28. $\begin{array}{r}\frac{1}{6}\\\frac{5}{6}\\\frac{3}{6}\\\hline\end{array}$	29. $\begin{array}{r}\frac{1}{8}\\\frac{3}{8}\\\frac{5}{8}\\\hline\end{array}$	30. $\begin{array}{r}\frac{7}{8}\\\frac{5}{8}\\\frac{4}{8}\\\hline\end{array}$
31. $\begin{array}{r}\frac{6}{8}\\\frac{7}{8}\\\frac{5}{8}\\\hline\end{array}$	32. $\begin{array}{r}\frac{3}{8}\\\frac{6}{8}\\\frac{5}{8}\\\hline\end{array}$	33. $\begin{array}{r}\frac{7}{10}\\\frac{9}{10}\\\frac{1}{10}\\\hline\end{array}$	34. $\begin{array}{r}\frac{3}{10}\\\frac{7}{10}\\\frac{4}{10}\\\hline\end{array}$	35. $\begin{array}{r}\frac{1}{10}\\\frac{8}{10}\\\frac{6}{10}\\\hline\end{array}$
36. $\begin{array}{r}\frac{5}{12}\\\frac{7}{12}\\\frac{1}{12}\\\hline\end{array}$	37. $\begin{array}{r}\frac{9}{12}\\\frac{7}{12}\\\frac{11}{12}\\\hline\end{array}$	38. $\begin{array}{r}\frac{5}{16}\\\frac{7}{16}\\\frac{10}{16}\\\hline\end{array}$	39. $\begin{array}{r}\frac{3}{16}\\\frac{15}{16}\\\frac{8}{16}\\\hline\end{array}$	40. $\begin{array}{r}\frac{9}{16}\\\frac{11}{16}\\\frac{12}{16}\\\hline\end{array}$

B

Find the sums.

1. $\begin{array}{r}\frac{5}{12}\\\frac{1}{6}\\\hline\end{array}$	2. $\begin{array}{r}\frac{5}{8}\\\frac{1}{2}\\\hline\end{array}$	3. $\begin{array}{r}\frac{7}{8}\\\frac{3}{4}\\\hline\end{array}$	4. $\begin{array}{r}\frac{5}{16}\\\frac{1}{4}\\\hline\end{array}$	5. $\begin{array}{r}\frac{1}{6}\\\frac{2}{3}\\\hline\end{array}$
6. $\begin{array}{r}\frac{7}{12}\\\frac{2}{3}\\\hline\end{array}$	7. $\begin{array}{r}\frac{11}{12}\\\frac{3}{4}\\\hline\end{array}$	8. $\begin{array}{r}\frac{3}{16}\\\frac{1}{2}\\\hline\end{array}$	9. $\begin{array}{r}\frac{7}{16}\\\frac{1}{4}\\\hline\end{array}$	10. $\begin{array}{r}\frac{3}{4}\\\frac{9}{16}\\\hline\end{array}$
11. $\begin{array}{r}\frac{1}{2}\\\frac{1}{3}\\\hline\end{array}$	12. $\begin{array}{r}\frac{2}{3}\\\frac{1}{2}\\\hline\end{array}$	13. $\begin{array}{r}\frac{3}{4}\\\frac{1}{3}\\\hline\end{array}$	14. $\begin{array}{r}\frac{1}{4}\\\frac{2}{3}\\\hline\end{array}$	15. $\begin{array}{r}\frac{1}{6}\\\frac{1}{4}\\\hline\end{array}$
16. $\begin{array}{r}\frac{1}{2}\\\frac{1}{3}\\\frac{2}{3}\\\hline\end{array}$	17. $\begin{array}{r}\frac{2}{3}\\\frac{2}{3}\\\frac{1}{2}\\\hline\end{array}$	18. $\begin{array}{r}\frac{3}{4}\\\frac{1}{2}\\\frac{1}{2}\\\hline\end{array}$	19. $\begin{array}{r}\frac{3}{4}\\\frac{2}{3}\\\frac{1}{3}\\\hline\end{array}$	20. $\begin{array}{r}\frac{1}{2}\\\frac{1}{3}\\\frac{1}{4}\\\hline\end{array}$

21. $\frac{1}{2}$ $\frac{2}{3}$ $\frac{3}{4}$	22. $\frac{3}{4}$ $\frac{5}{12}$ $\frac{1}{6}$	23. $\frac{1}{4}$ $\frac{2}{3}$ $\frac{7}{12}$	24. $\frac{11}{12}$ $\frac{5}{6}$ $\frac{2}{3}$	25. $\frac{7}{12}$ $\frac{5}{6}$ $\frac{3}{4}$

C

Find the sums.

1. $2\frac{1}{2}$ $3\frac{1}{4}$	2. $5\frac{1}{6}$ $4\frac{1}{6}$	3. $6\frac{5}{8}$ $2\frac{1}{4}$	4. $9\frac{1}{3}$ $7\frac{1}{6}$	5. $8\frac{3}{4}$ $3\frac{1}{8}$
6. $7\frac{15}{12}$ $9\frac{1}{4}$	7. $8\frac{17}{12}$ $4\frac{1}{3}$	8. $6\frac{2}{3}$ $3\frac{11}{12}$	9. $1\frac{13}{10}$ $2\frac{3}{5}$	10. $4\frac{11}{10}$ $6\frac{1}{5}$
11. $3\frac{1}{2}$ $4\frac{2}{3}$	12. $7\frac{1}{2}$ $6\frac{3}{4}$	13. $6\frac{5}{6}$ $9\frac{1}{2}$	14. $8\frac{5}{8}$ $7\frac{3}{4}$	15. $12\frac{3}{8}$ $9\frac{3}{4}$
16. $11\frac{5}{6}$ $12\frac{11}{12}$	17. $16\frac{3}{4}$ $14\frac{5}{6}$	18. $14\frac{15}{16}$ $16\frac{3}{4}$	19. $14\frac{3}{8}$ $9\frac{11}{16}$	20. $21\frac{5}{8}$ $20\frac{2}{3}$
21. $6\frac{1}{2}$ $5\frac{1}{3}$ $2\frac{1}{2}$	22. $5\frac{1}{2}$ $6\frac{1}{4}$ $3\frac{1}{4}$	23. $7\frac{1}{2}$ $9\frac{3}{4}$ $10\frac{1}{4}$	24. $1\frac{3}{4}$ $4\frac{1}{2}$ $7\frac{3}{4}$	25. $10\frac{2}{3}$ $5\frac{1}{2}$ $6\frac{1}{2}$
26. $2\frac{3}{4}$ $5\frac{1}{2}$ $4\frac{1}{2}$	27. $10\frac{1}{3}$ $7\frac{2}{3}$ $9\frac{1}{6}$	28. $5\frac{3}{8}$ $7\frac{1}{2}$ $2\frac{1}{4}$	29. $6\frac{7}{8}$ $4\frac{3}{4}$ $9\frac{1}{4}$	30. $1\frac{5}{6}$ $7\frac{1}{2}$ $8\frac{2}{3}$
31. $16\frac{3}{4}$ $10\frac{1}{2}$ $15\frac{1}{3}$	32. $7\frac{3}{8}$ $9\frac{3}{4}$ $6\frac{1}{2}$	33. $14\frac{1}{4}$ $2\frac{3}{4}$ $8\frac{1}{8}$	34. $11\frac{5}{6}$ $12\frac{2}{3}$ $7\frac{1}{2}$	35. $1\frac{1}{6}$ $2\frac{3}{4}$ $6\frac{1}{2}$
36. $11\frac{1}{4}$ $5\frac{5}{6}$ $12\frac{3}{4}$	37. $12\frac{1}{2}$ $6\frac{1}{6}$ $14\frac{3}{4}$	38. $17\frac{1}{2}$ $14\frac{13}{16}$ $12\frac{5}{8}$	39. $24\frac{2}{3}$ $27\frac{3}{4}$ $16\frac{5}{12}$	40. $26\frac{7}{8}$ $31\frac{7}{16}$ $28\frac{1}{2}$

Unit 2 Subtraction

Section 1 Subtraction of Whole Numbers and Decimals

Example 1 Subtract 65 from 83.

Solution

$$\begin{array}{r} 83 \\ -65 \\ \hline \end{array} = \begin{array}{cc} 7 & \\ \not{8} & {}^{1}3 \\ 6 & 5 \\ \hline 1 & 8 \end{array}$$

Rename as 70 + 13

Subtract 6 from 7

Subtract 8 from 13

Example 2 Subtract $5.98 from $10.

Solution

Insert decimal point

$$\begin{array}{r} \$10.00 \\ 5.98 \\ \hline \$\ 4.02 \end{array}$$

Attach two zeros

Exercises A

Find the differences.

1.	$\begin{array}{r} 4 \\ \underline{0} \end{array}$	2.	$\begin{array}{r} 7 \\ \underline{2} \end{array}$	3.	$\begin{array}{r} 6 \\ \underline{4} \end{array}$	4.	$\begin{array}{r} 7 \\ \underline{5} \end{array}$	5.	$\begin{array}{r} 6 \\ \underline{6} \end{array}$
6.	$\begin{array}{r} 8 \\ \underline{1} \end{array}$	7.	$\begin{array}{r} 6 \\ \underline{0} \end{array}$	8.	$\begin{array}{r} 9 \\ \underline{3} \end{array}$	9.	$\begin{array}{r} 8 \\ \underline{6} \end{array}$	10.	$\begin{array}{r} 9 \\ \underline{5} \end{array}$
11.	$\begin{array}{r} 7 \\ \underline{4} \end{array}$	12.	$\begin{array}{r} 3 \\ \underline{0} \end{array}$	13.	$\begin{array}{r} 7 \\ \underline{1} \end{array}$	14.	$\begin{array}{r} 9 \\ \underline{4} \end{array}$	15.	$\begin{array}{r} 8 \\ \underline{8} \end{array}$
16.	$\begin{array}{r} 9 \\ \underline{7} \end{array}$	17.	$\begin{array}{r} 1 \\ \underline{1} \end{array}$	18.	$\begin{array}{r} 5 \\ \underline{0} \end{array}$	19.	$\begin{array}{r} 9 \\ \underline{1} \end{array}$	20.	$\begin{array}{r} 8 \\ \underline{7} \end{array}$

21.	8	22.	9	23.	3	24.	9	25.	5
	5		0		3		8		5
26.	9	27.	8	28.	2	29.	7	30.	9
	9		2		2		0		6
31.	1	32.	4	33.	8	34.	9	35.	7
	0		2		3		2		3

B

Find the differences.

1.	12	2.	14	3.	11	4.	10	5.	11
	8		5		4		2		9
6.	18	7.	14	8.	13	9.	11	10.	12
	9		8		4		8		7
11.	16	12.	10	13.	12	14.	13	15.	14
	9		6		5		7		9
16.	15	17.	13	18.	10	19.	12	20.	15
	8		5		7		4		6
21.	13	22.	17	23.	16	24.	15	25.	17
	8		9		7		9		8

C

Find the differences.

1.	32	2.	41	3.	53	4.	22	5.	64
	24		15		44		16		26

6. 50 23	7. 54 35	8. 61 19	9. 73 48	10. 62 56
11. 87 78	12. 92 37	13. 83 29	14. 67 48	15. 58 39
16. 624 407	17. 737 418	18. 856 429	19. 583 247	20. 954 318
21. 524 367	22. 401 269	23. 500 127	24. 652 467	25. 741 582
26. 306 209	27. 534 148	28. 763 468	29. 873 585	30. 695 587
31. 478 299	32. 657 378	33. 504 278	34. 710 395	35. 683 486

D

Find the differences.

1. \$2.46 .35	2. \$4.39 1.26	3. \$5.88 2.65	4. \$3.69 1.27
5. \$8.27 3.04	6. \$5.75 2.35	7. \$6.58 4.17	8. \$7.31 2.10
9. \$5.14 3.91	10. \$2.26 1.84	11. \$9.57 4.62	12. \$8.39 1.43
13. \$6.44 3.72	14. \$7.06 2.46	15. \$9.23 5.71	16. \$7.54 6.83
17. \$7.00 3.58	18. \$2.00 .69	19. \$3.00 2.63	20. \$9.00 6.34

21.	22.	23.	24.
\$8.43 3.66	\$7.12 2.58	\$4.67 3.88	\$5.24 2.98

25.	26.	27.	28.
\$6.04 3.95	\$5.07 2.89	\$7.18 5.99	\$9.65 7.96

Section 2 Subtraction of Fractions

Example 1 Subtract 5 3/4 from 7 3/8.

Solution

Rename 7 as 6 + 1

Rename 1 as $\frac{8}{8}$

$$\begin{array}{rcrcrcrcr} 7\frac{3}{8} & = & 7\frac{3}{8} & = & \boxed{6+1}+\frac{3}{8} & = & 6+\frac{8}{8}+\frac{3}{8} & = & 6\frac{11}{8} \\ -5\frac{3}{4} & = & 5\frac{6}{8} & = & 5\frac{6}{8} & = & 5\frac{6}{8} & = & 5\frac{6}{8} \\ \hline & & & & & & & & 1\frac{5}{8} \end{array}$$

Rename $\frac{3}{4}$ as $\frac{6}{8}$

Exercises

A

Find the differences.

1.	2.	3.	4.	5.
$\frac{3}{5}$ $\frac{1}{5}$	$\frac{5}{7}$ $\frac{2}{7}$	$\frac{3}{4}$ $\frac{2}{4}$	$\frac{9}{10}$ $\frac{6}{10}$	$\frac{11}{12}$ $\frac{10}{12}$

6.	7.	8.	9.	10.
$\frac{3}{4}$ $\frac{1}{4}$	$\frac{7}{8}$ $\frac{3}{8}$	$\frac{1}{2}$ $\frac{1}{2}$	$\frac{5}{8}$ $\frac{1}{8}$	$\frac{7}{10}$ $\frac{3}{10}$

11.	12.	13.	14.	15.
$\frac{7}{16}$ $\frac{5}{16}$	$\frac{13}{16}$ $\frac{1}{16}$	$\frac{11}{12}$ $\frac{5}{12}$	$\frac{11}{12}$ $\frac{7}{12}$	$\frac{15}{16}$ $\frac{7}{16}$

B

Find the differences.

1.	2.	3.	4.	5.
$\frac{1}{2}$ $\frac{1}{4}$	$\frac{1}{2}$ $\frac{1}{6}$	$\frac{1}{3}$ $\frac{1}{6}$	$\frac{1}{2}$ $\frac{1}{8}$	$\frac{1}{4}$ $\frac{1}{8}$

6. $\begin{array}{r}\frac{1}{3}\\ \underline{\frac{1}{9}}\end{array}$	7. $\begin{array}{r}\frac{2}{3}\\ \underline{\frac{1}{9}}\end{array}$	8. $\begin{array}{r}\frac{3}{4}\\ \underline{\frac{5}{8}}\end{array}$	9. $\begin{array}{r}\frac{7}{8}\\ \underline{\frac{3}{4}}\end{array}$	10. $\begin{array}{r}\frac{3}{4}\\ \underline{\frac{1}{2}}\end{array}$
11. $\begin{array}{r}\frac{3}{4}\\ \underline{\frac{2}{3}}\end{array}$	12. $\begin{array}{r}\frac{2}{3}\\ \underline{\frac{1}{2}}\end{array}$	13. $\begin{array}{r}\frac{3}{4}\\ \underline{\frac{1}{6}}\end{array}$	14. $\begin{array}{r}\frac{5}{6}\\ \underline{\frac{3}{4}}\end{array}$	15. $\begin{array}{r}\frac{1}{2}\\ \underline{\frac{1}{5}}\end{array}$
16. $\begin{array}{r}\frac{3}{5}\\ \underline{\frac{1}{2}}\end{array}$	17. $\begin{array}{r}\frac{3}{4}\\ \underline{\frac{2}{5}}\end{array}$	18. $\begin{array}{r}\frac{4}{5}\\ \underline{\frac{3}{4}}\end{array}$	19. $\begin{array}{r}\frac{3}{4}\\ \underline{\frac{7}{10}}\end{array}$	20. $\begin{array}{r}\frac{9}{10}\\ \underline{\frac{1}{4}}\end{array}$

C

Find the differences.

1. $\begin{array}{r}6\frac{1}{2}\\ \underline{4\frac{1}{2}}\end{array}$	2. $\begin{array}{r}8\frac{1}{4}\\ \underline{5\frac{1}{4}}\end{array}$	3. $\begin{array}{r}9\frac{4}{5}\\ \underline{6\frac{1}{5}}\end{array}$	4. $\begin{array}{r}7\frac{5}{6}\\ \underline{4\frac{1}{6}}\end{array}$	5. $\begin{array}{r}5\frac{7}{8}\\ \underline{3\frac{3}{8}}\end{array}$
6. $\begin{array}{r}18\frac{1}{2}\\ \underline{3\frac{1}{5}}\end{array}$	7. $\begin{array}{r}16\frac{5}{16}\\ \underline{4\frac{1}{4}}\end{array}$	8. $\begin{array}{r}17\frac{1}{4}\\ \underline{6\frac{3}{16}}\end{array}$	9. $\begin{array}{r}15\frac{7}{8}\\ \underline{2\frac{3}{4}}\end{array}$	10. $\begin{array}{r}12\frac{5}{6}\\ \underline{9\frac{2}{3}}\end{array}$
11. $\begin{array}{r}13\frac{15}{16}\\ \underline{7\frac{5}{8}}\end{array}$	12. $\begin{array}{r}11\frac{7}{8}\\ \underline{2\frac{1}{4}}\end{array}$	13. $\begin{array}{r}15\frac{3}{4}\\ \underline{6\frac{3}{5}}\end{array}$	14. $\begin{array}{r}12\frac{5}{8}\\ \underline{8\frac{5}{16}}\end{array}$	15. $\begin{array}{r}14\frac{9}{16}\\ \underline{6\frac{1}{2}}\end{array}$

D

Find the differences.

1. $\begin{array}{r}7\\ \underline{5\frac{1}{2}}\end{array}$	2. $\begin{array}{r}9\\ \underline{4\frac{1}{4}}\end{array}$	3. $\begin{array}{r}6\\ \underline{5\frac{3}{8}}\end{array}$	4. $\begin{array}{r}8\\ \underline{6\frac{2}{5}}\end{array}$	5. $\begin{array}{r}7\\ \underline{6\frac{1}{3}}\end{array}$
6. $\begin{array}{r}9\\ \underline{2\frac{3}{4}}\end{array}$	7. $\begin{array}{r}12\\ \underline{5\frac{7}{8}}\end{array}$	8. $\begin{array}{r}14\\ \underline{2\frac{5}{8}}\end{array}$	9. $\begin{array}{r}23\\ \underline{1\frac{4}{5}}\end{array}$	10. $\begin{array}{r}17\\ \underline{2\frac{1}{6}}\end{array}$
11. $\begin{array}{r}18\\ \underline{5\frac{2}{3}}\end{array}$	12. $\begin{array}{r}16\\ \underline{2\frac{5}{12}}\end{array}$	13. $\begin{array}{r}14\\ \underline{8\frac{7}{16}}\end{array}$	14. $\begin{array}{r}11\\ \underline{7\frac{9}{16}}\end{array}$	15. $\begin{array}{r}10\\ \underline{2\frac{11}{12}}\end{array}$

E

Find the differences.

1. $\begin{array}{r} 9\frac{3}{10} \\ \underline{2\frac{7}{10}} \end{array}$ 2. $\begin{array}{r} 9\frac{1}{10} \\ \underline{8\frac{9}{10}} \end{array}$ 3. $\begin{array}{r} 6\frac{5}{12} \\ \underline{2\frac{7}{12}} \end{array}$ 4. $\begin{array}{r} 7\frac{1}{12} \\ \underline{3\frac{11}{12}} \end{array}$ 5. $\begin{array}{r} 8\frac{5}{16} \\ \underline{5\frac{7}{16}} \end{array}$

6. $\begin{array}{r} 8\frac{1}{4} \\ \underline{5\frac{1}{2}} \end{array}$ 7. $\begin{array}{r} 9\frac{1}{2} \\ \underline{4\frac{3}{4}} \end{array}$ 8. $\begin{array}{r} 12\frac{3}{8} \\ \underline{1\frac{3}{4}} \end{array}$ 9. $\begin{array}{r} 16\frac{2}{3} \\ \underline{2\frac{5}{6}} \end{array}$ 10. $\begin{array}{r} 12\frac{1}{2} \\ \underline{1\frac{2}{3}} \end{array}$

11. $\begin{array}{r} 17\frac{2}{3} \\ \underline{5\frac{3}{4}} \end{array}$ 12. $\begin{array}{r} 18\frac{1}{5} \\ \underline{4\frac{7}{10}} \end{array}$ 13. $\begin{array}{r} 16\frac{2}{5} \\ \underline{3\frac{1}{2}} \end{array}$ 14. $\begin{array}{r} 19\frac{1}{4} \\ \underline{6\frac{2}{3}} \end{array}$ 15. $\begin{array}{r} 15\frac{1}{6} \\ \underline{4\frac{1}{3}} \end{array}$

16. $\begin{array}{r} 11\frac{5}{12} \\ \underline{7\frac{1}{2}} \end{array}$ 17. $\begin{array}{r} 12\frac{5}{8} \\ \underline{5\frac{3}{4}} \end{array}$ 18. $\begin{array}{r} 14\frac{7}{8} \\ \underline{6\frac{15}{16}} \end{array}$ 19. $\begin{array}{r} 13\frac{5}{12} \\ \underline{7\frac{5}{6}} \end{array}$ 20. $\begin{array}{r} 10\frac{3}{4} \\ \underline{7\frac{5}{6}} \end{array}$

21. $\begin{array}{r} 21\frac{2}{3} \\ \underline{12\frac{5}{6}} \end{array}$ 22. $\begin{array}{r} 20\frac{1}{3} \\ \underline{16\frac{3}{4}} \end{array}$ 23. $\begin{array}{r} 24\frac{3}{10} \\ \underline{9\frac{4}{5}} \end{array}$ 24. $\begin{array}{r} 25\frac{1}{4} \\ \underline{17\frac{5}{16}} \end{array}$ 25. $\begin{array}{r} 28\frac{1}{8} \\ \underline{19\frac{3}{16}} \end{array}$

Multiplication Unit 3

Section 1 Multiplication of Whole Numbers and Decimals

Example 1 Find the product of 2.35 and .6 .

Solution

$$\begin{array}{r} 2.35 \\ \times\ \ .6 \\ \hline 1.410 \end{array}$$

2.35 ← **Number of digits to right of decimal point: 2**

.6 ← **Number of digits to right of decimal point: 1**

1.410 ← **Number of digits to right of decimal point: 2 + 1 = 3**

Exercises A

Find the products.

1. $\begin{array}{r}4\\ \underline{1}\end{array}$	**2.** $\begin{array}{r}0\\ \underline{8}\end{array}$	**3.** $\begin{array}{r}3\\ \underline{5}\end{array}$	**4.** $\begin{array}{r}4\\ \underline{4}\end{array}$	**5.** $\begin{array}{r}7\\ \underline{1}\end{array}$
6. $\begin{array}{r}0\\ \underline{7}\end{array}$	**7.** $\begin{array}{r}2\\ \underline{4}\end{array}$	**8.** $\begin{array}{r}3\\ \underline{7}\end{array}$	**9.** $\begin{array}{r}6\\ \underline{5}\end{array}$	**10.** $\begin{array}{r}8\\ \underline{2}\end{array}$
11. $\begin{array}{r}2\\ \underline{2}\end{array}$	**12.** $\begin{array}{r}1\\ \underline{8}\end{array}$	**13.** $\begin{array}{r}5\\ \underline{5}\end{array}$	**14.** $\begin{array}{r}6\\ \underline{3}\end{array}$	**15.** $\begin{array}{r}9\\ \underline{0}\end{array}$
16. $\begin{array}{r}4\\ \underline{0}\end{array}$	**17.** $\begin{array}{r}3\\ \underline{6}\end{array}$	**18.** $\begin{array}{r}5\\ \underline{1}\end{array}$	**19.** $\begin{array}{r}8\\ \underline{4}\end{array}$	**20.** $\begin{array}{r}4\\ \underline{5}\end{array}$
21. $\begin{array}{r}1\\ \underline{6}\end{array}$	**22.** $\begin{array}{r}5\\ \underline{4}\end{array}$	**23.** $\begin{array}{r}3\\ \underline{8}\end{array}$	**24.** $\begin{array}{r}6\\ \underline{6}\end{array}$	**25.** $\begin{array}{r}7\\ \underline{2}\end{array}$
26. $\begin{array}{r}5\\ \underline{3}\end{array}$	**27.** $\begin{array}{r}0\\ \underline{4}\end{array}$	**28.** $\begin{array}{r}6\\ \underline{1}\end{array}$	**29.** $\begin{array}{r}7\\ \underline{5}\end{array}$	**30.** $\begin{array}{r}4\\ \underline{6}\end{array}$
31. $\begin{array}{r}0\\ \underline{0}\end{array}$	**32.** $\begin{array}{r}2\\ \underline{8}\end{array}$	**33.** $\begin{array}{r}9\\ \underline{3}\end{array}$	**34.** $\begin{array}{r}6\\ \underline{7}\end{array}$	**35.** $\begin{array}{r}5\\ \underline{9}\end{array}$
36. $\begin{array}{r}1\\ \underline{4}\end{array}$	**37.** $\begin{array}{r}4\\ \underline{7}\end{array}$	**38.** $\begin{array}{r}0\\ \underline{1}\end{array}$	**39.** $\begin{array}{r}3\\ \underline{9}\end{array}$	**40.** $\begin{array}{r}7\\ \underline{3}\end{array}$
41. $\begin{array}{r}2\\ \underline{9}\end{array}$	**42.** $\begin{array}{r}5\\ \underline{6}\end{array}$	**43.** $\begin{array}{r}7\\ \underline{4}\end{array}$	**44.** $\begin{array}{r}8\\ \underline{3}\end{array}$	**45.** $\begin{array}{r}9\\ \underline{1}\end{array}$
46. $\begin{array}{r}4\\ \underline{8}\end{array}$	**47.** $\begin{array}{r}2\\ \underline{7}\end{array}$	**48.** $\begin{array}{r}1\\ \underline{1}\end{array}$	**49.** $\begin{array}{r}4\\ \underline{3}\end{array}$	**50.** $\begin{array}{r}6\\ \underline{4}\end{array}$
51. $\begin{array}{r}1\\ \underline{2}\end{array}$	**52.** $\begin{array}{r}0\\ \underline{9}\end{array}$	**53.** $\begin{array}{r}7\\ \underline{6}\end{array}$	**54.** $\begin{array}{r}8\\ \underline{5}\end{array}$	**55.** $\begin{array}{r}9\\ \underline{3}\end{array}$

56. $\begin{array}{r} 2 \\ \underline{3} \end{array}$	**57.** $\begin{array}{r} 7 \\ \underline{8} \end{array}$	**58.** $\begin{array}{r} 0 \\ \underline{5} \end{array}$	**59.** $\begin{array}{r} 9 \\ \underline{5} \end{array}$	**60.** $\begin{array}{r} 3 \\ \underline{0} \end{array}$
61. $\begin{array}{r} 6 \\ \underline{8} \end{array}$	**62.** $\begin{array}{r} 0 \\ \underline{3} \end{array}$	**63.** $\begin{array}{r} 4 \\ \underline{9} \end{array}$	**64.** $\begin{array}{r} 8 \\ \underline{7} \end{array}$	**65.** $\begin{array}{r} 9 \\ \underline{9} \end{array}$
66. $\begin{array}{r} 8 \\ \underline{9} \end{array}$	**67.** $\begin{array}{r} 8 \\ \underline{0} \end{array}$	**68.** $\begin{array}{r} 7 \\ \underline{7} \end{array}$	**69.** $\begin{array}{r} 9 \\ \underline{4} \end{array}$	**70.** $\begin{array}{r} 2 \\ \underline{1} \end{array}$
71. $\begin{array}{r} 9 \\ \underline{7} \end{array}$	**72.** $\begin{array}{r} 5 \\ \underline{0} \end{array}$	**73.** $\begin{array}{r} 8 \\ \underline{6} \end{array}$	**74.** $\begin{array}{r} 5 \\ \underline{8} \end{array}$	**75.** $\begin{array}{r} 9 \\ \underline{6} \end{array}$

B

Find the products.

1. $\begin{array}{r} 23 \\ \underline{2} \end{array}$	**2.** $\begin{array}{r} 41 \\ \underline{3} \end{array}$	**3.** $\begin{array}{r} 15 \\ \underline{4} \end{array}$	**4.** $\begin{array}{r} 24 \\ \underline{5} \end{array}$	**5.** $\begin{array}{r} 16 \\ \underline{3} \end{array}$
6. $\begin{array}{r} 18 \\ \underline{4} \end{array}$	**7.** $\begin{array}{r} 39 \\ \underline{2} \end{array}$	**8.** $\begin{array}{r} 47 \\ \underline{3} \end{array}$	**9.** $\begin{array}{r} 53 \\ \underline{6} \end{array}$	**10.** $\begin{array}{r} 42 \\ \underline{4} \end{array}$
11. $\begin{array}{r} 45 \\ \underline{2} \end{array}$	**12.** $\begin{array}{r} 36 \\ \underline{4} \end{array}$	**13.** $\begin{array}{r} 25 \\ \underline{6} \end{array}$	**14.** $\begin{array}{r} 28 \\ \underline{3} \end{array}$	**15.** $\begin{array}{r} 39 \\ \underline{5} \end{array}$
16. $\begin{array}{r} 43 \\ \underline{6} \end{array}$	**17.** $\begin{array}{r} 50 \\ \underline{7} \end{array}$	**18.** $\begin{array}{r} 40 \\ \underline{9} \end{array}$	**19.** $\begin{array}{r} 60 \\ \underline{8} \end{array}$	**20.** $\begin{array}{r} 51 \\ \underline{7} \end{array}$
21. $\begin{array}{r} 82 \\ \underline{7} \end{array}$	**22.** $\begin{array}{r} 81 \\ \underline{8} \end{array}$	**23.** $\begin{array}{r} 85 \\ \underline{6} \end{array}$	**24.** $\begin{array}{r} 87 \\ \underline{9} \end{array}$	**25.** $\begin{array}{r} 89 \\ \underline{5} \end{array}$
26. $\begin{array}{r} 93 \\ \underline{4} \end{array}$	**27.** $\begin{array}{r} 95 \\ \underline{6} \end{array}$	**28.** $\begin{array}{r} 96 \\ \underline{7} \end{array}$	**29.** $\begin{array}{r} 94 \\ \underline{8} \end{array}$	**30.** $\begin{array}{r} 98 \\ \underline{9} \end{array}$

C

Find the products.

1.	36	2.	39	3.	40	4.	40	5.	44
	27		26		29		32		35
6.	47	7.	53	8.	58	9.	61	10.	64
	38		39		46		48		51
11.	69	12.	70	13.	72	14.	75	15.	78
	56		58		64		65		67
16.	79	17.	76	18.	83	19.	88	20.	89
	69		72		73		76		78
21.	92	22.	94	23.	96	24.	97	25.	99
	80		83		85		87		98

D

Find the products.

1.	203	2.	104	3.	406	4.	507	5.	201
	4		5		3		6		8
6.	905	7.	703	8.	802	9.	809	10.	907
	2		4		8		7		6
11.	430	12.	560	13.	920	14.	740	15.	670
	5		4		6		8		9
16.	291	17.	403	18.	509	19.	870	20.	706
	60		80		70		60		20

21. 902 30	22. 804 50	23. 840 80	24. 760 60	25. 908 90

E

Find the products.

1. 57 .02	2. 95 .01	3. 76 .08	4. 83 .09	5. 96 .06
6. 1.05 23	7. 1.06 41	8. 1.08 34	9. 1.09 17	10. 1.02 25
11. 1.03 62	12. 1.04 48	13. 1.01 93	14. 1.06 85	15. 1.08 97
16. 250 .035	17. 470 .045	18. 640 .055	19. 800 .065	20. 750 .025

Section 2 Multiplication of Fractions and Mixed Numbers

Example 1 Find the product of 2/3 and 5/7.

Solution

$$\frac{2}{3} \times \frac{5}{7} = \frac{10}{21} \quad \begin{matrix} \leftarrow 2 \times 5 \\ \leftarrow 3 \times 7 \end{matrix}$$

Example 2 Find the product of 2/3 and 3/4.

Solution

6 ÷ 6

$$\frac{2}{3} \times \frac{3}{4} = \frac{6}{12} = \frac{1}{2}$$

12 ÷ 6

Example 3 Find the product of 6 and 2 1/4.

Solution

$$6 \times 2\tfrac{1}{4} = \tfrac{6}{1} \times \tfrac{9}{4} = \tfrac{54}{4} = \tfrac{27}{2} = 13\tfrac{1}{2}$$

$54 \div 2$

$4 \div 2$

Exercises

A

Find the products.

1.	$\frac{1}{2} \times \frac{3}{5}$	= ________	2.	$\frac{1}{2} \times \frac{5}{7}$	= ________
3.	$\frac{1}{3} \times \frac{4}{5}$	= ________	4.	$\frac{2}{3} \times \frac{4}{5}$	= ________
5.	$\frac{2}{3} \times \frac{5}{7}$	= ________	6.	$\frac{3}{4} \times \frac{1}{2}$	= ________
7.	$\frac{1}{4} \times \frac{3}{5}$	= ________	8.	$\frac{1}{4} \times \frac{1}{5}$	= ________
9.	$\frac{3}{4} \times \frac{3}{5}$	= ________	10.	$\frac{4}{5} \times \frac{2}{5}$	= ________
11.	$\frac{1}{6} \times \frac{1}{3}$	= ________	12.	$\frac{5}{7} \times \frac{2}{3}$	= ________
13.	$\frac{1}{8} \times \frac{5}{6}$	= ________	14.	$\frac{3}{8} \times \frac{3}{5}$	= ________
15.	$\frac{5}{8} \times \frac{1}{6}$	= ________	16.	$\frac{5}{6} \times \frac{5}{8}$	= ________

B

Find the products and reduce each answer to its lowest terms.

1.	$\frac{1}{2} \times \frac{2}{5}$	= ________	2.	$\frac{1}{3} \times \frac{3}{5}$	= ________
3.	$\frac{2}{3} \times \frac{6}{7}$	= ________	4.	$\frac{2}{3} \times \frac{5}{8}$	= ________
5.	$\frac{3}{4} \times \frac{1}{6}$	= ________	6.	$\frac{3}{4} \times \frac{2}{9}$	= ________
7.	$\frac{1}{4} \times \frac{2}{5}$	= ________	8.	$\frac{4}{5} \times \frac{1}{2}$	= ________
9.	$\frac{1}{5} \times \frac{5}{6}$	= ________	10.	$\frac{4}{5} \times \frac{7}{10}$	= ________
11.	$\frac{2}{5} \times \frac{5}{8}$	= ________	12.	$\frac{2}{3} \times \frac{9}{10}$	= ________
13.	$\frac{5}{6} \times \frac{2}{5}$	= ________	14.	$\frac{5}{8} \times \frac{4}{5}$	= ________
15.	$\frac{3}{10} \times \frac{5}{9}$	= ________	16.	$\frac{5}{12} \times \frac{2}{5}$	= ________

C

Find the products.

1.	$400 \times \frac{7}{200}$	= ________	2.	$200 \times \frac{9}{25}$	= ________
3.	$300 \times \frac{90}{270}$	= ________	4.	$800 \times \frac{60}{400}$	= ________
5.	$720 \times \frac{50}{360}$	= ________	6.	$900 \times \frac{40}{360}$	= ________
7.	$1{,}500 \times \frac{120}{360}$	= ________	8.	$2{,}400 \times \frac{150}{480}$	= ________
9.	$4{,}200 \times \frac{75}{300}$	= ________	10.	$6{,}900 \times \frac{36}{240}$	= ________

D

Find the products and reduce each answer to its lowest terms.

1.	$4 \times 1\frac{1}{2}$	= ________	2.	$6 \times 2\frac{1}{2}$	= ________
3.	$10 \times 5\frac{1}{2}$	= ________	4.	$3\frac{1}{2} \times 6$	= ________
5.	$4\frac{1}{2} \times 8$	= ________	6.	$7\frac{1}{2} \times 12$	= ________
7.	$6 \times 1\frac{1}{3}$	= ________	8.	$12 \times 2\frac{2}{3}$	= ________
9.	$9 \times 3\frac{1}{3}$	= ________	10.	$3 \times 6\frac{2}{3}$	= ________
11.	$15 \times 1\frac{2}{3}$	= ________	12.	$21 \times 1\frac{1}{7}$	= ________
13.	$2\frac{2}{5} \times 10$	= ________	14.	$3\frac{4}{5} \times 5$	= ________
15.	$5\frac{1}{4} \times 8$	= ________	16.	$3\frac{1}{4} \times 12$	= ________

E

Find the products and reduce each answer to its lowest terms.

1.	$1\frac{1}{2} \times 2\frac{1}{2}$	= ________	2.	$1\frac{1}{4} \times 2\frac{1}{2}$	= ________
3.	$1\frac{1}{3} \times 2\frac{1}{4}$	= ________	4.	$2\frac{1}{2} \times 2\frac{1}{4}$	= ________
5.	$2\frac{1}{3} \times 2\frac{1}{2}$	= ________	6.	$2\frac{2}{3} \times 1\frac{1}{4}$	= ________
7.	$3\frac{1}{2} \times 1\frac{1}{3}$	= ________	8.	$3\frac{1}{3} \times 1\frac{1}{5}$	= ________
9.	$2\frac{3}{4} \times 1\frac{1}{3}$	= ________	10.	$2\frac{3}{4} \times 3\frac{1}{2}$	= ________

Section 3 Multiplication of a Whole Number and a Mixed Number

Solution

```
   14 8
 × 12.5  ←———— 12 1/2 renamed as 12.5
 ------
   74 0
  296
 148
 ------
 1850.0
```

Exercises

Find the products.

1.	$59 \times 7\frac{1}{2}$	= ________	2.	$157 \times 3\frac{1}{2}$	= ________
3.	$24\frac{1}{2} \times 12$	= ________	4.	$68\frac{1}{2} \times 10$	= ________
5.	$35\frac{1}{2} \times 16$	= ________	6.	$125\frac{1}{2} \times 46$	= ________
7.	$137\frac{1}{2} \times 23$	= ________	8.	$168\frac{1}{2} \times 31$	= ________

9.	$24 \times 6\frac{1}{4}$	= ________	10.	$36 \times 5\frac{1}{4}$	= ________
11.	$64 \times 10\frac{1}{4}$	= ________	12.	$76 \times 18\frac{1}{4}$	= ________
13.	$129 \times 20\frac{1}{4}$	= ________	14.	$158 \times 25\frac{1}{4}$	= ________
15.	$58\frac{1}{4} \times 12$	= ________	16.	$71\frac{1}{4} \times 40$	= ________
17.	$85\frac{1}{4} \times 28$	= ________	18.	$116\frac{1}{4} \times 20$	= ________
19.	$243\frac{1}{4} \times 22$	= ________	20.	$250\frac{1}{4} \times 37$	= ________
21.	$56 \times 7\frac{3}{4}$	= ________	22.	$92 \times 6\frac{3}{4}$	= ________

Unit 4 Division

Section 1 Division of Whole Numbers and Decimals

Example 1 Divide 2.028 by 2.6

Explanation

Step 1: Move decimal point to the right of the last digit in the divisor.

2⊙6 . — **Move 1 place**

Step 2: Move decimal point the same number of places in dividend.

2⊙0 . 2 8 — **Move 1 place**

Solution

Align decimal points

```
        .78
    ______
26 ) 20.28
     18 2
     ----
      2 08
      2 08
      ----
```

Exercises A

Find the quotients mentally.

1.	$5 \div 1$	= ________	2.	$12 \div 2$	= ________
3.	$20 \div 4$	= ________	4.	$15 \div 5$	= ________
5.	$6 \div 3$	= ________	6.	$10 \div 2$	= ________
7.	$21 \div 3$	= ________	8.	$4 \div 2$	= ________
9.	$8 \div 4$	= ________	10.	$18 \div 3$	= ________

11.	16 ÷ 2	= ______	12.	8 ÷ 1	= ______
13.	20 ÷ 5	= ______	14.	18 ÷ 6	= ______
15.	14 ÷ 7	= ______	16.	15 ÷ 3	= ______
17.	32 ÷ 4	= ______	18.	24 ÷ 3	= ______
19.	9 ÷ 1	= ______	20.	45 ÷ 5	= ______
21.	24 ÷ 4	= ______	22.	14 ÷ 2	= ______
23.	3 ÷ 1	= ______	24.	25 ÷ 5	= ______
25.	12 ÷ 6	= ______	26.	21 ÷ 7	= ______
27.	12 ÷ 3	= ______	28.	16 ÷ 4	= ______
29.	7 ÷ 1	= ______	30.	18 ÷ 2	= ______
31.	9 ÷ 3	= ______	32.	4 ÷ 1	= ______
33.	42 ÷ 6	= ______	34.	30 ÷ 5	= ______
35.	4 ÷ 4	= ______	36.	30 ÷ 6	= ______
37.	24 ÷ 8	= ______	38.	6 ÷ 6	= ______
39.	6 ÷ 1	= ______	40.	35 ÷ 5	= ______
41.	28 ÷ 4	= ______	42.	42 ÷ 7	= ______
43.	40 ÷ 8	= ______	44.	24 ÷ 6	= ______
45.	3 ÷ 3	= ______	46.	10 ÷ 5	= ______
47.	36 ÷ 6	= ______	48.	56 ÷ 7	= ______
49.	40 ÷ 5	= ______	50.	28 ÷ 7	= ______
51.	2 ÷ 1	= ______	52.	12 ÷ 4	= ______
53.	48 ÷ 6	= ______	54.	32 ÷ 8	= ______
55.	27 ÷ 9	= ______	56.	5 ÷ 5	= ______
57.	35 ÷ 7	= ______	58.	16 ÷ 8	= ______
59.	1 ÷ 1	= ______	60.	63 ÷ 9	= ______
61.	48 ÷ 8	= ______	62.	7 ÷ 1	= ______
63.	56 ÷ 8	= ______	64.	36 ÷ 9	= ______
65.	7 ÷ 7	= ______	66.	64 ÷ 8	= ______
67.	63 ÷ 7	= ______	68.	72 ÷ 8	= ______
69.	49 ÷ 7	= ______	70.	45 ÷ 9	= ______
71.	54 ÷ 9	= ______	72.	8 ÷ 8	= ______
73.	27 ÷ 3	= ______	74.	18 ÷ 9	= ______
75.	2 ÷ 2	= ______	76.	72 ÷ 9	= ______
77.	36 ÷ 4	= ______	78.	54 ÷ 6	= ______
79.	8 ÷ 2	= ______	80.	81 ÷ 9	= ______

B

Find the quotients.

1.	92 ÷ 4	= ______	2	84 ÷ 6	= ______
3.	96 ÷ 6	= ______	4.	98 ÷ 7	= ______
5.	135 ÷ 3	= ______	6.	192 ÷ 3	= ______
7.	207 ÷ 3	= ______	8.	184 ÷ 4	= ______
9.	675 ÷ 5	= ______	10.	590 ÷ 5	= ______
11.	324 ÷ 6	= ______	12.	504 ÷ 6	= ______
13.	744 ÷ 6	= ______	14.	828 ÷ 6	= ______
15.	161 ÷ 7	= ______	16.	245 ÷ 7	= ______
17.	336 ÷ 7	= ______	18.	469 ÷ 7	= ______
19.	256 ÷ 8	= ______	20.	360 ÷ 8	= ______

21. 504 ÷ 8 = ______ 22. 768 ÷ 8 = ______
23. 198 ÷ 9 = ______ 24. 387 ÷ 9 = ______
25. 666 ÷ 9 = ______ 26. 954 ÷ 9 = ______

C

Find the quotients.

1. 384 ÷ 24 = ______ 2. 480 ÷ 32 = ______
3. 525 ÷ 25 = ______ 4. 414 ÷ 23 = ______
5. 351 ÷ 27 = ______ 6. 805 ÷ 35 = ______
7. 608 ÷ 38 = ______ 8. 943 ÷ 41 = ______
9. 1,058 ÷ 46 = ______ 10. 1,813 ÷ 49 = ______
11. 1,508 ÷ 52 = ______ 12. 1,512 ÷ 54 = ______
13. 322 ÷ 14 = ______ 14. 592 ÷ 16 = ______
15. 846 ÷ 18 = ______ 16. 663 ÷ 17 = ______
17. 3,712 ÷ 64 = ______ 18. 2,275 ÷ 65 = ______
19. 5,628 ÷ 67 = ______ 20. 3,456 ÷ 72 = ______
21. 4,056 ÷ 78 = ______ 22. 5,644 ÷ 83 = ______
23. 6,192 ÷ 86 = ______ 24. 7,644 ÷ 91 = ______

D

Find the quotients.

1. 4,500 ÷ 125 = ______ 2. 3,289 ÷ 137 = ______
3. 7,290 ÷ 162 = ______ 4. 10,672 ÷ 184 = ______
5. 4,738 ÷ 103 = ______ 6. 7,004 ÷ 206 = ______
7. 34,036 ÷ 508 = ______ 8. 58,546 ÷ 802 = ______
9. 9,933 ÷ 231 = ______ 10. 9,288 ÷ 258 = ______
11. 8,424 ÷ 324 = ______ 12. 32,376 ÷ 568 = ______
13. 48,766 ÷ 659 = ______ 14. 62,264 ÷ 724 = ______
15. 61,539 ÷ 843 = ______ 16. 54,834 ÷ 962 = ______
17. 25,235 ÷ 245 = ______ 18. 75,235 ÷ 367 = ______
19. 267,208 ÷ 526 = ______ 20. 447,811 ÷ 637 = ______

E

Find the quotients.

1. .54 ÷ .6 = ______ 2. 4.2 ÷ .7 = ______
3. 56 ÷ .8 = ______ 4. .24 ÷ .3 = ______
5. .02 ÷ .2 = ______ 6. .2 ÷ .02 = ______
7. 8.4 ÷ .06 = ______ 8. .85 ÷ .5 = ______
9. 81 ÷ .09 = ______ 10. .72 ÷ .08 = ______
11. 3.5 ÷ .1 = ______ 12. .64 ÷ .04 = ______
13. 156 ÷ 1.2 = ______ 14. 18.2 ÷ .13 = ______
15. .675 ÷ .25 = ______ 16. 1.116 ÷ 3.6 = ______

17. 1.458 ÷ .54 = ________ 18. .2294 ÷ .062 = ________
19. 45.75 ÷ 7.5 = ________ 20. 1.29 ÷ .86 = ________
21. 6.448 ÷ 1.04 = ________ 22. 826.8 ÷ 1.06 = ________
23. 89.64 ÷ 1.08 = ________ 24. 1.0355 ÷ 1.09 = ________

F

Find the quotients.

1. 245.61 ÷ 10 = ________ 2. 3,876.39 ÷ 10 = ________
3. 398.36 ÷ 100 = ________ 4. 4.83 ÷ 100 = ________
5. 527.64 ÷ 1,000 = ________ 6. 9.61 ÷ 1,000 = ________
7. 86 ÷ 10 = ________ 8. 4 ÷ 10 = ________
9. 293 ÷ 100 = ________ 10. 6 ÷ 100 = ________
11. 73 ÷ 1,000 = ________ 12. 9 ÷ 1,000 = ________

Section 2 Division of Fractions

Example 1 Divide 2/3 by 5/6.

Solution

Change division to multiplication

$$\frac{2}{3} \div \frac{5}{6} = \frac{2}{3} \times \frac{6}{5} = \frac{12}{15} = \frac{4}{5}$$

Replace $\frac{5}{6}$ with inverse

Example 2 Divide 4/5 by 8.

Solution

Rename 8 as $\frac{8}{1}$

$$\frac{4}{5} \div 8 = \frac{4}{5} \div \frac{8}{1} = \frac{4}{5} \times \frac{1}{8} = \frac{}{40} = \frac{}{10}$$

Example 3 Divide 3 1/2 by 2 1/4.

Solution

$$3\tfrac{1}{2} \div 2\tfrac{1}{4} = \frac{7}{2} \div \frac{9}{4} = \frac{7}{2} \times \frac{4}{9} = \frac{28}{18} = \frac{14}{19} = 1\tfrac{5}{9}$$

Exercises

A

Find the quotients.

1. $\frac{3}{5} \div \frac{1}{2}$ = ______
2. $\frac{1}{2} \div \frac{3}{7}$ = ______
3. $\frac{5}{4} \div \frac{1}{3}$ = ______
4. $\frac{1}{4} \div \frac{3}{5}$ = ______
5. $\frac{2}{3} \div \frac{5}{7}$ = ______
6. $\frac{3}{4} \div \frac{1}{5}$ = ______
7. $\frac{4}{5} \div \frac{2}{5}$ = ______
8. $\frac{1}{6} \div \frac{2}{3}$ = ______
9. $\frac{5}{6} \div \frac{1}{3}$ = ______
10. $\frac{3}{8} \div \frac{5}{6}$ = ______
11. $\frac{5}{8} \div \frac{5}{6}$ = ______
12. $\frac{1}{6} \div \frac{7}{10}$ = ______
13. $\frac{7}{8} \div \frac{3}{10}$ = ______
14. $\frac{3}{8} \div \frac{9}{16}$ = ______
15. $\frac{2}{3} \div \frac{5}{6}$ = ______
16. $\frac{3}{5} \div \frac{9}{10}$ = ______
17. $\frac{3}{8} \div \frac{15}{16}$ = ______
18. $\frac{5}{8} \div \frac{7}{16}$ = ______
19. $\frac{3}{10} \div \frac{4}{15}$ = ______
20. $\frac{8}{9} \div \frac{8}{15}$ = ______
21. $\frac{7}{10} \div \frac{21}{25}$ = ______
22. $\frac{5}{9} \div \frac{20}{21}$ = ______

B

Find the quotients.

1. $8 \div \frac{1}{2}$ = ______
2. $10 \div \frac{1}{3}$ = ______
3. $6 \div \frac{1}{4}$ = ______
4. $12 \div \frac{3}{4}$ = ______
5. $16 \div \frac{4}{5}$ = ______
6. $10 \div \frac{2}{5}$ = ______
7. $\frac{4}{5} \div 2$ = ______
8. $\frac{3}{5} \div 6$ = ______
9. $\frac{1}{2} \div 2$ = ______
10. $\frac{5}{9} \div 10$ = ______
11. $\frac{5}{8} \div 15$ = ______
12. $\frac{3}{8} \div 24$ = ______

C

Find the quotients.

1. $1\frac{1}{2} \div 2$ = ______
2. $3\frac{3}{4} \div 3$ = ______
3. $4\frac{1}{2} \div 3$ = ______
4. $3\frac{1}{5} \div 8$ = ______
5. $4\frac{2}{3} \div 7$ = ______
6. $5\frac{1}{4} \div 7$ = ______
7. $6 \div 1\frac{1}{2}$ = ______
8. $10 \div 2\frac{1}{2}$ = ______
9. $3 \div 3\frac{1}{5}$ = ______
10. $24 \div 5\frac{1}{3}$ = ______
11. $18 \div 4\frac{1}{2}$ = ______
12. $14 \div 2\frac{5}{8}$ = ______

D

Find the quotients.

1. $2\frac{1}{2} \div 1\frac{1}{2}$ = ______
2. $2\frac{3}{4} \div 1\frac{1}{4}$ = ______
3. $3\frac{1}{2} \div 2\frac{1}{2}$ = ______
4. $1\frac{3}{4} \div 1\frac{1}{4}$ = ______
5. $2\frac{1}{3} \div 3\frac{1}{2}$ = ______
6. $3\frac{1}{4} \div 1\frac{1}{4}$ = ______
7. $4\frac{3}{4} \div 2\frac{3}{8}$ = ______
8. $1\frac{3}{4} \div 3\frac{1}{2}$ = ______
9. $5\frac{1}{4} \div 2\frac{1}{2}$ = ______
10. $6\frac{1}{2} \div 2\frac{1}{4}$ = ______

Percent Unit 5

Section 1 Renaming Percent Numerals as Decimal Numerals and Decimal Numerals as Percent Numerals

Example 1 Rename 6.5% as a decimal.

Solution

$$6.5\% = .06\odot 5$$

Move decimal point two places to the *left*.

Example 2 Rename 15% as a decimal.

Solution

$$15\% = 15.\% = .15\odot$$

Insert decimal point.

Move decimal point two places to the *left*.

Example 3 Rename $4\frac{1}{2}\%$ as a decimal.

Solution

$$4\tfrac{1}{2}\% = 4.5\% = .04\odot 5$$

Rename $4\frac{1}{2}$ as 4.5

Move decimal point two places to the *left*.

Example 4 Rename .085 as a percent numeral.

Solution

$$.085 = \odot 08.5\%, \text{ or } 8.5\%$$

Move decimal point two places to the *right*.

Exercises

A

Rename each percent numeral as a decimal.

1. 15% ________	2. 26% ________	3. 53% ________
4. 72% ________	5. 93% ________	6. 31% ________
7. 49% ________	8. 67% ________	9. 85% ________
10. 14% ________	11. 60% ________	12. 70% ________
13. 80% ________	14. 90% ________	15. 10% ________
16. 2% ________	17. 8% ________	18. 3% ________
19. 5% ________	20. 7% ________	21. 124% ________
22. 136% ________	23. 158% ________	24. 235% ________
25. 375% ________	26. 150% ________	27. 250% ________
28. 140% ________	29. 320% ________	30. 460% ________
31. 100% ________	32. 200% ________	33. 300% ________
34. 400% ________	35. 500% ________	36. 26.5% ________
37. 47.8% ________	38. 52.3% ________	39. 34.6% ________
40. 367% ________	41. 3.5% ________	42. 6.5% ________
43. 4.2% ________	44. 5.7% ________	45. 9.3% ________
46. 105.4% ________	47. 207.8% ________	48. 184.7% ________
49. 169.1% ________	50. 300.2% ________	51. 14.65% ________
52. 29.37% ________	53. 51.04% ________	54. 62.71% ________
55. 84.39% ________	56. 2.54% ________	57. 4.65% ________
58. 3.75% ________	59. 5.25% ________	60. 6.35% ________
61. .045% ________	62. .058% ________	63. .074% ________
64. .0956% ________	65. .0682% ________	66. .0904% ________

B

Write each percent numeral as a decimal.

1. $14\frac{1}{2}\%$ ________	2. $16\frac{1}{2}\%$ ________	3. $21\frac{1}{2}\%$ ________
4. $53\frac{1}{2}\%$ ________	5. $67\frac{1}{2}\%$ ________	6. $8\frac{1}{2}\%$ ________
7. $4\frac{1}{2}\%$ ________	8. $3\frac{1}{2}\%$ ________	9. $6\frac{1}{2}\%$ ________
10. $1\frac{1}{2}\%$ ________	11. $128\frac{1}{2}\%$ ________	12. $157\frac{1}{2}\%$ ________
13. $220\frac{1}{2}\%$ ________	14. $325\frac{1}{2}\%$ ________	15. $450\frac{1}{2}\%$ ________
16. $27\frac{1}{4}\%$ ________	17. $86\frac{1}{4}\%$ ________	18. $37\frac{1}{4}\%$ ________
19. $72\frac{1}{4}\%$ ________	20. $93\frac{1}{4}\%$ ________	21. $5\frac{1}{4}\%$ ________
22. $7\frac{1}{4}\%$ ________	23. $8\frac{1}{4}\%$ ________	24. $1\frac{1}{4}\%$ ________
25. $3\frac{1}{4}\%$ ________	26. $158\frac{1}{4}\%$ ________	27. $194\frac{1}{4}\%$ ________
28. $237\frac{1}{4}\%$ ________	29. $205\frac{1}{4}\%$ ________	30. $370\frac{1}{4}\%$ ________
31. $32\frac{3}{4}\%$ ________	32. $47\frac{3}{4}\%$ ________	33. $61\frac{3}{4}\%$ ________
34. $80\frac{3}{4}\%$ ________	35. $76\frac{3}{4}\%$ ________	36. $6\frac{3}{4}\%$ ________

C

Write each decimal as a percent numeral.

1. .54 ________%	2. .76 ________%	3. .42 ________%
4. .85 ________%	5. .93 ________%	6. .27 ________%

7. .19	_______%	8. .30	_______%	9. .40	_______%
10. .67	_______%	11. .04	_______%	12. .05	_______%
13. .06	_______%	14. .09	_______%	15. .01	_______%
16. .07	_______%	17. .02	_______%	18. .08	_______%
19. .03	_______%	20. .10	_______%	21. 1.25	_______%
22. 2.50	_______%	23. 1.75	_______%	24. 1.50	_______%
25. 1.46	_______%	26. 3.47	_______%	27. 3.10	_______%
28. 2.05	_______%	29. 1.07	_______%	30. 1.01	_______%
31. .145	_______%	32. .267	_______%	33. .394	_______%
34. .876	_______%	35. .105	_______%	36. .283	_______%
37. .504	_______%	38. .309	_______%	39. .324	_______%
40. .207	_______%	41. .065	_______%	42. .073	_______%
43. .058	_______%	44. .061	_______%	45. .027	_______%

D

Find the products.

1. 500 × 2%	= _______	2. 350 × 4%	= _______
3. 740 × 5%	= _______	4. 960 × 6%	= _______
5. 2,300 × 8%	= _______	6. 4,500 × 9%	= _______
7. 400 × 5.5%	= _______	8. 700 × 6.5%	= _______
9. 3,700 × 8.5%	= _______	10. 4,000 × 10.5%	= _______
11. 2,000 × 6.25%	= _______	12. 5,000 × 6.75%	= _______
13. 600 × .02841%	= _______	14. 700 × .01874%	= _______
15. 1,300 × .0561%	= _______	16. 2,500 × .03142%	= _______

E

Find the products.

1. 50 × $3\frac{1}{2}$%	= _______	2. 160 × $1\frac{1}{2}$%	= _______
3. 240 × $7\frac{1}{2}$%	= _______	4. 96 × $14\frac{1}{2}$%	= _______
5. 500 × $5\frac{1}{4}$%	= _______	6. 1,200 × $6\frac{1}{4}$%	= _______
7. 180 × $1\frac{1}{4}$%	= _______	8. 230 × $10\frac{1}{4}$%	= _______
9. 2,100 × $6\frac{3}{4}$%	= _______	10. 3,600 × $3\frac{3}{4}$%	= _______
11. 4,500 × $15\frac{3}{4}$%	= _______	12. 5,800 × $20\frac{3}{4}$%	= _______
13. 245 × $\frac{1}{2}$%	= _______	14. 392 × $\frac{1}{4}$%	= _______
15. 581 × $\frac{1}{4}$%	= _______	16. 624 × $\frac{3}{4}$%	= _______
17. 6.50 × $4\frac{1}{2}$%	= _______	18. 7.20 × $3\frac{1}{2}$%	= _______
19. 9.10 × $12\frac{1}{2}$%	= _______	20. 24.30 × $31\frac{1}{2}$%	= _______
21. 8.40 × $7\frac{1}{4}$%	= _______	22. 12.70 × $2\frac{1}{4}$%	= _______
23. 112.80 × $5\frac{1}{4}$%	= _______	24. 136.50 × $3\frac{1}{4}$%	= _______
25. 10.30 × $4\frac{3}{4}$%	= _______	26. 28.35 × $2\frac{3}{4}$%	= _______
27. 102.50 × $7\frac{3}{4}$%	= _______	28. 270.08 × $10\frac{3}{4}$%	= _______
29. 143.70 × $\frac{1}{2}$%	= _______	30. 57.65 × $\frac{1}{2}$%	= _______
31. 304.50 × $\frac{1}{4}$%	= _______	32. 621.35 × $\frac{3}{4}$%	= _______

F

Find the products.

1. 800 × 5% × $\frac{3}{4}$ = ________
2. 900 × 8% × $\frac{1}{2}$ = ________
3. 1,200 × 9% × $\frac{1}{3}$ = ________
4. 1,600 × 10% × $\frac{3}{4}$ = ________
5. 2,000 × 8% × $\frac{2}{5}$ = ________
6. 500 × .325% × 20 = ________
7. 900 × .461% × 40 = ________
8. 6,000 × .0356% × 100 = ________
9. 8,000 × .0541% × 80 = ________

Unit 6 Rounding Numbers

The concept of rounding numbers in itself is quite simple. The difficulty arises over the many different ways in which we may be asked to round a number. Let us begin by examining the following number.

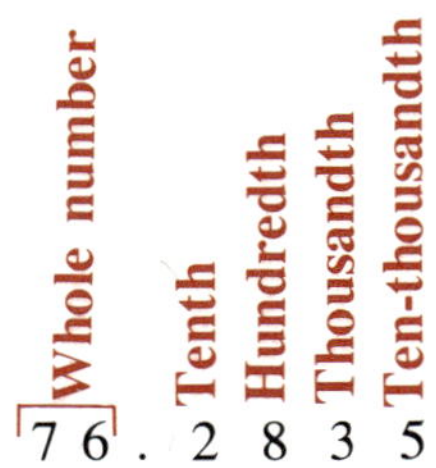

Whenever it is necessary to round a number to any particular digit, we must examine the digit to its immediate right. If we were asked to round the above number to the nearest hundredth, then

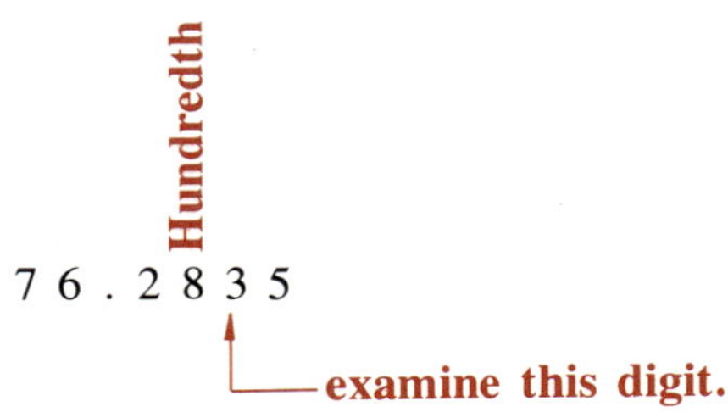

If that digit is:
0, 1, 2, 3, or 4,
drop that digit and all digits to its right.

If that digit is:
5, 6, 7, 8, or 9,
add 1 to the hundredths digit and drop all digits to its right.

Example 1 Round 76.2835 to the nearest tenth.

Explanation

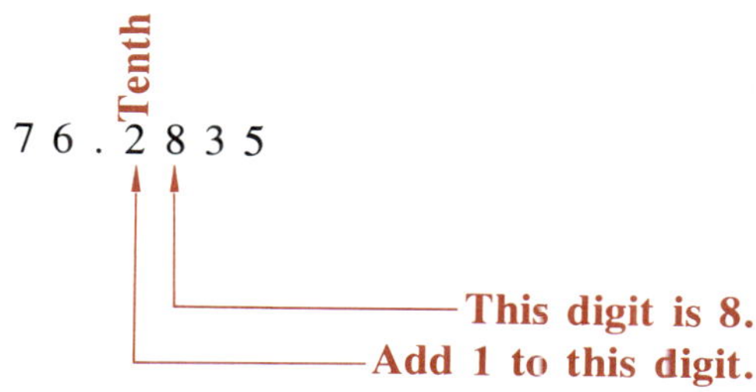

Solution

76.2835 rounded to nearest tenth is 76.3.

Example 2 Round 76.2835 to the nearest whole number.

Explanation

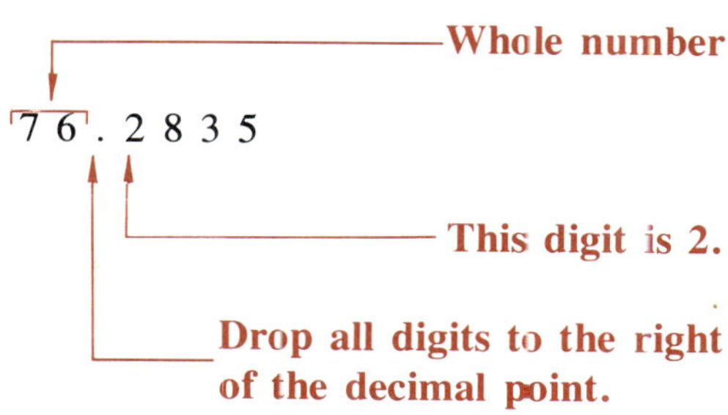

Solution

76.2835 rounded to the nearest whole number is 76.

If the number is used to name money, the rounding process is done in the same manner as above; the digits, however, have different names. The names most important to us are shown in the number below.

Dollars Dimes Cents

$ 4 5 . 6 7 2

Example 3 Round $45.872 to the nearest cent.

Explanation $45.872

This digit is 2.

Drop all digits to the right of 7.

Solution

$45.872 rounded to the nearest cent is $45.87.

Example 4 Round $45.872 to the nearest dime.

Explanation $45.872

This digit is 7.

Add 1 to this digit.

Solution

$45.872 rounded to the nearest dime is $45.90.

Example 5 Round $39.73 to the nearest dollar.

Explanation $39.73

This digit is 7.

Add 1 to this number.

Solution

$39.73 rounded to the nearest dollar is $40.

Example 6 Round \$184.62 to the next higher dime.

Explanation \$ 1 8 4 . 6 2

└**The next dime after 6 is 7.**

Solution

\$184.62 rounded to the next dime is \$184.70.

Exercises

A

Round each of the following numbers to the nearest tenth.

1.	23.41	________	2.	35.82	________	3.	47.68	________
4.	64.17	________	5.	58.05	________	6.	42.96	________
7.	87.98	________	8.	79.97	________	9.	69.05	________

B

Round each of the following numbers to the nearest hundredth.

1.	41.672	________	2.	58.593	________	3.	34.876	________
4.	42.935	________	5.	60.197	________	6.	21.399	________
7.	72.998	________	8.	89.996	________	9.	49.095	________

C

Round each of the following numbers to the nearest whole number.

1.	67.21	________	2.	84.73	________	3.	95.04	________
4.	68.54	________	5.	123.75	________	6.	146.49	________
7.	181.39	________	8.	117.92	________	9.	329.82	________
10.	439.59	________	11.	619.49	________	12.	499.61	________

D

Round each of the following numbers to the nearest cent.

1. \$4.672 \$________ 2. \$5.837 \$________ 3. \$2.835 \$________

4. $6.027 ______ 5. $12.291 ______ 6. $16.396 ______
7. $27.898 ______ 8. $35.993 ______ 9. $46.996 ______
10. $59.991 ______ 11. $59.998 ______ 12. $99.999 ______

E

Round each of the following numbers to the nearest tenth of a cent.

1. 5.72¢ $______ 2. 9.61¢ $______ 3. 28.34¢ $______
4. 6.47¢ ______ 5. 7.85¢ ______ 6. 36.39¢ ______
7. 4.06¢ ______ 8. 12.09¢ ______ 9. 16.92¢ ______
10. 38.96¢ ______ 11. 59.96¢ ______ 12. 99.99¢ ______

F

Round each number to the next higher cent.

1. $6.341 $______ 2. $1.542 $______ 3. $.623 $______
4. $.836 ______ 5. $27.752 ______ 6. $35.849 ______
7. $62.291 ______ 8. $49.992 ______ 9. $29.905 ______

G

Round each number to the nearest dime.

1. $2.34 $______ 2. $4.61 $______ 3. $5.86 $______
4. $9.12 ______ 5. $16.43 ______ 6. $27.02 ______
7. $84.07 ______ 8. $53.55 ______ 9. $67.93 ______
10. $85.96 ______ 11. $39.91 ______ 12. $89.97 ______

H

Round each number to the next higher dime.

1. $3.21 $______ 2. $4.57 $______ 3. $6.03 $______
4. $8.15 ______ 5. $35.89 ______ 6. $43.01 ______
7. $51.45 ______ 8. $69.91 ______ 9. $39.96 ______

I

Round each number to the nearest dollar.

1.	$7.23	$______	2.	$5.48	$______	3.	$6.87	$______
4.	$2.16	______	5.	$32.59	______	6.	$44.80	______
7.	$53.49	______	8.	$68.50	______	9.	$157.08	______
10.	$126.46	______	11.	$109.63	______	12.	$129.40	______
13.	$349.75	______	14.	$299.20	______	15.	$399.50	______

J

Round each number as indicated.

1. 12.347 to the nearest tenth ______
2. 16.482 to the nearest hundredth ______
3. 11.271 to the nearest tenth ______
4. 24.693 to the nearest whole number ______
5. 58.529 to the nearest hundredth ______
6. 69.891 to the nearest tenth ______
7. 40.497 to the nearest whole number ______
8. 39.952 to the nearest tenth ______
9. 49.967 to the nearest hundredth ______
10. 69.991 to the nearest hundredth ______
11. 79.992 to the nearest tenth ______
12. 99.593 to the nearest whole number ______

K

Round each number as indicated.

1. $14.673 to the nearest cent $______
2. $18.226 to the nearest cent $______
3. $24.61 to the nearest dime $______
4. $34.52 to the next higher dime $______
5. $41.273 to the next higher penny $______
6. $67.41 to the nearest dollar $______
7. $59.236 to the nearest cent $______
8. $29.92 to the nearest dime $______
9. $29.92 to the next higher dime $______
10. $59.50 to the nearest dollar $______
11. $31.295 to the nearest penny $______
12. $31.295 to the next higher penny $______
13. $49.63 to the nearest dollar $______
14. $99.91 to the next higher dime $______
15. $99.91 to the nearest dollar $______

Index